Lecture Notes in Computer Science 8453

Commenced Publication in 1973
Founding and Former Series Editors:
Gerhard Goos, Juris Hartmanis, and Jan van Leeuwen

T0224061

Anne Remke   Mariëlle Stoelinga (Eds.)

Stochastic Model Checking

Rigorous Dependability Analysis Using Model
Checking Techniques for Stochastic Systems

International Autumn School, ROCKS 2012
Vahrn, Italy, October 22-26, 2012
Advanced Lectures

 Springer

Volume Editors

Anne Remke
University of Twente
Faculty of Electrical Engineering, Mathematics and Computer Science
Design and Analysis of Communication Systems
P.O. Box 217, 7500 AE Enschede, The Netherlands
E-mail: anne@cs.utwente.nl

Mariëlle Stoelinga
University of Twente
Department of Computer Science
Formal Methods and Tools
P.O. Box 217, 7500 AE Enschede, The Netherlands
E-mail: marielle@cs.utwente.nl

ISSN 0302-9743   e-ISSN 1611-3349
ISBN 978-3-662-45488-6   e-ISBN 978-3-662-45489-3
DOI 10.1007/978-3-662-45489-3
Springer Heidelberg New York Dordrecht London

Library of Congress Control Number: 2014953970

LNCS Sublibrary: SL 1 – Theoretical Computer Science and General Issues

Typesetting: Camera-ready by author, data conversion by Scientific Publishing Services, Chennai, India

Printed on acid-free paper

Springer is part of Springer Science+Business Media (www.springer.com)

Preface

Stochastic models are widely used in the modeling and analysis of a wide range of phenomena, ranging from psychology, speech recognition, to political coalition forming, particle behavior, and many more applications. Their use in computer science is also wide-spread, for instance, in performance modeling, analysis of randomized algorithms, and communication protocols that form the structure of the Internet. *Stochastic model checking* is an important field in stochastic analysis. It has rapidly gained popularity, due to its powerful and systematic methods for modeling and analyzing stochastic systems.

In order to inform young researchers about the fundamentals and state of the art in stochastic model checking, an Autumn School was organized by the ROCKS project, funded by the Dutch NWO and German DFG. The school was held during Ocotber 22-26, 2012, in Vahrn, Italy. Leading scientists from the field gave lectures on foundations as well as state-of-the-art research.

The seven chapters of this tutorial were initiated at the ROCKS Autumn School, summarizing the state of the art in the field, centered around the three areas of stochastic models, abstraction techniques, and stochastic model checking. All submissions were thoroughly reviewed in a two-stage review process by at least three Program Committee members and in the end the committee decided to accept all seven papers.

Stochastic model checking is a rich field, which provides powerful and systematic methods for modeling and analyzing stochastic systems. A wide variety of stochastic models exist, depending on probabilistic choices that are used (discrete, continuous, or both), on whether nondeterminism is present (Markov models versus decision models) and the state space of the models (discrete versus continuous). These models allow for a wide variety of analysis methods to investigate their behavior and properties.

Stochastic models. The analysis of stochastic behavior starts with the choice of a suitable modeling framework. This volume contains an introduction in three important frameworks.

- Interactive Markov chains (IMCs) combine continuous time Markov chains with labeled transition systems, thus leveraging the power of stochastic choices with interaction. The paper "On Interactive Markov Chains" by Arnold, Gebler, Guck, and Hatefi provides an in-depth treatment of their stochastic model-checking techniques, scheduler classes, behavioral equivalences, and case studies.
- The paper "A Theory for the Semantics of Stochastic and Non-deterministic Continuous Systems" by Budde, D'Argenio, Sànchez, Terraf, and Wolovick considers models with uncountable state spaces, emerging when modeling continuous quantities such as time, distance, temperature. The paper studies equivalences for these models, as well as logical characterizations.

- Authors Gouberman and Siegle discuss in their paper "Markov Reward Models and Markov Decision Processes in Discrete and Continuous Time: Performance Evaluation and Optimization" stochastic models containing reward structures and nondeterminism. They present techniques to compute rewards and cost-optimal strategies and apply these to performance models.

Abstraction techniques. Since complex models are often too large to be analyzed, a simpler model is used that contains fewer states, but preserves the relevant properties.

- The paper "On Abstraction of Probabilistic Systems" by Dehnert, Gebler, Volpato, and Jansen presents three important abstraction techniques: multi-valued abstraction, counterexample-guided abstraction-refinement, and game-based abstraction. The authors show how these abstraction preserve probabilistic properties formulated in PCTL.
- In the paper "Computing Behavioral Relations for Probabilistic Concurrent Systems," Daniel Gebler, Vahid Hashemi, and Andrea Turrini study behavioral equivalences and preorders. In particular, they discuss strong and weak probabilistic (bi-)simulation and provide efficient algorithms to compute these relations. Finally, the paper presents approximate relations in terms of metrics.

Advanced analysis techniques. Model checking can be an alternative to classic analysis in terms of differential equations. This volume discusses two important directions in this regard.

- Mean-field approximation is a powerful technique to analyze systems with a large number of identical components. The paper by Kolesnichenko, Pourranjbar, Senni, and Remke, titled "Applying Mean-field Approximation to Continuous Time Markov Chains," provides, through a number of real-life examples, a rigorous introduction to classic mean-field analysis, as well as model checking approaches.
- Oscillatory behavior, that is, periodically re-occurrent behavior, is an important property in population models. The paper "Analyzing Oscillatory Behavior with Formal Methods" by Andreychenko, Krüger and Spieler presents an overview of analysis methods for systems with cyclic behavior. The authors discuss deterministic, stochastic, and mixed models and elucidate how model checking complements traditional analysis via differential equations.

We would like to once more thank the invited speakers at the ROCKS Autumn School, who helped to turn this school into a succes:

- Christel Baier
- Jean-Yves Le Boudec
- Martin Fränzle
- Thomas Henzinger
- Jane Hillston

- Jozef Hooman
- Joost-Pieter Katoen
- Marta Kwiatkowska
- John Lygeros
- Gethin Norman

We would also like to thank our co-organizers of the autumn school: Erika Ábrahám, Christel Baier, David Jansen, Markus Siegle, and Verena Wolf. Finally, we thank our sponsors, NWO and DFG, for their generous support of the ROCKS research collaboration.

We hope that this volume is enjoyable to read, and helps to give you a good overview and understanding of the inspiring field of stochastic model checking.

June 2014

Anne Remke
Mariëlle Stoelinga

Organization

Program Committee

Christel Baier	TU Dresden, Germany
David N. Jansen	Radboud University Nijmegen, The Netherlands
Anne Remke	University of Twente, The Netherlands
Markus Siegle	Bundeswehr University Munich, Germany
Marielle I.A. Stoelinga	University of Twente, The Netherlands
Verena Wolf	Saarland University, Germany
Erika Ábrahám	RWTH Aachen, Germany

Additional Reviewers

Ahmad, Waheed	Kolesnischenko, Anna
Arnold, Florian	Krčál, Jan
Gouberman, Alexander	Ruijters, Enno
Guck, Dennis	Spieler, David
Jansen, Nils	Timmer, Mark

Table of Contents

Analyzing Oscillatory Behavior
with Formal Methods

Alexander Andreychenko, Thilo Krüger, and David Spieler

Saarland University
{makedon,thilo,spieler}@mosi.uni-saarland.de

Abstract. An important behavioral pattern that can be witnessed in many systems is periodic re-occurrence. For example, most living organisms that we know are governed by a 24 hours rhythm that determines whether they are awake or not. On a larger scale, also whole population numbers of organisms fluctuate in a cyclic manner as in predator-prey relationships. When treating such systems in a deterministic way, i.e., assuming that stochastic effects are negligible, the analysis is a well-studied topic. But if those effects play an important role, recent publications suggest that at least a part of the system should be modeled stochastically. However, in that case, one quickly realizes that characterizing and defining oscillatory behavior is not a straightforward task, which can be solved once and for all. Moreover, efficiently checking whether a given system oscillates or not and if so determining the amplitude of the fluctuations and the time in-between is intricate. This paper shall give an overview of the existing literature on different modeling formalisms for oscillatory systems, definitions of oscillatory behavior, and the respective analysis methods.

1 Introduction

In the years 1926 respectively 1924, V. Volterra [102] and A. Lotka [70] independently from each other studied the dynamics of predator and prey populations. Their key insight was that the amount of both species showed regular *oscillatory fluctuations*, where the predator population followed the prey population. But the phenomenon of oscillation is also present at various granularities and forms in many other systems. Examples are the 24 hour day/night rhythm of living organisms on this planet [13] and calcium ion transport between membranes in cells [91]. But oscillation can also be found at macroscopic levels as e.g. in whole ecospheres like Savanna patches [76].

The tool-set that was used by Lotka and Volterra were deterministic models, where the amount of each species was computed over time. Those models and the respective analysis techniques are well-understood and are still used widely today to reason about reaction networks in the fields of chemistry and systems biology [33,60,91]. However, recent insights suggest that a completely deterministic approach might not be appropriate in all of the cases since *noise* resulting from small populations plays an important role. For example, *circadian clocks*,

A. Remke and M. Stoelinga (Eds.): ROCKS Autumn School 2012, LNCS 8453, pp. 1–25, 2014.

the basic mechanism behind the 24 hour rhythm, rely on stochastic effects to maintain an oscillatory pattern and to prevent being trapped in an equilibrium state [13]. Results like that and others [6,74] clearly speak in favor of stochastic modeling. The framework of (continuous-time) Markov chains is rigorously justified [38] for those systems from the area of biology and chemistry and it is also used in the area of formal methods [8]. The aim of this paper is to give an overview of how to model and analyze systems that exhibit an oscillatory behavioral pattern by the use of formal methods. There is a great variety of different aspects, definitions, notions, and techniques used in the literature and not all can be handled in this paper. We therefore restrict to selected examples characteristic for each of the three major abstraction levels, i.e., deterministic, fully stochastic and their combination in the form of hybrid models. Further, there is no common formal definition of terms like *oscillation* or *attractor* and we will refer to the mostly informal and varying interpretations as presented in the recpective references.

The structure of the paper is as follows. In Section 2, we discuss the respective modeling formalisms dependent on whether no, some, or all components rely on a stochastic representation. Section 3 compares different notions of oscillatory behavior depending on the model type and describes the corresponding approaches to analyze the models with respect to the various definitions. We also give a series of examples for those approaches in Section 4. One of these examples is the two-species predator-prey model, which also serves as our running example. Finally, we conclude our paper in Section 5.

2   Modeling of Oscillatory Systems

At first, we would like to clarify the notation of this paper. Vectors are written like $\mathbf{x} = [x_1 \ldots x_N]$ and $\mathbf{e}$ denotes the column vector of ones. Analogously, matrices are written like $\mathbf{Q} = [q_{ij}]$. We will assume suitable enumeration schemes for vectors $\mathbf{x} \in \mathbb{N}^N$ such that matrices and vectors can be indexed by them, e.g. as in $q_{\mathbf{xy}}$ for $\mathbf{x}, \mathbf{y} \in \mathbb{N}^N$.

Next, we will give an overview of how oscillatory systems are modeled. We will distinguish the modeling approaches by the amount of influence stochasticity has on the behavior. In general we will treat three main classes, which are completely deterministic models, fully stochastic models and hybrid models which have deterministic as well as stochastic parts.

2.1   Population Structure and Chemical Reaction Networks

In order to base these models on a common framework, we first observe that regardless of the exact semantics, they share a *population structure*. By population structure, we mean that a system consists of N different *populations*, i.e., distinct quantities whose amounts determine the (possible) future behavior. Consequently, those quantities make up the *state* of the system, are usually non-negative, and might all be discrete, real valued, or mixed. Also they might be bounded or unbounded. Moreover, the state of the system usually changes over time, where time can be treated as discrete or continuous. In this overview, we will focus on *continuous-time* models. We can summarize the common mathematical model structure by defining that a system $X(t)$ is in a state $\mathbf{x} = [x_1 \ \ldots \ x_N] \in \mathbb{R}_{\geq 0}^N$ for each $t \in \mathbb{R}_{\geq 0}$. The precise definition of $X(t)$ is a function in the deterministic setting, and a *stochastic process* in the stochastic and hybrid setting as described later on.

All of the example models described in this summary paper originate from a biologically motivated application area. Consequently, the populations described in the previous paragraph mainly correspond to *chemical species* and their evolutions are usually described by *chemical reactions* forming a *chemical reaction network* (CRN). For a detailed overview of a structural analysis of CRN, we refer to [5]. In short, a system consists of N chemical species $S_1, \ldots, S_N$ and R different chemical reactions, where each reaction is of the form

$$u_{r1} \cdot S_1 + \cdots + u_{rN} \cdot S_N \xrightarrow{c_r} w_{r1} \cdot S_1 + \cdots + w_{rN} \cdot S_N,$$

with *stoichiometric coefficients* given by vectors $\mathbf{u_r} = [u_{r1}, \ldots, u_{rN}]^T \in \mathbb{N}^N$ and $\mathbf{w_r} = [w_{r1}, \ldots, w_{rN}]^T \in \mathbb{N}^N$ and a *reaction rate* $c_r \in \mathbb{R}_{\geq 0}$ for $r \in \{1, \ldots, R\}$. The intuition of a chemical reaction is that the reactant molecules (u_{ri} molecules of species i) can be transformed into the product molecules (w_{rj} molecules of species j) with the corresponding likelihood and speed determined by the reaction rate c_r. But the exact physical interpretation and units of c_r both depend on the model type and the order of reaction [104]. Symbol $\emptyset$ denotes that $\mathbf{u_r}$ respectively $\mathbf{w_r}$ is the zero vector.

Example 1 (Predator-Prey CRN). This representation is not limited to chemical reactions but can also be used for example to characterize population dynamics of animals. An example is the predator-prey model proposed independently by Lotka and Volterra [70,102]. The CRN is described by

$$S_1 \xrightarrow{c_1} 2 \cdot S_1, \quad S_1 + S_2 \xrightarrow{c_2} 2 \cdot S_2, \quad S_2 \xrightarrow{c_3} \emptyset$$

where the first reaction describes reproduction of a prey species S_1, the second reaction encodes the reproduction of predators S_2 consuming prey (food) and the last reaction relates to the natural death of the predator species.

2.2 Deterministic Semantics

The traditional way of treating the behavior of chemical reaction networks is the deterministic approach justified by the *law of chemical mass action* as for example described in [51,96]. For that, the amount of molecules per chemical species is not modeled as such but as a *concentration*, i.e., molecules per volume, commonly measured in mol/L. The state space in that case is continuous, i.e., $X : \mathbb{R}_{\geq 0} \to \mathbb{R}_{\geq 0}^N$. We define the *deterministic propensity* $\alpha_r^{det}(\mathbf{x})$ of a reaction r for state $\mathbf{x} = [x_1 \ldots x_N]$ as

$$\alpha_r^{det}(\mathbf{x}) = c_r^{det} \cdot \prod_{i=1}^{N} x_i^{u_{ri}}.$$

Note that c_r differs depending on the type of the model [104]. Here we have to use the deterministic version. The exponent u_{ri} is the number of occurences of the species in the reaction. It follows that if the species does not take part in the reaction, it will not occur in the equation. Given an *initial concentration* $X(0)$, the future behavior is completely determined by a set of *ordinary differential equations* (ODE) that is given for each species i by

$$\dot{X}_i(t) = \sum_{r=1}^{R} (\mathbf{w_{r_i}} - \mathbf{u_{r_i}}) \cdot \alpha_r^{det}(X(t)). \qquad (1)$$

A deeper mathematical analysis of the law of mass action can be found in [23,78]). There exist other ways to derive the ODE describing the concentration of certain species in time [105]. In order to model the feedback mechanisms in physiological systems delay differential equations can be exploited. They are capable of modeling and formally reasoning about the effect of one or several feedback loops with different delays that control many processes in living organisms [40,69]. Please also notice that the given deterministic framework does not account for possible measurement errors neither for uncertainties in states. Taking those into consideration leads to a stochastic differential equations model [17,38,104].

Example 2 (Predator-Prey ODE). The ODE for the predator-prey model from Example 1 as obtained from the CRN is

$$\dot{X}(t) = [1 \ 0]^T \cdot \alpha_1^{det}(X(t)) + [-1 \ 1]^T \cdot \alpha_2^{det}(X(t)) + [0 \ -1]^T \cdot \alpha_3^{det}(X(t))$$

with $\alpha_1^{det}(\mathbf{x}) = c_1 \cdot x_1$, $\alpha_2^{det}(\mathbf{x}) = c_2 \cdot x_1 \cdot x_2$, $\alpha_3^{det}(\mathbf{x}) = c_3 \cdot x_2$.

2.3   Stochastic Semantics

In the fully stochastic approach, the granularity of the modeling is the molecule level. Thus, instead of keeping track of the concentration as in the deterministic setting, individual molecules are treated as such. The stochastic semantics of a chemical reaction network is given by a *(homogeneous) continuous-time Markov chain (CTMC)* [38,36,37]. In detail, the behavior of the system is determined by a stochastic process $\{X(t)\}_{t\in\mathbb{R}_{\geq 0}}$ which is a family of random variables indexed by time. These processes satisfy the *Markov* and *homogeneity* properties, i.e., the possible future behavior only depends on the current state and does not change over time, as formalized by

$$\mathbf{Pr}[X(t_n) = s_n \mid X(t_{n-1}) = s_{n-1}, \ldots, X(t_0) = s_0]$$
$$= \mathbf{Pr}[X(t_n) = s_n \mid X(t_{n-1}) = s_{n-1}] = \mathbf{Pr}[X(t_n - t_{n-1}) = s_n \mid X(0) = s_{n-1}].$$

The main reason to exploit the stochastic model when analyzing biological processes comes from the fact that given low molecular counts, the corresponding stochastic effects can not be ignored. It is often referred to as a *noisy* process, where the noise can be decoupled into two components, i.e., intrinsic and extrinsic noise [31]. In contrast to the deterministic setting, there is not a determined solution, but the system is in a state $\mathbf{x} \in \mathbb{N}^N$ with a certain *probability* $\pi_{\mathbf{x}}(t)$. Distribution $\pi(t)$ is called the *transient probability distribution* at time t and is defined as a row vector $\pi(t)$ such that $\pi_{\mathbf{x}}(t) = \mathbf{Pr}[X(t) = \mathbf{x}] \in [0,1]$ and $\pi(t) \cdot \mathbf{e} = 1$. Due to the above constraints, the behavior of the CTMC is fully described by an *initial distribution* $\pi(0)$ and an *infinitesimal generator matrix* $\mathbf{Q} = [q_{\mathbf{xy}}]$ where all elements $q_{\mathbf{x} \neq \mathbf{y}} \in \mathbb{R}_{\geq 0}$ and diagonal entries are defined as $q_{\mathbf{xx}} = -\sum_{\mathbf{y} \neq \mathbf{x}} q_{\mathbf{xy}}$. More precisely, the transient distribution satisfies the *Kolmogorov differential equations* [58]

$$\dot{\pi}(t) = \pi(t) \cdot \mathbf{Q} \tag{2}$$

also called the *master equation* [38,56] in the natural sciences. In order to define the matrix entries q_{xy}, we define the (stochastic) *propensity* $\alpha^{st}(\mathbf{x})$ of a reaction r for state $\mathbf{x} = [x_1 \ldots x_N]$ as

$$\alpha_r^{st}(\mathbf{x}) = \begin{cases} c_r^{st} \cdot \prod_{i=1}^{N} \binom{x_i}{u_{ri}}, & \text{if } x_i \geq u_{ri}, \\ 0, & \text{otherwise} \end{cases} \tag{3}$$

and get

$$q_{\mathbf{xy}} = \begin{cases} \sum_{\{r \mid \mathbf{x} - \mathbf{u}_r + \mathbf{w}_r = \mathbf{y}\}} \alpha_r^{st}(\mathbf{x}), & \mathbf{x} \neq \mathbf{y}, \\ -\sum_{\mathbf{z} \neq \mathbf{x}} q_{\mathbf{xz}}, & \mathbf{x} = \mathbf{y}. \end{cases}$$

The reaction rate constant c_r^{st} encodes physical properties of reaction r and its units depend on the type of reaction. The actual value can be computed using the information about the rate constant c_r^{det} used in the deterministic semantic as described in [61,75]. The binomial coefficients refer to the number of possible combinations of the reactants. We assume that the Kolmogorov equations (2) have a unique solution under a given initial condition $\pi(0)$. To solve the system of ODE (2) in practice, one can exploit generally applicable analytical and numerical methods for solving ODE systems (like Runge-Kutta method) as well as those specially designed for Markov chain analysis as for example *simulation* [17,18,35,37,39], *abstraction* [1,27,57,64], *uniformization* [30,44,46,55] and others [16,54].

Example 3 (Predator-Prey CTMC). A stochastic treatment of the model from Example 1 results in a CTMC with state space $\mathbb{N}^2$ and an infinitesimal generator $\mathbf{Q}$ defined by

$$q_{\mathbf{xy}} = \begin{cases} c_1 \cdot x_1 & \text{if } x_1 > 0, y_1 = x_1 + 1, y_2 = x_2, \\ c_2 \cdot x_1 \cdot x_2 & \text{if } x_1 > 0, x_2 > 0, y_1 = x_1 - 1, y_2 = x_2 + 1, \\ c_3 \cdot x_2 & \text{if } x_2 > 0, y_1 = x_1, y_2 = x_2 - 1, \\ -\sum_{\mathbf{z} \neq \mathbf{x}} q_{\mathbf{xz}} & \text{if } \mathbf{x} = \mathbf{y}, \text{and} \\ 0 & \text{otherwise.} \end{cases}$$

In order to simplify the description of stochastic models, tools like PRISM [65] use *guarded-commands*, i.e., rules that for each state describe the possible transitions (consisting of successor state and rate) that are possible depending on the validity of a boolean predicate on the current state (*guard*). Moreover, the chemical species and reactions can be encoded as *modules* that are composed via *synchronization* [65] as done for example in [11,12]. If there is no a-priori knowledge about the bounds on molecular counts as in the previous example, the corresponding Markov chain has a *countably infinite state space*. In those cases, the stochastic simulation or specially designed numerical methods like [15,30,48,49,80] are needed. A tool specifically tailored to analyze such systems numerically is SHAVE [66]. Sophisticated techniques like inexact matrix vector multiplications exploiting a truncation of the state space to focus on the significant states, i.e., states with a probability mass greater than a threshold, and automated hybridization of populations help to cope with large or even infinite state spaces. Another way to treat such systems approximately is to take the limit of the number of individuals per volume and describe the system solely by the mean population numbers, arriving at the *mean-field limit* of the system. This mean-field limit coincides with the deterministic approach modeled by the ODE of Equation (1) but only under certain conditions [47,62].

2.4 Hybrid Semantics

Analysis of real-life case studies from biology reveals that there exist systems where some species posses low molecular counts and others are present in extremely high amounts. Treating the whole system as discrete using the fully stochastic approach described in the previous section, results in a very large state space. Consequently, the direct solution of Equation (2) is inefficient. On the other hand, we can treat the dynamics of the whole system deterministically thus omitting the stochastic effects as described in Section 2.2, where solving the system is computationally far less demanding. In order to still profit from the computational speed of the deterministic model and be able to account for the underlying stochasticity, the hybrid approach can be used. Here, the process $X(t)$ is split into two parts, i.e., $X(t) = (M(t), Y(t))$, where the discrete stochastic component $M(t)$ accounts for the small populations and the deterministic continuous component $Y(t)$ deals with large populations. We assume that the dimension of $M(t)$ is $\hat{n}$ and the dimension of $Y(t)$ is $\tilde{n}$ with $N = \hat{n} + \tilde{n}$. The dynamics of the stochastic component $M(t) \in \mathbb{N}^{\hat{n}}$ is described by a CTMC where the rates not only depend on the current state $\mathbf{m}$ but also on the state $\mathbf{y}$ of the deterministic counterpart $Y(t) \in \mathbb{R}^{\tilde{n}}$. Consequently, the entries of the generator matrix $\mathbf{Q}(t)$ are described by

$$q_{\mathbf{mm'}}(t) = \begin{cases} \sum_{\{r \mid \mathbf{m}-\mathbf{u}_r+\mathbf{w}_r=\mathbf{m'}\}} \alpha_r^{hyb}(\mathbf{m}, \mathbf{y}(t)), & \text{if } m \neq m' \\ -\sum_{\mathbf{m''} \neq \mathbf{m}} q_{\mathbf{mm''}}(t), & \text{if } m = m' \end{cases} \quad (4)$$

with propensity functions $\alpha_r^{hyb}(\mathbf{m}, \mathbf{y}(t))$ defined as

$$\alpha_r^{hyb}(\mathbf{m}, \mathbf{y}(t)) = \begin{cases} c_r^{st} \cdot \prod_{i=1}^{\hat{n}} \binom{m_i}{u_{ri}} \cdot \prod_{j=1}^{\tilde{n}} y_j^{u_{rj}}(t), & \text{if } m_i \geq u_{ri} \\ 0, & \text{otherwise} \end{cases}$$

The transient probability distribution for $M(t)$ follows the time-dependent Kolmogorov equation $\dot{\pi}(t) = \pi(t) \cdot \mathbf{Q}(t)$. As before, we assume that this system of equations has a unique solution under the given initial condition $\pi(0)$. To solve this equation system the above mentioned methods can be applied as well as methods designed to analyze time-dependent continuous-time Markov chains [7,79,100] and hybrid systems [3,26,75,88]. The behavior of the continuous process $Y(t)$ is defined by the system of ODE similar to Equation (1)

$$\dot{Y}_i(t) = \sum_{r=1}^{R'} (\mathbf{w_{r_i}} - \mathbf{u_{r_i}}) \cdot \alpha_r^{det}(X(t)),$$

where R' is the number of reactions, where only species whose concentration is treated as a continuous quantity, take part. Such a representation is especially useful if the Markov chain $M(t)$ only encodes the states (on/off) of a set of genes that control the production of proteins. Then each state variable $m_1, m_2, \ldots, m_{\hat{n}}$ can take values 1 or 0 and the concentration of the correspondent proteins is controlled by the deterministic process $Y(t)$. We can refer to states of $M(t)$ as to *modes* [92] of the hybrid system. In each mode the components of $Y(t)$ follow the system of ODE as in Equation (1). For a rigorous derivation of the corresponding theory, we refer to [52].

The presented fixed decomposition of the state space vector onto two parts is based on the molecular counts. There exist other approaches to divide the system into subsystems. They allow for dynamical repartitioning by exploiting the model structure. For details, we refer to [2,10,17,26,45,86,89,103].

Example 4 (Predator-Prey Hybrid). The predator-prey model from Example 1 is often used to test the applicability of hybrid solution techniques since under certain parameter values it possesses high stiffness. In this example we assume that the reaction rate constants are such that species S_2 (predator) has low molecular counts and that species S_1 (prey) is present in high counts. A hybrid approach leads to a definition of the process as $X(t) = (S_1(t), S_2(t))$, where S_2 is treated stochastically and its behavior is determined by the time-dependent generator matrix $Q(t)$ with entries $q_{s_1,s_1'}(t)$ according to Equation (4), where $\alpha_2^{hyb}(s_1(t), \mathbf{s}_2) = c_2^{st} \cdot s_1(t) \cdot \mathbf{s}_2$, $\alpha_3^{hyb}(s_1(t), \mathbf{s}_2) = c_3^{st} \cdot \mathbf{s}_2$. The evolution of species S_1 is deterministic and follows the ODE $\dot{S}_1(t) = c_1^{det} s_1(t) - c_2^{det} s_1(t) \mathbf{s}_2$.

2.5 Other Model Semantics

In this tutorial we address only a selection of formalisms among many used to describe biological and chemical systems. Other notable formalisms are more *structure* oriented, where individual molecules even of the same type can be distinguished as e.g. in [22]. Another way to model biological systems and signaling pathways [90] is to represent them as timed automata [4]. It allows to treat deterministic and hybrid systems [21,59,71,84,97] and formally verify their properties using existing tools [82]. The interested reader can find more detailed information about the variety of formalisms in [29,47].

3   Defining and Analyzing Oscillatory Behavior

Prior to describing how *oscillatory behavior* is defined in the literature, we first have to agree on the *quantity* that shall be analyzed. This question is directly related to what measures about a system are observable. In order to treat all presented approaches under a common framework and to be able to compare them, we assume that in general, the current number of individuals of each population type at any moment in time can be observed. In detail, this means that given the system is in state $\mathbf{x} = [x_1 \ldots x_N]$ at time t, each value x_i for $1 \leq i \leq N$ is visible to the analysis. In order to further simplify the presentation, we focus on studying the behavior of single population types as the quantities of interest. We would like to note that in all presented approaches, the analysis of more complex observations like the mean or sum of two or more populations is possible either directly or with little effort.

For approaches based on model checking like [11,12,22], no information about the current state $X(t)$ is a priori visible to the underlying logic. Consequently, the observable information must be stated explicitly. For that, a model with state space S is augmented with a *labeling function* $L : S \to 2^{AP}$, which assigns to each state $s \in S$ a set of *atomic propositions* $L(s)$ from the set AP that hold in that state. In [11,12] for example, each state $[x_1 \ldots x_N]$ is labeled by propositions like $X_i = x_i$ for each $1 \leq i \leq N$. In order to allow for more complex observation queries, also inequalities like $X_1 \neq 1$ and $X_1 \geq 3$ are for example used to describe a state with $x_1 = 4$.

3.1   General and Mathematical Notions of Oscillatory Behavior

At first we would like to discuss the difficulty of choosing a *general definition* of oscillatory behavior. When considering the term *oscillation* from an intuitive point of view, it usually refers to a repeated fluctuation pattern in time of some observable quantity. Before going further, we would like to differentiate the intended notion from *switching behavior*, a phenomenon that is also studied very often in the literature and that might be called oscillatory as well. Some stochastic models like the *genetic toggle switches* [34,53] possess two or more regions in the state space where the steady-state probability mass concentrates (around global maxima). Those regions are called *attracting regions* and due to the high probability mass, the system most of the time is in one of those regions. Transitions between those regions are always possible in principle but they might be very unlikely depending on the exact parameters. Consequently, switching between those regions becomes a rare event. Due to the long time between significant fluctuations, we will not treat those systems and multi-stable systems with low switching-rate in general as oscillatory in this overview paper.

In order to approach more formal definitions, we would like to continue with a discussion of the notions of periodic and oscillatory behavior in the strict mathematical sense. In general, a function $f(t)$ is called *periodic* with *period length p* if for all times $t \in \mathbb{R}_{\geq 0}$ holds that $f(t + p) = f(t)$ and *oscillatory* with *frequency* θ and *amplitude a* if $f(t) = a \cdot \sin(\theta \cdot t + \omega)$. While these definitions provide very precise notions of the behavior, their use in our setting is severely limited [95]. In detail, they are too strict in the sense that hardly any system adheres to them over the complete time horizon. Where their use for deterministic models might be justified, e.g. when altering the definitions to cope with phase shifts or more complex fluctuation patterns, their use in the stochastic setting becomes complicated. For example, the probability measure of all paths of a CTMC based on a biological model that follow such a predefined trajectory pattern is trivially zero due to the inherent stochastic noise [95]. In addition, periodicity also would not cover the full intended behavior where fluctuations are required, as a constant function $f(t) = c$ for some $c \in \mathbb{R}$ would be periodic for any period length $p \in \mathbb{R}$. When comparing the literature, it is therefore not surprising to see several definitions of the intended *oscillatory* behavior specifically tailored to the respective applications like an inclusion of noise corridors [11,12] in order to relax the strict definitions.

3.2 Oscillatory Behavior in the Deterministic Setting

The most common approach to model and analyze oscillatory behavior is the deterministic modeling via ODEs. Since this general approach is the most established, it is reasonable to start with a short historical overview of the development of these deterministic methods. Since this is done in detail by Tyson et al. in 2008 [99] for the biological context, we limit ourselves to a brief overview. First observations of oscillatory behavior in biology were made in macroscopic biological systems. Thus the first models, that were discussed by Lotka [70] and Volterra [102], are models of predator and prey relationships of animals. The well-known Lotka-Volterra mechanism in CRN form and the corresponding equations [50] are shown in Example 2. Around 1950 Beloussov for the first time observed oscillating behavior in a homogeneous chemical reaction, the Beloussov-Zhabotinsky-Reaction [99]. Another much simpler reaction scheme, which was analyzed in the sixties was the so-called Brusselator, which was used as a theoretical system to show that oscillatory behavior is possible in homogeneous chemical reaction systems. First larger developments in analyzing oscillatory behavior were made in the early sixties and were summarized by a paper of Higgins [50]. He gave a classification of oscillations that allowed to distinguish *damped* oscillations, *sustained* oscillations and general oscillations, and gave very general equations for oscillating ODEs. He also summarizes general thermodynamical and kinetic considerations from different papers with the result, that it is generally impossible to achieve sustained oscillatory behavior in closed chemical systems due to thermodynamic reasons. For kinetic reasons, it is not possible to achieve sustained oscillatory behavior with first order reactions exclusively, but with pseudo first order reactions or reactions of higher order. One important part of the work of Higgins is the classification of oscillatory mechanisms in back-activation and forward inhibition mechanisms. This classification is still valid today but is now known by the terms *positive* and *negative feedback loops* [91]. For both mechanisms, Higgins gave possible descriptions by chemical reaction systems and ODEs.

Also today, there is no possibility to generally verify oscillations in a system of ODEs. The prevalent approach used to detect them is to integrate the equations numerically and to manually classify the solutions. If sustained oscillations are detected, it is easy to get the frequency or the amplitude by inspecting the solutions. To analyze the oscillatory behavior more deeply, Goldbeter [42] proposes five steps. Two of these steps deal with analyzing the structure of the ODE system. The third step is the evaluation of the steady state. A steady state of a system is a state vector that satisfies $\dot{X}(t) = 0$. Consequently, the steady states can be computed either analytically or numerically by solving the respective (possibly non-linear) equation system. To analyze the retrieved steady states, *linear stability analysis* is used [33]. Here, the aim is to find out whether a steady state is *stable*, i.e., whether after a small perturbation of the steady state, the system comes back. The key insight is that only in cases where a steady state is unstable, oscillations can occur. Technically, the Jacobian matrix for the ODE is set up and evaluated at the steady state(s) (cf. [33], page 881 for details). Finally, analysis of the eigenvalues $\lambda_1, \ldots, \lambda_K$ of the resulting matrix gives information about the stability. In general, if at least one eigenvalue λ_i has a positive real part, the steady state is unstable. If all eigenvalues have a negative real part, the steady state is stable. In all other cases, e.g. if some of the real parts are zero, further analysis is necessary. With the help of linear stability analysis it is possible to compute some general properties of the oscillating system. For example, *nullclines* are functions of one species that show under which conditions the value of the species does not change ($\dot{X}_n(t) = 0$). They can be illustrated by phase-plots of two or three species. It is now also possible to obtain the Hopf bifurcation points of the system [94]. In these points, at least one eigenvalue has to have a real part of zero. Based on Hopf bifurcation points, it is possible to obtain bifurcation diagrams. These are plots of a system parameter, that show under which conditions a steady state is stable or unstable. In the unstable region, the bifurcation diagram can also give insight into the amplitudes of obtained oscillations. For examples of bifurcation diagrams, we refer to [99]. We want to illustrate some of these analysis techniques in the following example.

Example 5 (Predator-Prey ODE). We continue with the predator-prey model from the previous examples. To obtain the steady states, we have to solve the equation system

$$0 = x_1 \cdot (c_1 - c_2 \cdot x_2), \quad 0 = x_2 \cdot (c_2 \cdot x_1 - c_3)$$

which can be done analytically. One steady state is $x_1 = c_3/c_2, x_2 = c_1/c_2$. The Jacobian of this steady state is $\mathbf{J} = [0 \ -c_3, c_1 \ 0]$, which leads to the eigenvalues $\lambda_{1,2} = \pm i \cdot (c_1 \cdot c_3)^{0.5}$. Since there are imaginary parts of the eigenvalues, the system oscillates. Since the real parts of the eigenvalues are zero for all parameters c_1, c_2, and c_3, these oscillations are sustained and are not ending in an attracting limit circle and it is not reasonable to set up a bifurcation diagram because it is not possible to compute Hopf bifurcation points. It is indeed necessary to have either an ODE system with more than two equations or at least one reaction of 3rd order in the reaction system to obtain an attracting limit cycle for an oscillation [91].

3.3 Oscillatory Behavior in the Stochastic Setting

One principal approach that is widely used when analyzing stochastic systems with respect to periodic and oscillatory behavior is *model checking*. The underlying mechanism is that the property of interest is encoded as a formula in some temporal logic or as some kind of (finite state) automaton and the model checking routine efficiently decides whether the model satisfies the property or not. Model checking based approaches reason about the structure of the models and allow precise statements that hold for sure, unlike simulative approaches which can not give such strong guarantees [12].

Temporal Logic Based Model Checking Approaches. In an early model checking approach used to analyze biochemical systems [22], the authors make use of *computation tree-logic* (CTL) [24]. For a detailed introduction to CTL, we refer to [9]. In the context of reasoning about qualitative aspects of biological models, the idea of requiring an infinite repetition of cyclic (non-constant) behavior was first described in [22]. The authors capture that property via the CTL formula

$$\exists \Box \left((P \Rightarrow \exists \Diamond \neg P) \wedge (\neg P \Rightarrow \exists \Diamond P) \right). \tag{5}$$

Formula (5) demands that there exists at least one path such that whenever some predicate P is satisfied it will be invalid later on, and vice versa whenever it is not satisfied it will become valid again in the future. Intuitively, there should exist at least one path such that the validity of P alternates forever. CTL model checking was originally applied to *labeled transition systems* (LTS), but a CTMC can be interpreted as a LTS as well by assuming a state transition whenever there is a positive rate between a state and its successor. Using such a construction, one can reason qualitatively about the possible behavior of a CTMC as done in [12] for example. Here, the authors propose the CTL formula

$$\forall \Box (((X_i = k) \Rightarrow \exists \Diamond (X_i \neq k)) \wedge ((X_i \neq k) \Rightarrow \exists \Diamond (X_i = k))), \tag{6}$$

in order to query whether a system shows *permanent oscillations*. This formula is similar to Formula (5) where the inner formula is strengthened by exchanging the outer $\exists$ by a $\forall$ operator and predicate P is instantiated with $X_i = k$.

The intuitive meaning of this formula is that *all* paths should cross a level of k molecules infinitely often.

The authors are concerned that noise could affect the fluctuations. Thus, they change the previous formula slightly to

$$\forall\square(((X_i = k) \Rightarrow \exists\lozenge\phi_n) \wedge (\phi_n \Rightarrow \exists\lozenge(X_i = k))) \tag{7}$$

with $\phi_n := (X_i > k + n) \vee (X_i < k - n)$. They call Formula (7) *noise filtered oscillations permanence* and intuitively, in contrast to the previous formula, the required crossing of molecule level k is extended to a crossing of the interval $[k - n, k + n]$ which resembles a noise band of size n around the desired level k.

So far, we have discussed how qualitative aspects like the permanence of oscillatory behavior are analyzed in the literature. The logic CTL proved to be expressive enough for this task but in order to be able to reason about quantities like the time needed for a fluctuation or the probability of a peak, CTL is ill-equipped. Consequently, approaches like [11,12] make use of the *continuous stochastic logic* (CSL) which lifts CTL to the continuous-time probabilistic setting, i.e., to continuous-time Markov chains. For a detailed introduction to CSL, we refer the reader to [8]. In [12], Ballarini et al. extend their qualitative approach by incorporating time and probability bounds.

$$\mathcal{P}_{=?}[\lozenge\,(\mathcal{P}_{\geq 1}[\lozenge(X_i = k)] \wedge \mathcal{P}_{\geq 1}[\lozenge(X_i \neq k)])] \tag{8}$$

More precisely, they use Formula (8) to compute for each state s, the probability p_s that oscillations will *not* terminate in the respective state. With the rest of the probability mass $1 - p_s$, the oscillatory pattern will end in (initial) state s since either a level of $X_i = k$ can not be reached or left any more with sufficient probability. Note that Formula (8) does not explicitly encode oscillatory behavior but is used implicitly to compute those states, where no oscillatory behavior will be possible any more. The authors further analyze the specific model structure of their case study (the 3-way oscillator discussed in Section 4.2) to define short-cut predicates $X_i = INV$ holding in states where oscillations terminate in species i. This way, the probability of termination of oscillation for any species within time T is captured by the formula[1]

$$\mathcal{P}_{=?}[\lozenge^{[0,T]}\,(X_1 = INV \vee X_2 = INV \vee \cdots \vee X_N = INV)]. \tag{9}$$

Moreover, computing the satisfaction probability of formula $\mathcal{S}_{=?}[X_i = k]$ for every molecule level k allows the authors to reason about the long-run probability distribution of molecules of type i. Finally, they use *rewards* introduced in CSLR [25,63], an extension of CSL, to query the expected perimeter of k molecules around the initial state resembling the *amplitude* of oscillation. We will not elaborate the construction in detail, since the described approach is tailored to the specific case study and can not be used for arbitrary models.

---

[1] In [12], only three short-cut predicates were used, we generalized the formula to N species.

In a follow-up paper [11], the authors base their qualitative characterization of oscillatory behavior on several notions of monotonicity. More precisely, a chemical species i is either monotonically increasing or decreasing indicated by boolean flags inc_i respectively dec_i, or nothing thereof. In order to relax the strict sense of monotonicity, increasing (decreasing) behavior might include up to a maximum of ns steps without an increase (decrease) in the number of molecules of species i, where ns denotes the chosen maximum extent of the noise band. For that, an additional model variable keeps track of the center of the current noise band. This additional information is provided by the augmentation of the model by a step counting automaton that keeps track of the noise band and the boolean flags. Technically, the PRISM model is composed in parallel with a module encoding the automaton and synchronizing on the chemical reactions to detect changes in the molecule counts. Finally, a CSL formula to query the probability that some species i increases monotonically (modulo noise) until some level k is reached is given by

$$\mathcal{P}_{=?}[inc\_i \ U \ (X_i = k)]. \tag{10}$$

Oscillatory behavior of a species i, more precisely, a single period, is then characterized by a monotonic increase from a current level j to some level $k > j$, followed by a monotonic decrease back to level j, where the noise band ns is used. The authors finally use (probabilistic) *linear temporal logic* (LTL) to formalize such an oscillation pattern of amplitude $j - 1$ via[2]

$$\mathcal{P}_{=?} \left[inc\_i \ U \ (X_i = k \wedge (dec\_i \ U \ X_i = j)) \right]. \tag{11}$$

The major difference between LTL and the branching time logics CTL and CSL is that it is a *linear time* logic, i.e., the semantics is based on the paths of a model in contrast to the states as in CTL/CSL. More precisely, although there is a path operator in CTL/CSL, the final judgment, whether a path formula is satisfied is done per state (by validating the path formula satisfaction probability against the probability bounds). As a consequence, path formulae can not be nested. Since the semantics of LTL is based on paths, nesting is possible. We will not give a full introduction to LTL but will discuss the intuitive meaning of the presented formulae and refer to [85] for details. Formula (11) queries the probability measure of all paths where species i is monotonically increasing until a level of k molecules is reached, followed by a monotonic decrease until level j.

The authors further relax the requirements of oscillatory behavior by allowing the fluctuation to stay at the peak, i.e., at around a molecule level of k (module noise band ns), for an unlimited amount of time as described by Formula (12)[3].

$$\mathcal{P}_{=?} \left[inc\_i \ U \ (X_i = k \wedge ((k - ns \leq X_i \leq k + ns) \ U \ (dec\_i \ U \ X_i = j))) \right] \tag{12}$$

---

[2] The original formula in [11] uses the W operator which behaves like the U operator but is also satisfied if the first sub-formula holds forever.

[3] Again, the original formula used the W operator instead of the U operator.

Automata Specification Based Model Checking Approaches. A different idea to capture the essence of oscillatory behavior is to encode the desired behavior as an automaton. One such approach is described in [77], where CTMCs with population structure and potentially infinite state spaces are model checked against single-clock *deterministic timed automata* (DTA) specifications. In the paper, DTAs are used to describe *linear-time* properties, i.e., the desired behavior corresponds to the set of paths that are accepted by the DTA. Consequently, the presented model checking approach computes the probability measure of all CTMC paths that are accepted by the DTA. In the paper, a DTA encoding oscillatory behavior is given. This DTA describes all trajectories, that cross a lower molecule threshold L, followed by reaching a higher threshold H and finally returning to the lower threshold L again. This fluctuation of *amplitude* $H - L$ must happen within a *period length* of t time units with $t \in [T_0^{\min}, T_0^{\max}]$. The real valued interval bounds $T_0^{\min} < T_0^{\max}$ can be chosen freely. We note that the requirements on oscillatory behavior are relaxed in the sense that not a specific period length but a range of possible period lengths is allowed. Accordingly, only a minimal but not maximal amplitude is specified. These relaxations were done in order to be able to deal with stochastic noise. Note that in contrast to the usual model checking based approaches, here, the state space does not necessarily has to be finite. The reason is that the model checking problem is reduced to the truncation based transient analysis [30,48] based on inexact vector matrix multiplication, which allows the routine to concentrate on *significant* states, i.e., states with a transient probability greater than some threshold $\delta > 0$.

3.4   Oscillatory Behavior in the Hybrid Setting

The hybrid approach is aimed at combining the advantages of the deterministic and noisy stochastic approaches. When we are dealing with pathways where there are species both with large and small molecular counts, the hybrid simulation or numerical solution methods are the methods of choice. Although the analysis is usually aimed at oscillations of species whose concentration is represented by continuous variables, we still can reason formally about the variety of complex behaviors that can arise. To do so we can represent the hybrid system as a timed automaton and apply the corresponding techniques [14]. It is important to notice that often we can apply deterministic methods to describe the mean behavior of large quantities modelled stochastically. However for some systems the effect of the noise can be so intense that the average behavior of stochastic components differs from the deterministic solution [47].

4   Applications

In this section, we would like to give various examples of oscillatory models with different underlying semantics and show how they are analyzed in the literature.

4.1   The Predator-Prey Model

At first, we will have a look at the predator-prey model from Example 1. In the work of Dayar et al. [28], the authors compare numerical analysis methods applied to the mentioned model. More precisely, they solve the system deterministically, stochastically and in a hybrid fashion and manually compare the results. The ODE corresponding to the deterministic approach (cf. Example 2) proves the system to possesses a never-ending oscillatory character according to the curve

$$c_2 \cdot S_1 - c_3 \cdot \log S_1 + c_2 \cdot S_2 - c_1 \cdot \log S_2 = \text{const}$$

in the phase space, assuming $c_1, c_2, c_3 > 0$ and an initial condition different from the equilibrium $[c_3/c_2 \; c_1/c_2]$. Intuitively, a peak in the prey population is followed by a peak in the predator population decreasing the prey and ultimately the predator population again, and the cycle repeats. However, when considering the system to be stochastic (cf. Example 3), i.e., treating predator and prey counts as such and not as concentrations, simulation shows that the mean of the resulting transient distribution behaves differently from the solution of the deterministic approach. Here, for the parameter set $c_1 = c_3 = 1$, $c_2 = 0.01$ and initial condition $\pi_{[20\ 20]}(0) = 1$, the oscillatory pattern already breaks after the first period. The number of predators goes to zero and the number of prey individuals grows unboundedly. The reason is that in general, the predators will die out eventually with probability one due to stochastic effects, allowing the prey population to proliferate. That case suggests that depending on the granularity, with which a system shall be analyzed, an appropriate semantics must be chosen. In contrast, when assuming the *thermo-dynamic limit*, i.e., taking the limit of the number of predator and prey individuals and of the system's volume to infinity (such that a fixed concentration is reached), the deterministic approach is justified. Intuitively, the probability of extinction approaches zero. If such an assumption can not be made, stochastic effects need to be considered. However, a problem of the fully stochastic approach is the high run time and memory requirements due to the large amount of individual states that need to be considered. Consequently, the authors suggest to use a hybrid approach as described in Section 2.4 with the additional property that populations are *dynamically* treated as individuals as long as they do not surpass a certain threshold. As soon as that threshold is reached, the respective population is treated deterministically, with the mean and covariances of S_i being computed for each time step. This helps to decrease run time and memory requirements by up to two magnitudes while still being able to capture the relevant stochastic effects with negligible error.

4.2   The 3-Way Oscillator Model

While the previous model has been analyzed mainly by solving the systems over time and manually reasoning about the resulting oscillatory behavior, another model, the *3-way oscillator* [19,20] has in addition been studied using fully automated model checking. The 3-way oscillator is a CRN which consists of three chemical species S_1, S_2, and S_3 which form a positive feedback loop as described by the chemical reactions

$$S_1 + S_2 \xrightarrow{\tau_1} 2 \cdot S_2, \quad S_2 + S_3 \xrightarrow{\tau_2} 2 \cdot S_3, \quad S_3 + S_1 \xrightarrow{\tau_3} 2 \cdot S_1$$

which resemble a cyclic predator-prey scheme, where species S_1 is eaten by species S_2, which is eaten by S_3, which itself is consumed by species S_1. For simplicity, we assume $\tau_1 = \tau_2 = \tau_3 = 1$ as in [12]. The corresponding ODE is

$$\dot{X}(t) = \alpha_1(\mathbf{x}) \cdot [-1 \ 1 \ 0]^T + \alpha_2(\mathbf{x}) \cdot [0 \ -1 \ 1]^T + \alpha_3(\mathbf{x}) \cdot [1 \ 0 \ -1]^T$$

with $\alpha_1(\mathbf{x}) = x_1 \cdot x_2$, $\alpha_2(\mathbf{x}) = x_2 \cdot x_3$, and $\alpha_3(\mathbf{x}) = x_1 \cdot x_2$. In [12], the authors use an initial concentration of $X(0) = [100 \ 200 \ 10]^T$ and conclude that the system oscillates forever. Using Matlab [73] to solve the ODE, one can compute the center of oscillation to be at around 133 with a peak-to-peak amplitude of 251 for each species. Also, the oscillation is regular corresponding to the mathematical notion of periodicity with a period length of around 0.05. However, when started in equilibrium, i.e. $x_1 = x_2 = x_3 = 10$, the system stays there forever [12].

In contrast to the deterministic semantics, a stochastic treatment results in different characteristics. First observations resulting from stochastic simulations starting in an initial state with $x_1 = x_2 = x_3 = 10$ show that in contrast to the ODE approach, fluctuations are present, although there is no strict regularity due to stochastic noise. Moreover, the authors argue that eventually, a terminal/absorbing state, i.e., a state with no outgoing transitions, corresponding to a depletion of two of the species will be reached. Consequently, the system will stop to oscillate since the state can not be left any more invalidating Formula (6) since either no state with $X_i = k$ or $X_i \neq k$ can be reached any more. That effect can be witnessed in the predator-prey model as well. Alongside to this argumentation, automated model checking reveals that the permanent oscillation property as stated in Formula (6) is not satisfied for that system. Using probabilistic model checking, further characteristics like the probability of depletion corresponding to the termination of oscillatory behavior within T time units as specified by Formula (9) are investigated. In order to be able to recover from a depletion, the authors suggest to allow the direct transformation of one species into another

$$S_1 \xrightarrow{\tau_1'} S_2, \quad S_2 \xrightarrow{\tau_2'} S_3, \quad S_3 \xrightarrow{\tau_3'} S_1,$$

usually with $\tau_i' \ll \tau_i$ for all $i \in \{1, 2, 3\}$ which results in no terminal states anymore. They call that augmentation *doping* and show that it indeed suffices to finally satisfy Formula (6), i.e., to make the oscillatory character permanent.

4.3   Circadian Clocks

An interesting phenomenon common to many living organisms is the day-night rhythm where the period length of 24 hours is kept nearly constant [87]. External stimuli like light and temperature might help to adjust the underlying mechanism called *circadian clock*. But in order to maintain a stable period length [13], the living organism should also cope with the external perturbations and internal stochastic noise (induced by species presented in small molecule numbers). In the paper [101], the authors model circadian clocks via the chemical reaction network described by

$$
\begin{array}{lll}
G_A \xrightarrow{\alpha_A} G_A + M_A & G_R \xrightarrow{\alpha_R} G_R + M_R & A \xrightarrow{\delta_A} \emptyset \\[4pt]
G_A \circ A \xrightarrow{\alpha'_A} G_A \circ A + M_A & G_R \circ A \xrightarrow{\alpha'_R} G_R \circ A + M_R & R \xrightarrow{\delta_R} \emptyset \\[4pt]
G_A + A \underset{\theta_A}{\overset{\gamma_A}{\rightleftharpoons}} G_A \circ A & G_R + A \underset{\theta_R}{\overset{\gamma_R}{\rightleftharpoons}} G_R \circ A & A + R \xrightarrow{\gamma_C} C \\[4pt]
M_A \xrightarrow{\beta_A} M_A + A & M_R \xrightarrow{\beta_R} M_R + R & C \xrightarrow{\delta_A} R \\[4pt]
M_A \xrightarrow{\delta_{M_A}} \emptyset & M_R \xrightarrow{\delta_{M_R}} \emptyset &
\end{array}
$$

where initially both genes are active ($G_A = G_R = 1$) and molecular count for the rest of species is 0. The parameters are set to α_A=50, α'_A=500, β_A=50, δ_{M_A}=10, δ_A=1, γ_A=1, θ_A=50, α_R=0.01, α'_R=50, β_R=5, δ_{M_R}=0.5, δ_R=0.2, γ_R=1, θ_R=100, γ_C=2. In that model, genes G_A and G_R are transcribed to mRNA molecules M_A and M_R which are finally translated into activator proteins A and repressor proteins R. The repressor protein R can bind to protein A to form a complex protein C which then degrades to a single R protein. The activator protein can bind to the promotor region of both genes to boost their transcription rate. The authors compare the integration of the corresponding ODE with stochastic simulations (cf. Figure 2 in [101]). The simulations show random fluctuations (noise) with respect to period length and amplitude in contrast to the ODE solution, where every period behaves the same. Using quasi-steady state assumptions [81], the deterministic system is reduced to the slow species R and C and it is possible to use limit cycle and stability analysis (cf. Section 3.2) to show that the system oscillates permanently. But, those results turn out not to be applicable to the stochastic setting. More precisely, the set of parameters giving rise to a stable steady state (fixed point) for the ODE semantics produce oscillations in the stochastic setting just like for the 3-way oscillator model (cf. Section 4.2). Here, the stochastic noise intuitively pushes the system away from the equilibrium to initiate a new cycle resulting in fluctuations not present in the deterministic limit. In addition, we would like to refer to the PRISM tutorial [83], which deals with stochastic simulation of the described model to show the oscillatory character.

Another example for circadian rhythms is the oscillatory character of the period protein (PER) in living organisms like *Drosophila*. Here, oscillations are induced by the negative feedback of PER on the transcription of the period gene and by the fact that PER has multiple phosphorylation sites. The minimal deterministic model capable of generating oscillations is provided in [41] by the system of 5 non-linear ODEs. The exact values of parameters and a detailed description of the reaction network are omitted here. Numerical integration shows that the system follows limit cycle oscillations for PER. It allows to investigate the influence of different factors onto the period length. The analysis shows that the sustained oscillations occur only when the maximum degradation rate of

PER and the rate of PER transport both fall into a certain region. This model is further developed in [67,68] to analyze circadian rhythms in *Neurospora* and *Drosophila* and incorporates the influence of external stimuli. The authors investigate how day-night light cycles as well as light pulses change the oscillatory properties, where the models account for different effects of light in both organisms. The goal is to study the relative contribution of various molecular processes to the oscillatory behavior. The possibility of chaotic behavior was revealed in both systems for a certain range of parameter values, however the corresponding physiological aspects are not clear yet. In [67], the authors enrich the model by a collection of additional components in order to describe the mammalian circadian clock. The resulting system is capable of generating the oscillatory cycles in the complete absence of light. The obtained results help to explain certain human physiological disorders associated with sleep. It is also noted that in such a complex system the mechanism giving rise to oscillatory behavior can be non-unique. Other refinements of this model and a comparison with experimental data for another biological species (*Arabidopsis*) can be found in [93,105].

An important comparison with the results of stochastic simulation was conducted in [43] under low copy numbers of mRNA and protein molecules. The comparison includes the initial deterministic model and two newly developed stochastic models. It studies the influence of molecular noise (controlled through the volume of the compartment) on circadian oscillations. These oscillations revealed to be quite robust with respect to the noise in both developed models. If the influence of stochastic fluctuations is small (when the volume of the modeled compartment is large) stochastic and deterministic models provide similar predictions. This fact motivates the usage of a deterministic model whenever it is a-priori known that molecular noise does not play an important role.

4.4 Calcium Oscillations

A well-known oscillatory process is the cellular calcium oscillation. Here, we will present the analysis of a deterministic model of this process, for a stochastic treatment, we refer to [61]. There are many more or less complex ODE models of calcium oscillations but we chose a simple model by Somogyi and Stucky [94]. The populations in this model are the calcium ions in different parts of a cell, i.e., S_1 denotes the ions in the endoplasm and S_2 the ions in the cytoplasm. x_1 and x_2 are the concentrations of S_1 and S_2. The model mostly describes transport processes from one of these regions to the other and the reactions are

$$\emptyset \xrightarrow{c_1} S_2, \quad S_2 \xrightarrow{c_2} S_1, \quad S_1 \xrightarrow{c_3 \cdot f(x_2)} S_2, \quad S_2 \xrightarrow{c_4} \emptyset$$

which describe the flow into the cytoplasm, transfer from cytoplasm to endoplasm, production of Ca^{2+} during the transport from endoplasm to the cytoplasm, and the outflow of ions from the cytoplasm[4]. The ODE system corresponding to the reactions is

---

[4] With $f(x_2) = x_2^2$. These reactions are similar to the reactions of the Brusselator [99] and the model in [94].

$$[\dot{x}_1, \quad \dot{x}_2] = [c_2 \cdot x_2 - c_3 \cdot f(x_2) \cdot x_1, \quad c_1 + c_3 \cdot f(x_2) \cdot x_1 - c_2 \cdot x_2 - c_4 \cdot x_2].$$

In order to analyze this system, the steady state has to be obtained. Here, we first compute the nullclines for both species as in

$$x_1 = \frac{c_2 \cdot x_2}{c_3 \cdot f(x_2)} \text{ for } (\dot{x}_1 = 0) \text{ and } x_1 = \frac{(c_2 + c_4) \cdot x_2 - c_1}{c_3 \cdot f(x_2)} \text{ for } (\dot{x}_2 = 0).$$

The intersection of both nullclines is the steady states of the system

$$x_1 = c_1 \cdot c_2 / (c_3 \cdot c_4 \cdot f(x_2)), \quad x_2 = c_1 / c_4.$$

The next steps are the computation of the Jacobian matrix in the steady state and the calculation of the trace of the matrix to get the real part of the eigenvalue which is

$$Re(\lambda) = c_2 \cdot c_4 / c_1 \cdot f'(x_2) - c_3 \cdot f(x_2) - c_2 - c_4.$$

With $Re(\lambda) = 0$ it is now possible to obtain the Hopf-Bifurcation points and to draw the bifurcation diagrams. For the Brusselator model such a diagram is given in Figure 1b in [99].

4.5 Other Applications

The presented examples are only a selection of models that exhibit oscillatory behavioral patterns. Another model is the *repressilator* [32], a chemical reaction network consisting of three chemical species mutually repressing each other, which has been studied mainly stochastically and which has already been implemented *in vivo*. Another biological system, which is analyzed in detail is the NF-κB signalling system [60]. This system is involved in several cellular processes and most of the analysis is based on ODEs. As shown above, the calcium oscillations can be used for intra-cellular communication. Another way to encode the information in extracellular communication as well is the cyclic AMP (cAMP) signaling. An example of such system is *Dictyostelium discoideum* studied in [72,98].

5   Conclusion

We have presented an overview of the literature on modeling and analyzing oscillatory population models. While the problem of characterizing such behavior for deterministic systems has been thoroughly studied, a unifying approach for stochastic models is not present so far. More precisely, there is no common formal agreement on what defines oscillatory behavior, neither how to define the respective period length and amplitude. Nevertheless, several approaches dealing with qualitative and quantitative aspects using manual and automated techniques have been discussed. Current work in progress by the authors tries to give a foundational definition of oscillatory character for stochastic systems and develops a fast numerical algorithm to approximate the distributions of period length for a chosen amplitude. A problem of the fully stochastic treatment is the high computational demand from the runtime and memory perspectives. However, several recent evidences indicate that stochastic effects are essential for certain systems to maintain a sustained oscillatory pattern. Consequently, hybrid techniques treating one part of the system stochastically while approximating the rest deterministically, have been developed to increase the scalability and the efficiency of the analysis. We finally illustrated the presented approaches by a large collection of examples.

Acknowledgments. This research has been partially funded by the Graduate School of Computer Science at Saarland University and the German Research Council (DFG) as part of the Cluster of Excellence on Multimodal Computing and Interaction at Saarland University and the Transregional Collaborative Research Center "Automatic Verification and Analysis of Complex Systems" (SFB/TR 14 AVACS).

References

1. de Alfaro, L., Roy, P.: Magnifying-lens abstraction for Markov decision processes. In: Damm, W., Hermanns, H. (eds.) CAV 2007. LNCS, vol. 4590, pp. 325–338. Springer, Heidelberg (2007)
2. Alfonsi, A., Cancès, E., Turinici, G., Di Ventura, B., Huisinga, W.: Exact simulation of hybrid stochastic and deterministic models for biochemical systems. Research Report RR-5435, INRIA (2004)
3. Alfonsi, A., Cancès, E., Turinici, G., Ventura, B.D., Huisinga, W.: Adaptive simulation of hybrid stochastic and deterministic models for biochemical systems. ESAIM: Proc., 14:1–14:13 (2005)
4. Alur, R., Dill, D.L.: A theory of timed automata. Theoretical Computer Science 126(2), 183–235 (1994)
5. Aris, R.: Prolegomena to the rational analysis of systems of chemical reactions. Archive for Rational Mechanics and Analysis 19, 81–99 (1965)
6. Arkin, A., Ross, J., McAdams, H.H.: Stochastic kinetic analysis of developmental pathway bifurcation in phage lambda-infected escherichia coli cells. Genetics 149(4), 1633–1648 (1998)
7. Arns, M., Buchholz, P., Panchenko, A.: On the numerical analysis of inhomogeneous continuous-time Markov chains. INFORMS Journal on Computing 22(3), 416–432 (2009)
8. Baier, C., Haverkort, B., Hermanns, H., Katoen, J.-P.: Model-checking algorithms for continuous-time Markov chains. IEEE Transactions on Software Engineering 29(6), 524–541 (2003)
9. Baier, C., Katoen, J.-P.: Principles of Model Checking. The MIT Press (2008)
10. Ball, K., Kurtz, T.G., Popovic, L., Rempala, G.: Asymptotic analysis of multiscale approximations to reaction networks. The Annals of Applied Probability 16(4), 1925–1961 (2006)
11. Ballarini, P., Guerriero, M.L.: Query-based verification of qualitative trends and oscillations in biochemical systems. Theoretical Computer Science 411(20), 2019–2036 (2010)

12. Ballarini, P., Mardare, R., Mura, I.: Analysing biochemical oscillation through probabilistic model checking. ENTCS 229(1), 3–19 (2009)
13. Barkai, N., Leibler, S.: Biological rhythms: Circadian clocks limited by noise. Nature 403, 267–268 (2000)
14. Bartocci, E., Corradini, F., Merelli, E., Tesei, L.: Model checking biological oscillators. ENTCS 229(1), 41–58 (2009)
15. Burrage, K., Hegland, M., Macnamara, S., Sidje, R.: A Krylov-based finite state projection algorithm for solving the chemical master equation arising in the discrete modeling of biological systems. In: Langville, A.N., Stewart, W.J. (eds.) Markov Anniversary Meeting 2006: An International Conference to Celebrate the 150th Anniversary of the Birth of A. A. Markov, pp. 21–38. Boston Books, Charleston (2006)
16. Burrage, K., Tian, T.: Poisson Runge-Kutta methods for chemical reaction systems. In: Lu, Y., Sun, W., Tang, T. (eds.) Advances in Scientific Computing and Applications, pp. 82–96. Science Press, Beijing (2004)
17. Burrage, K., Tian, T., Burrage, P.: A multi-scaled approach for simulating chemical reaction systems. Progress in Biophysics and Molecular Biology 85(2-3), 217–234 (2004)
18. Cao, Y., Gillespie, D.T., Petzold, L.R.: The slow-scale stochastic simulation algorithm. The Journal of Chemical Physics 122(1), 014116 (2005)
19. Cardelli, L.: Artificial biochemistry. Technical report, Microsoft Research (2006)
20. Cardelli, L.: Artificial biochemistry. In: Algorithmic Bioproceses. LNCS. Springer (2008)
21. Casagrande, A., Mysore, V., Piazza, C., Mishra, B.: Independent dynamics hybrid automata in systems biology. In: Proceedings of the First International Conference on Algebraic Biology, pp. 61–73. Universal Academy Press, Tokyo (2005)
22. Chabrier-Rivier, N., Chiaverini, M., Danos, V., Fages, F., Schächter, V.: Modeling and querying biomolecular interaction networks. Theoretical Computer Science 325, 25–44 (2003)
23. Chellaboina, V., Bhat, S., Haddad, W., Bernstein, D.: Modeling and analysis of mass-action kinetics. IEEE Control Systems Magazine 29(4), 60–78 (2009)
24. Clarke, E.M., Emerson, E.A.: Design and synthesis of synchronization skeletons using branching-time temporal logic. In: Logic of Programs, pp. 52–71. Springer, London (1982)
25. Cloth, L., Katoen, J.-P., Khattri, M., Pulungan, R.: Model-checking Markov reward models with impulse rewards. In: DSN, Yokohama (2005)
26. Crudu, A., Debussche, A., Radulescu, O.: Hybrid stochastic simplifications for multiscale gene networks. BMC Systems Biology 3(1), 89 (2009)
27. D'Argenio, P.R., Jeannet, B., Jensen, H.E., Larsen, K.G.: Reachability analysis of probabilistic systems by successive refinements. In: de Luca, L., Gilmore, S. (eds.) PAPM-PROBMIV 2001. LNCS, vol. 2165, pp. 39–56. Springer, Heidelberg (2001)
28. Dayar, T., Mikeev, L., Wolf, V.: On the numerical analysis of stochastic Lotka-Volterra models. In: IMCSIT, pp. 289–296 (2010)
29. de Jong, H.: Modeling and simulation of genetic regulatory systems: A literature review. Journal of Computational Biology 9(1), 67–103 (2002)
30. Didier, F., Henzinger, T.A., Mateescu, M., Wolf, V.: Fast adaptive uniformization of the chemical master equation. In: Proc., HIBI 2009, pp. 118–127. IEEE Computer Society, Washington, DC (2009)
31. Elowitz, M.B.: Stochastic gene expression in a single cell. Science 297(5584), 1183–1186 (2002)

32. Elowitz, M.B., Leibler, S.: A synthetic oscillatory network of transcriptional regulators. Nature 403(6767), 335–338 (2000)
33. Ferrell, J.E., Tsai, T.Y.-C., Yang, Q.: Modeling the cell cycle: Why do certain circuits oscillate? Cell 144(6), 874–885 (2011)
34. Gardner, T.S., Cantor, C.R., Collins, J.J.: Construction of a genetic toggle switch in Escherichia coli. Nature 403(6767), 339–342 (2000)
35. Gibson, M.A., Bruck, J.: Efficient exact stochastic simulation of chemical systems with many species and many channels. The Journal of Physical Chemistry A 104(9), 1876–1889 (2000)
36. Gillespie, D.T.: A general method for numerically simulating the stochastic time evolution of coupled chemical reactions. Journal of Computational Physics 22(4), 403–434 (1976)
37. Gillespie, D.T.: Exact stochastic simulation of coupled chemical reactions. The Journal of Physical Chemistry 81(25), 2340–2361 (1977)
38. Gillespie, D.T.: A rigorous derivation of the chemical master equation. Physica A 188, 404–425 (1992)
39. Gillespie, D.T.: Approximate accelerated stochastic simulation of chemically reacting systems. The Journal of Chemical Physics 115(4), 1716 (2001)
40. Glass, L., Beuter, A., Larocque, D.: Time delays, oscillations, and chaos in physiological control systems. Mathematical Biosciences 90(1-2), 111–125 (1988)
41. Goldbeter, A.: A model for circadian oscillations in the drosophila period protein (PER). Proceedings of the Royal Society B: Biological Sciences 261(1362), 319–324 (1995)
42. Goldbeter, A.: Computational approaches to cellular rhythms. Nature 420(6912), 238–245 (2002)
43. Gonze, D., Halloy, J., Goldbeter, A.: Deterministic versus stochastic models for circadian rhythms. Journal of Biological Physics 28(4), 637–653 (2002)
44. Grassmann, W.: Finding transient solutions in Markovian event systems through randomization. In: The First International Conference on the Numerical Solution of Markov Chains, pp. 375–385 (1990)
45. Griffith, M., Courtney, T., Peccoud, J., Sanders, W.H.: Dynamic partitioning for hybrid simulation of the bistable HIV-1 transactivation network. Bioinformatics 22(22), 2782–2789 (2006)
46. Gross, D., Miller, D.R.: The randomization technique as a modeling tool and solution procedure for transient Markov processes. Operations Research 32(2), 343–361 (1984)
47. Guerriero, M.L., Heath, J.K.: Computational modeling of biological pathways by executable biology. Methods in Enzymology 487, 217–251 (2011)
48. Henzinger, T.A., Mateescu, M., Wolf, V.: Sliding window abstraction for infinite Markov chains. In: Bouajjani, A., Maler, O. (eds.) CAV 2009. LNCS, vol. 5643, pp. 337–352. Springer, Heidelberg (2009)
49. Henzinger, T.A., Mikeev, L., Mateescu, M., Wolf, V.: Hybrid numerical solution of the chemical master equation. In: Proc., CMSB 2010, pp. 55–65. ACM, New York (2010)
50. Higgins, J.: The theory of oscillating reactions. Industrial and Engineering Chemistry 59, 18–62 (1967)
51. Horn, F., Jackson, R.: General mass action kinetics. ARMA 47, 81–116 (1972)
52. Horton, G., Kulkarni, V.G., Nicol, D.M., Trivedi, K.S.: Fluid stochastic Petri nets: Theory, applications, and solution techniques. European Journal of Operational Research 105(1), 184–201 (1998)

53. Lohmueller, J., et al.: Progress toward construction and modelling of a tri-stable toggle switch in e. coli. IET Synthetic Biology 1(1.2), 25–28 (2007)
54. Jahnke, T., Huisinga, W.: Solving the chemical master equation for monomolecular reaction systems analytically. Journal of Mathematical Biology 54(1), 1–26 (2006)
55. Jensen, A.: Markoff chains as an aid in the study of Markoff processes. Scandinavian Actuarial Journal 1953(suppl. 1), 87–91 (1953)
56. Kampen, N.V.: Stochastic processes in physics and chemistry. North Holland (2007)
57. Katoen, J.-P., Klink, D., Leucker, M., Wolf, V.: Three-valued abstraction for continuous-time Markov chains. In: Damm, W., Hermanns, H. (eds.) CAV 2007. LNCS, vol. 4590, pp. 311–324. Springer, Heidelberg (2007)
58. Kolmogoroff, A.: Über die analytischen Methoden in der Wahrscheinlichkeitsrechnung. Mathematische Annalen 104, 415–458 (1931)
59. Kowalewski, S., Engell, S., Stursberg, O.: On the generation of timed discrete approximations for continuous systems. MCMDS 6(1), 51–70 (2000)
60. Krishna, S.: Minimal model of spiky oscillations in NF-κb signaling. PNAS 103(29), 10840–10845 (2006)
61. Kummer, U., Krajnc, B., Pahle, J., Green, A.K., Dixon, C.J., Marhl, M.: Transition from stochastic to deterministic behavior in calcium oscillations. Biophysical Journal 89(3), 1603–1611 (2005)
62. Kurtz, T.G.: The Relationship between Stochastic and Deterministic Models for Chemical Reactions. The Journal of Chemical Physics 57(7), 2976–2978 (1972)
63. Kwiatkowska, M., Norman, G., Pacheco, A.: Model checking expected time and expected reward formulae with random time bounds. In: Proc. 2nd Euro-Japanese Workshop on Stochastic Risk Modelling for Finance, Insurance, Production and Reliability (2002)
64. Kwiatkowska, M., Norman, G., Parker, D.: Game-based abstraction for Markov decision processes. In: Proc. QEST, pp. 157–166. IEEE CS Press (2006)
65. Kwiatkowska, M., Norman, G., Parker, D.: PRISM 4.0: Verification of probabilistic real-time systems. In: Gopalakrishnan, G., Qadeer, S. (eds.) CAV 2011. LNCS, vol. 6806, pp. 585–591. Springer, Heidelberg (2011)
66. Lapin, M., Mikeev, L., Wolf, V.: SHAVE – Stochastic hybrid analysis of Markov population models. In: Proc. HSCC. ACM, New York (2011)
67. Leloup, J.C.: Toward a detailed computational model for the mammalian circadian clock. PNAS 100(12), 7051–7056 (2003)
68. Leloup, J.C., Gonze, D., Goldbeter, A.: Limit cycle models for circadian rhythms based on transcriptional regulation in drosophila and neurospora. Journal of Biological Rhythms 14(6), 433–448 (1999)
69. Lewis, J.: Autoinhibition with transcriptional delay: A simple mechanism for the zebrafish somitogenesis oscillator. Current Biology 13(16), 1398–1408 (2003)
70. Lotka, A.: Elements of mathematical biology. Dover Publications (1956); Reprinted from Lotka, A.J. Elements of physical biology (1924)
71. Maler, O., Batt, G.: Approximating continuous systems by timed automata. In: Fisher, J. (ed.) FMSB 2008. LNCS (LNBI), vol. 5054, pp. 77–89. Springer, Heidelberg (2008)
72. Martiel, J.-L., Goldbeter, A.: A model based on receptor desensitization for cyclic AMP signaling in dictyostelium cells. Biophysical Journal 52(5), 807–828 (1987)
73. MATLAB. Version 7.11.0.584 (R2010b). The MathWorks Inc., Natick, Massachusetts (2010)

74. McAdams, H.H., Arkin, A.: Stochastic mechanisms in gene expression. Proceedings of the National Academy of Sciences of the United States of America 94(3), 814–819 (1997)
75. Menz, S., Latorre, J.C., Schtte, C., Huisinga, W.: Hybrid stochastic–deterministic solution of the chemical master equation. MMS 10(4), 1232–1262 (2012)
76. Meyer, K.M., Wiegand, K., Ward, D., Moustakas, A.: SATCHMO: A spatial simulation model of growth, competition, and mortality in cycling savanna patches. Ecological Modelling 209(24), 377–391 (2007)
77. Mikeev, L., Neuhäußer, M.R., Spieler, D., Wolf, V.: On-the-fly verification and optimization of DTA-properties for large Markov chains. FMSD, 1–25 (2012)
78. Mincheva, M.: Oscillations in non-mass action kinetics models of biochemical reaction networks arising from pairs of subnetworks. Journal of Mathematical Chemistry 50(5), 1111–1125 (2011)
79. van Moorsel, A.P.A., Wolter, K.: Numerical solution of non-homogeneous Markov processes through uniformization. In: Proceedings of the 12th European Simulation Multiconference on Simulation - Past, Present and Future, pp. 710–717. SCS Europe (1998)
80. Munsky, B., Khammash, M.: The finite state projection algorithm for the solution of the chemical master equation. The Journal of Chemical Physics 124(4), 044104 (2006)
81. Murray, J.D.: Mathematical Biology. Springer, New York (1993)
82. Nakano, S., Yamaguchi, S.: Two modeling methods for signaling pathways with multiple signals using uppaal. Proc. BioPPN, 87–101 (2011)
83. Parker, D.: PRISM Tutorial - Circadian Clock, http://www.prismmodelchecker.org/tutorial/circadian.php
84. Piazza, C., Antoniotti, M., Mysore, V., Policriti, A., Winkler, F., Mishra, B.: Algorithmic algebraic model checking I: Challenges from systems biology. In: Etessami, K., Rajamani, S.K. (eds.) CAV 2005. LNCS, vol. 3576, pp. 5–19. Springer, Heidelberg (2005)
85. Pnueli, A.: The temporal logic of programs. In: Proc., SFCS, pp. 46–57. IEEE Computer Society, Washington, DC (1977)
86. Rao, C.V., Arkin, A.P.: Stochastic chemical kinetics and the quasi-steady-state assumption: Application to the Gillespie algorithm. The Journal of Chemical Physics 118(11), 4999 (2003)
87. Reppert, S.M., Weaver, D.R.: Coordination of circadian timing in mammals. Nature 418(6901), 935–941 (2002)
88. Salis, H., Kaznessis, Y.: Accurate hybrid stochastic simulation of a system of coupled chemical or biochemical reactions. The Journal of Chemical Physics 122(5), 54103 (2005)
89. Sanft, K., Gillespie, D., Petzold, L.: Legitimacy of the stochastic Michaelis Menten approximation. IET Systems Biology 5(1), 58 (2011)
90. Schivo, S., et al.: Modelling biological pathway dynamics with timed automata. In: BIBE, pp. 447–453. IEEE (2012)
91. Schuster, S., Marhl, M., Höfer, T.: Modelling of simple and complex calcium oscillations. European Journal of Biochemistry 269(5), 1333–1355 (2002)
92. Singh, A., Hespanha, J.P.: Stochastic hybrid systems for studying biochemical processes. Philosophical Transactions of the Royal Society A: Mathematical, Physical and Engineering Sciences 368(1930), 4995–5011 (2010)
93. Smolen, P., Hardin, P.E., Lo, B.S., Baxter, D.A., Byrne, J.H.: Simulation of drosophila circadian oscillations, mutations, and light responses by a model with VRI, PDP-1, and CLK. Biophysical Journal 86(5), 2786–2802 (2004)

94. Somogyi, R., Stucki, J.W.: Hormone-induced calcium oscillations in liver cells can be explained by a simple one pool model. Journal of Biological Chemistry 266(17), 11068–11077 (1991)

95. Spieler, D.: Model checking of oscillatory and noisy periodic behavior in Markovian population models. Technical report, Saarland University (2009), Master thesis available at http://mosi.cs.uni-saarland.de/?page_id=93

96. Steinfeld, J., Francisco, J., Hase, W.: Chemical kinetics and dynamics. Prentice Hall (1989)

97. Stiver, J.A., Antsaklis, P.J.: State space partitioning for hybrid control systems. In: American Control Conference, pp. 2303–2304. IEEE (1993)

98. Tang, Y., Othmer, H.G.: Excitation, oscillations and wave propagation in a G-protein-based model of signal transduction in dictyostelium discoideum. Philosophical Transactions of the Royal Society B: Biological Sciences 349(1328), 179–195 (1995)

99. Tyson, J.J.: Biological switches and clocks. Journal of the Royal Society Interface 5, S1–S8 (2008)

100. van Dijk, N.M.: Uniformization for nonhomogeneous Markov chains. Operations Research Letters 12(5), 283–291 (1992)

101. Vilar, J., Kueh, H.-Y., Barkai, N., Leibler, S.: Mechanisms of noise-resistance in genetic oscillators. PNAS 99(9), 5988–5992 (2002)

102. Volterra, V.: Fluctuations in the abundance of a species considered mathematically. Nature 118, 558–560 (1926)

103. Wagner, H., Möller, M., Prank, K.: COAST: controllable approximative stochastic reaction algorithm. The Journal of Chemical Physics 125(17), 174104 (2006)

104. Wolkenhauer, O., Ullah, M., Kolch, W., Cho, K.-H.: Modeling and simulation of intracellular dynamics: Choosing an appropriate framework. IEEE Transactions on Nanobioscience 3(3), 200–207 (2004)

105. Zeilinger, M.N., Farr, E.M., Taylor, S.R., Kay, S.A., Doyle, F.J.: A novel computational model of the circadian clock in arabidopsis that incorporates PRR7 and PRR9. Molecular Systems Biology 2 (2006)

A Tutorial on Interactive Markov Chains

Florian Arnold[1], Daniel Gebler[2],
Dennis Guck[1], and Hassan Hatefi[3]

[1] Formal Methods and Tools Group, Department of Computer Science
University of Twente, P.O. Box 217, 7500 AE Enschede, The Netherlands
[2] Department of Computer Science, VU University Amsterdam,
De Boelelaan 1081a, NL-1081 HV Amsterdam, The Netherlands
[3] Department of Computer Science
Saarland University, 66123 Saarbrücken, Germany

Abstract. Interactive Markov chains (IMCs) constitute a powerful stochastic model that extends both continuous-time Markov chains and labelled transition systems. IMCs enable a wide range of modelling and analysis techniques and serve as a semantic model for many industrial and scientific formalisms, such as AADL, GSPNs and many more. Applications cover various engineering contexts ranging from industrial system-on-chip manufacturing to satellite designs. We present a survey of the state-of-the-art in modelling and analysis of IMCs.

We cover a set of techniques that can be utilised for compositional modelling, state space generation and reduction, and model checking. The significance of the presented material and corresponding tools is highlighted through multiple case studies.

1 Introduction

The increasing complexity of systems and software requires appropriate formal modelling and verification tools to gain insights into the system's possible behaviour and dependability. Imagine the construction of a satellite equipped with hardware components and software systems. Once sent into its orbit, the satellite has to work mostly autonomously. In case of any hardware or software component failure, the required maintenance work is time-consuming and cannot be executed immediately, leading to excessive costs and even complete system failures. To avoid such shortcomings, the system's components need to be highly dependable and any design mistakes must be identified as soon as possible. Rigorous modelling and analysis techniques can help significantly by accompanying the development process from the blue-print to the testing phase. They can answer quantitative questions like "what is the probability that the system fails within 3 years" by synthesising an abstract system model.

In the last years a plethora of formalisms [45, 25, 55, 47, 35, 23] and tools (PRISM [43], ETMCC [39], MRMC [42], YMER [58], VESTA [56] and MAMA [27]) have been developed for this purpose. The advent of large-scale, distributed, dependable systems requires formal specification and verification methods that

A. Remke and M. Stoelinga (Eds.): ROCKS Autumn School 2012, LNCS 8453, pp. 26–66, 2014.
© Springer-Verlag Berlin Heidelberg 2014

capture both qualitative and quantitative aspects of systems. Labelled transition systems (LTS) allow to capture qualitative aspects of software and hardware systems by defining the interaction between system components, but they lack quantitative predicates. On the other hand, continuous time Markov chains (CTMC) allow to model and reason over quantitative aspects of systems. However, CTMCs do not allow to model component dependencies and interaction with the environment.

A prominent formalism to remedy these drawbacks are interactive Markov Chains (IMCs) [35]. IMCs conservatively extend LTSs and CTMCs and thereby allow to accurately model system dependencies as well as quantitative aspects. IMCs strictly separate between nondeterministic choices, called interactive transitions, and exponentially distributed delays, called Markovian transitions. Hence, they can be considered as an extension of CTMCs with nondeterminism or, the other way around, as enriched labelled transition systems. Interactive transitions, denoted as $s \xrightarrow{\alpha} s'$, allow to model actions that are executed in zero time and account for nondeterministic choices by the system's environment. They allow very efficient bisimulation minimisation since quotienting can be done in a compositional fashion. A system's progress over time can be modelled by Markovian transitions, denoted by $s \xrightarrow{\lambda} s'$. They indicate that the system is moving from state s to s' after a delay exponentially distributed with parameter λ, and thereby account for time dependencies between system states.

IMCs are closely related to continuous-time Markov decision processes (CTMDPs), but they are strictly more expressive. CTMDPs closely entangle nondeterminism and stochastic behaviour in their transition relation. The separation of nondeterministic and stochastic choices allows for a well-behaved and natural notion of composition and hierarchy. Recently, IMCs were extended to Markov automata (MA) [23] by adding the possibility of random switching to interactive transitions.

Recent works on model checking opened the door for far-reaching industrial applications. IMCs provide a strict formal semantics of modelling and engineering formalisms such as generalised stochastic Petri nets (GSPNs) [50], Dynamic Fault Trees (DFTs) [12], the Architectural Analysis and Design Language (AADL) [13], and STATEMATE [10]. The powerful compositional design and verification approach of IMCs is applied for instance to Globally Asynchronous Locally Synchronous (GALS) designs [19], supervisory control [48, 49], state-of-the-art satellite design [24], water-treatment facilities [34] and train control systems [10].

This paper aims to give an extensive introduction to IMCs and survey recent developments in the field. Therefore, we present a detailed description of the fundamentals of the IMC formalism. Besides, we introduce related concepts such as CTMDPs and describe their relationship to IMCs. An important aspect of IMCs is that they can be analysed with respect to certain properties. Therefore, we introduce a logic that is capable of specifying important properties like "is the system running at least 99% of the time?". Furthermore, we provide a rich set of model checking algorithms to efficiently compute and thus check these properties.

Especially for time-bounded reachability, expected time and long-run average properties, we give an in-depth description of the algorithms with accompanying examples. Another challenge in a model like IMCs is the state space explosion problem. The prevention of this is a major research topic and covered by this paper in terms of bisimulation minimisation. Therefore, we present the notion of strong and weak bisimulation, and provide an algorithm for computing the bisimulation quotient.

Organisation of the Paper. Section 2 introduces the model, semantics and compositional construction methods of IMCs. A survey on model checking techniques is provided in Section 3 and behavioural equivalences and abstraction are discussed in Section 4. Section 5 shows extensions of IMCs, Section 6 provides a number of case studies and applications, and Section 7 concludes the paper.

2 Preliminaries

This section summarises the basic notions and definitions to provide a formal underpinning of the concept of interactive Markov chains [35, 14] and related concepts. The interested reader can find more details in the referred material throughout this section.

Before we describe interactive Markov chains, we give a brief introduction to two widely used models which are related to them. We start with a discrete time and nondeterministic model, namely *Markov Decision Processes* (MDPs). They extend Markov chains by adding nondeterministic decisions.

Definition 1 (Markov Decision Process). *A Markov decision process (MDP) is a tuple $\mathcal{M} = (S, Act, \mathbf{P}, s_0)$ where S is a finite set of states, Act a finite set of actions, s_0 the initial state, and $\mathbf{P}: S \times Act \times S \to [0,1]$ the transition probability function such that $\sum_{s' \in S} \mathbf{P}(s, \alpha, s') \in \{0,1\}$ for all $s \in S$ and $\alpha \in Act$.*

MDPs are a well studied model with a wide range of efficient algorithms [53] for various types of analysis. Later on in this survey, we exploit some of those algorithms to solve problems on interactive Markov chains.

Unsurprisingly, CTMDPs are the extension of MDPs to continuous time and are closely related to IMCs.

Definition 2 (Continuous Time Markov Decision Process). *A CTMDP is a tuple $\mathcal{C} = (S, Act, \mathbf{R}, s_0)$ where S is a finite set of states, Act a finite set of actions, s_0 the initial state, and $\mathbf{R}: S \times Act \times S \to \mathbb{R}_{>0}$ a three dimensional rate matrix.*

A CTMDP is a stochastic nondeterministic model that describes the behaviour of a system in continuous time. The delay of each transition (s_1, α, s_2) is exponentially distributed with rate $\mathbf{R}(s_1, \alpha, s_2)$ for $s_1, s_2 \in S$ and $\alpha \in Act$. IMCs extend CTMDPs by breaking the tight connection between nondeterministic and stochastic behaviour.

2.1   Interactive Markov Chains

The Syntax of an IMC. IMCs are finite transition systems with action-labelled interactive transitions, as well as Markovian transitions that are labelled with a positive real number identifying the rate of an exponential distribution. Hence, they strictly separate between interactive and Markovian behaviour. This enables for a wide range of modelling features. On the one hand, based on the action-labelled interactive transitions, IMCs can be used for compositional modelling with intermittent weak bisimulation [35]. On the other hand, the Markovian transitions allow to encode arbitrary distributions in terms of acyclic phase-type distributions [52]. An in depth discussion of the advantages of the IMC formalism is given in [14].

Definition 3 (Interactive Markov Chain). *An* interactive Markov chain *is a tuple* $\mathcal{I} = (S, Act, \rightarrow, \dashrightarrow, s_0)$ *where S is a nonempty, finite set of states with initial state $s_0 \in S$, Act is a finite set of actions, $\rightarrow \subseteq S \times Act \times S$ is a finite set of* interactive *transitions and* $\dashrightarrow \subseteq S \times \mathbb{R}_{>0} \times S$ *is a finite set of* Markovian *transitions.*

We abbreviate $(s, \alpha, s') \in \rightarrow$ by $s \xrightarrow{\alpha} s'$ and $(s, \lambda, s') \in \dashrightarrow$ by $s \overset{\lambda}{\dashrightarrow} s'$. Let:
- $IT(s) = \{s \xrightarrow{\alpha} s'\}$ be the set of interactive transitions that leave s, and
- $MT(s) = \{s \overset{\lambda}{\dashrightarrow} s'\}$ be the set of Markovian transitions that leave s.

We denote with $MS \subseteq S$ the set of Markovian states, with $IS \subseteq S$ the set of interactive states and with $HS \subseteq S$ the set of hybrid states of an IMC $\mathcal{I}$, where:
- $s \in MS$ iff $MT(s) \neq \emptyset$ and $IT(s) = \emptyset$,
- $s \in IS$ iff $MT(s) = \emptyset$ and $IT(s) \neq \emptyset$, and
- $s \in HS$ iff $MT(s) \neq \emptyset$ and $IT(s) \neq \emptyset$.

Further, we distinguish *external* actions from *internal* τ-actions. Note that a labelled transition system (LTS) is an IMC with $MS = \emptyset$ and $HS = \emptyset$. Further, a continuous-time Markov chain (CTMC) is an IMC with $IS = \emptyset$ and $HS = \emptyset$. Therefore, IMCs are a natural extension of LTSs as well as CTMCs.

The Semantics of an IMC. A *distribution* μ over a countable set S is a function $\mu \colon S \longmapsto [0, 1]$ such that $\sum_{s \in S} \mu(s) = 1$. If $\mu(s) = 1$ for some $s \in S$, μ is a *Dirac* distribution, and is denoted μ_s. Let $Distr(S)$ be the set of all distributions over a set S. The interpretation of a Markovian transition $s \overset{\lambda}{\dashrightarrow} s'$ is that the IMC moves from state s to s' within d time units with probability $\int_0^d \lambda e^{-\lambda t} dt = (1 - e^{-\lambda \cdot d})$. For a state $s \in MS$, let $\mathbf{R}(s, s') = \sum \{\lambda \mid s \overset{\lambda}{\dashrightarrow} s'\}$ be the total rate to move from state s to s', and let $E(s) = \sum_{s' \in S} \mathbf{R}(s, s')$ be the total outgoing rate of s. If s has multiple outgoing Markovian transitions to different successors, then we speak of a race between these transitions, known as the *race condition*. In this case, the probability to move from s to s' within d time units is $\frac{\mathbf{R}(s,s')}{E(s)} \cdot (1 - e^{-E(s)d})$, utilising that the IMC moves to a successor state s' after a delay of at most d time units with discrete branching probability $\mathbf{P}(s, s') = \frac{\mathbf{R}(s,s')}{E(s)}$. As defined on CTMDPs [6], uniformity can also be adapted to IMCs [40]. An IMC is called *uniform* iff there exists an $e \in \mathbb{R}_{\geq 0}$ such that

$\forall s \in MS$ it holds that $E(s) = e$. Thus, the distribution of the sojourn time is the same for all Markovian states if the IMC is uniform.

IMCs are compositional, i. e. if a system comprises several IMC components, then it can be assembled via *parallel composition* of the components. The components can communicate through external actions visible to all of them, while internal τ-actions are invisible and cannot communicate with any other action. Instead of communication, we say in the following that two IMCs synchronize on an action. Consider a state $s \in HS$ with a Markovian transition with rate λ and a τ-labelled transition. We assume that the τ-transition takes no time and is fired immediately since it is not subject to any interaction and cannot be delayed. On the other hand, the probability that the Markovian transition fires immediately is zero. Thus, internal interactive transitions always take precedence over Markovian transitions.

Assumption 1 (Maximal Progress) *In any IMC, internal interactive transitions take precedence over Markovian transitions.*

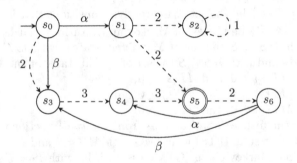

Fig. 1. An interactive Markov chain

Example 1. Let $\mathcal{I}$ be the IMC depicted in Figure 1. Then s_0 is a hybrid state with Markovian transition $s_0 \dashrightarrow^{2} s_3$ and interactive transitions $s_0 \xrightarrow{\alpha} s_1$ and $s_0 \xrightarrow{\beta} s_3$. We assume that all actions are no longer subject to any further synchronisation. W.l.o.g. we consider α and β as τ-transitions. Hence, we can apply the maximal progress assumption and obtain $s_0 \in IS$ with $s_0 \xrightarrow{\alpha} s_1$ and $s_0 \xrightarrow{\beta} s_3$. Therefore, in s_0 we have to choose between α and β. Since both transitions are fired without delay and take no precedence over each other, this choice has to be made nondeterministicly by a scheduler, see Section 2.3. The same holds for state s_6. If we choose β in s_0, then the successor state is s_3, which is a Markovian state with transition $s_3 \dashrightarrow^{3} s_4$ with rate $\lambda = 3$. The delay of this transition is exponentially distributed with parameter λ; thus, the transition fires in the next $z \in \mathbb{R}_{\geq 0}$ time units with probability $\int_0^z \lambda e^{-\lambda t} dt = (1 - e^{-3z})$. In case we choose α in s_0 we reach state s_1, which has two Markovian transitions.

We encounter a race condition, and the IMC moves along the transition whose delay expires first. Consequently, the sojourn time in s_1 is determined by the delay of the first transition that executes. The minimum of exponential distributions with parameters $\lambda_1, \lambda_2, \ldots$ is again exponentially distributed with the parameter $\lambda_1 + \lambda_2 + \cdots$. Thus, the sojourn time is determined by the exit rate, in our case we have $E(s_1) = 4$. The probability to move from a state $s \in MS$ to a successor state $s' \in S$ equals the probability that one of the outgoing Markovian transitions from s to s' wins the race. Therefore, the discrete branching probabilities for s_1 are given by $\mathbf{P}(s_1, s_2) = \mathbf{P}(s_1, s_5) = \frac{2}{4} = \frac{1}{2}$. ∎

2.2 Behavioural and Measurability Concepts

In this section we define fundamental concepts relating to the behaviour and the measurability of IMCs. We start with the definition of paths and then define the σ-algebra over the set of paths.

Paths. Like in other transition systems, an execution in an IMC is described by a path. We define finite and infinite paths and provide several useful notations and operators relating to paths. Before proceeding with the definition, for the uniformity of notation, we use a distinguished action $\bot \notin Act$ to indicate Markovian transitions and extend the set of actions to $Act_\bot = Act \cup \{\bot\}$. Formally, a finite path is an initial state followed by a finite sequence of interactive and Markovian transitions annotated with times, i. e.

$$\pi = s_0 \xrightarrow{t_0, \alpha_0} s_1 \xrightarrow{t_1, \alpha_1} \cdots s_{n-1} \xrightarrow{t_{n-1}, \alpha_{n-1}} s_n$$

with $\alpha_i \in Act_\bot$, $t_i \in \mathbb{R}_{\geq 0}$, $i = 0 \ldots n - 1$ and $s_0 \ldots s_n \in S$. Each step of a path π describes how the IMC evolves from one state to the next; with what action and after spending what sojourn time. For example, when the IMC is in an interactive state $s \in IS$ where only internal actions are enabled, it must immediately (in zero time) choose an enabled action α and go to state s'. This gives rise to the finite path $s \xrightarrow{0, \alpha} s'$. On the other hand, if $s \in MS$, the IMC stays in s for $t > 0$ time units and then moves to the next state s' based on the distribution $\mathbf{P}(s, \cdot)$ by $s \xrightarrow{t, \bot} s'$.

For a finite path π we use $|\pi| = n$ as the length of π and $\pi{\downarrow} = s_n$ as the last state of π. Assume $k \leq n$ is an index, then $\pi[k] = s_k$ is the $k + 1$-th state of π. Moreover, the time spent on π up to state $\pi[k]$ is denoted by $\Delta(\pi, k)$ which is zero if $k = 0$, and otherwise $\sum_{i=0}^{k-1} t_i$. We use $\Delta(\pi)$ as an abbreviation for $\Delta(\pi, |\pi|)$. For $t \leq \Delta(\pi)$, let $\pi@t$ denote the set of states that π occupies at time t. Note that $\pi@t$ is in general not a single state, but rather a set of states, as an IMC may exhibit immediate transitions and thus may occupy various states at the same time instant. Operator *Pref* extracts the prefix of length k from path π by $Pref(\pi, k) = s_0 \xrightarrow{t_0, \alpha_0} s_1 \cdots s_{k-1} \xrightarrow{t_{k-1}, \alpha_{k-1}} s_k$. By removing the sojourn time from transitions of path π, we obtain a time-abstract path denoted by $\mathsf{abs}(\pi) = s_0 \xrightarrow{\alpha_0} s_1 \xrightarrow{\alpha_1} \cdots s_{n-1} \xrightarrow{\alpha_{n-1}} s_n$. Furthermore, $Paths^n$ refers to the set of all paths with length n and $Paths^\star$ to the set of all finite paths. In this

Table 1. An example derivation of $\pi@t$ for IMCs

$t \leq \Delta(\pi, i)$	0	1	2	3	4	5	6	7	$\min j$	$\max j$	$\pi@t$
0	✓	✓	✓	✓	✓	✓	✓	✓	0	3	$\langle s_0 s_1 s_2 s_3 \rangle$
$t_3 - \epsilon$	×	×	×	×	✓	✓	✓	✓	4	-	$\langle s_3 \rangle$
t_3	×	×	×	×	✓	✓	✓	✓	4	5	$\langle s_4 s_5 \rangle$
$t_3 + \epsilon$	×	×	×	×	×	×	✓	✓	6	-	$\langle s_5 \rangle$
$t_3 + t_5$	×	×	×	×	×	×	✓	✓	6	7	$\langle s_6 s_7 \rangle$

context, we add subscript abs for the set of time-abstract paths i.e. $Paths^n_{\mathsf{abs}}$ and $Paths^\star_{\mathsf{abs}}$. A (possibly time-abstract) path could be infinite which means it is constructed by an infinite sequence of (time-abstract) transitions. Accordingly, we use $Paths^\omega$ ($Paths^\omega_{\mathsf{abs}}$) to refer to the set of all (time-abstract) infinite paths.

Example 2. Consider the path

$$\pi = s_0 \xrightarrow{0, \alpha_0} s_1 \xrightarrow{0, \alpha_1} s_2 \xrightarrow{0, \alpha_2} s_3 \xrightarrow{t_3, \perp} s_4 \xrightarrow{0, \alpha_4} s_5 \xrightarrow{t_5, \perp} s_6 \xrightarrow{0, \alpha_6} s_7.$$

Let $0 < \epsilon < \min\{t_3, t_5\}$. The derivations for the sequence $\pi@0$, $\pi@(t_3 - \epsilon)$, $\pi@(t_3)$, $\pi@(t_3 + \epsilon)$ and $\pi@(t_3 + t_5)$ are depicted in Table 1, where ✓ indicates that $t \leq \Delta(\pi, i)$, and × denotes the states where $t > \Delta(\pi, i)$. Further, $\min j$ describes the minimum path length and $\max j$ the maximum path length such that $t \leq \Delta(\pi, j)$. Hence, with $\min j$, $\pi[j]$ describes the first state on path π for the sequence $\pi@t$, respectively for $\max j$ the last state. ■

σ-*Algebra.* Here we recall the definition of σ-algebra for IMCs as described in [40, 50]. First we recapitulate the concept of *compound transitions*. A compound transition is a triple of (t, α, s), which describes the behaviour of the IMC when it waits in its current state for t time units then takes action α and finally evolves to the next state s. The set of all compound transitions over action space Act and state space S is denoted by $CT = \mathbb{R}_{\geq 0} \times Act_\perp \times S$. As a path in IMCs is composed of a sequence of compound transitions originating from an initial state, first we define a σ-algebra over compound transitions and then extend it over finite and infinite paths. Let $\mathfrak{F}_S = 2^S$ and $\mathfrak{F}_{Act_\perp} = 2^{Act_\perp}$ be σ-algebras over S and $Act_\perp$, respectively. We define the σ-algebra over compound transitions using the concept of Cartesian product of a collection of σ-algebras [4], as $\mathfrak{F}_{CT} = \sigma(\mathfrak{B}(\mathbb{R}_{\geq 0}) \times \mathfrak{F}_{Act_\perp} \times \mathfrak{F}_S)$, where $\mathfrak{B}(\mathbb{R}_{\geq 0})$ is the Borel σ-algebra over non-negative reals. Furthermore, it can be extended to the σ-algebra over finite paths using the same technique as follows. Let $\mathfrak{F}_{Paths^n} = \sigma(\mathfrak{F}_S \times \prod_{i=1}^n \mathfrak{F}_{CT})$ be the σ-algebra over finite paths of length n, then the σ-algebra over finite paths is defined as $\mathfrak{F}_{Paths^\star} = \cup_{i=0}^\infty \mathfrak{F}_{Paths^n}$. The σ-algebra over infinite paths is defined using the standard cylinder set construction [4]. We define the cylinder set of a given base $B \in \mathfrak{F}_{Paths^n}$ as $Cyl(B) = \{\pi \in Paths^\omega : Pref(\pi, n) \in B\}$. $Cyl(B)$ is measurable if its base B is measurable. The σ-algebra over infinite paths, $\mathfrak{F}_{Paths^\omega}$, is therefore the smallest σ-algebra over measurable cylinders. Finally the σ-algebra over the set of paths is the disjoint union of the σ-algebras over the finite paths and the infinite paths.

2.3   Schedulers

In states with more than one outgoing interactive transition the choice of the transition to be taken is nondeterministic, just as in the LTS setting. This nondeterminism is resolved by schedulers. Different classes of schedulers exist in order to resolve nondeterminism for different kinds of objectives. The most general scheduler class maps a finite path to a distribution over the set of interactive transitions that are enabled in the last state of the path:

Definition 4 (Generic Measurable Scheduler [40]). *A generic scheduler over IMC $\mathcal{I} = (S, Act, \longrightarrow, \dashrightarrow, s_0)$ is a function, $D\colon Paths^\star \rightarrowtail Distr(\longrightarrow)$, where the support of $D(\pi)$ is a subset of $(\{\pi\downarrow\} \times Act \times S) \cap \longrightarrow$ and $\pi\downarrow \in IS$. A generic scheduler is measurable iff for all $T \subseteq \longrightarrow$, $D(\cdot)(T)\colon Paths^\star \rightarrowtail [0,1]$ is measurable.*

For a finite path π ending in an interactive state, a scheduler specifies how to resolve nondeterminism by defining a distribution over the set of enabled transitions of $\pi\downarrow$. Measurability of scheduler D means that it never resolves nondeterminism in a way that induces a set of paths that is not measurable, i.e. $\{\pi \mid D(\pi)(T) \in B\} \in \mathfrak{F}_{Paths^\star}$ for all $T \subseteq \longrightarrow$ and $B \in \mathfrak{B}([0,1])$, where $\mathfrak{B}([0,1])$ is the Borel σ-algebra over interval $[0,1]$. We use the term $\mathcal{GM}$ to refer to the set of all generic schedulers. Since schedulers in IMCs are closely related to schedulers in CTMDPs, most of the concepts are directly applied from the latter to the former. A slight difference is that schedulers in IMCs resolve nondeterminism only for finite paths that end up in interactive states.

A variety of scheduler classes in CTMDPs [40, 50], which can also be employed in IMCs, has been proposed in order to resolve nondeterminism for different kinds of objectives. These schedulers are classified according to the level of time and history details they use to resolve nondeterminism. Another criterion is whether they are *deterministic*, i.e. the distribution over the set of target transitions is Dirac, or *randomised*. In *history-dependent* schedulers the resolution of nondeterminism on an interactive state may depend on the path is visited upto the state. A scheduler is *hop counting* if all finite paths with the same length lead to the same resolution of nondeterminism. It is *positional* if its decision for a given path is only made based on the last state of the path. On the other hand, schedulers can be *time-dependent*, *total time-dependent* or *time-abstract*. Time-dependent schedulers utilise the whole timing information of a path including the sojourn time of all intermediate states for resolution of nondeterminism, while total time-dependent schedulers only employ the total time that has elapsed to reach the current state for that purpose. No timing information is used by time-abstract schedulers and a path is thus considered time-abstract by them.

The most general class, $\mathcal{GM}$ schedulers, uses the complete trajectory up to the current interactive state to randomly determine the next state. Therefore, they are also called *time- and history-dependent randomised* (THR) schedulers. The class has an excessive power which is not necessary for most types of analysis. For example, for time-abstract criteria like expected reachability, long-run average and unbounded reachability, it suffices to consider *time-abstract positional deterministic* (TAPD) schedulers [27], which are also called *stationary deterministic*.

Table 2. Randomised scheduler classes for IMCs. The classification criteria are denoted by TA (Time-Abstract), TT (Total Time-dependent), T (Time-dependent), P (Positional), HOP (HOP counting) and H (History-dependent).

		Abbreviation	Scheduler Signature	Parameters of Scheduler for a given path π
TA	P	TAPR	$D : IS \rightarrowtail Distr(\longrightarrow)$	$\pi{\downarrow} \in IS$
	HOP	TAHOPR	$D : IS \times \mathbb{N} \rightarrowtail Distr(\longrightarrow)$	$\pi{\downarrow} \in IS, |\pi|$
	H	TAHR	$D : Paths^\star_{abs} \rightarrowtail Distr(\longrightarrow)$	$abs(\pi)$ with $\pi{\downarrow} \in IS$
TT	P	TTPR	$D : IS \times \mathbb{R}_{\geq 0} \rightarrowtail Distr(\longrightarrow)$	$\pi{\downarrow} \in IS, \Delta(\pi)$
	HOP	TTHOPR	$D : IS \times \mathbb{N} \times \mathbb{R}_{\geq 0} \rightarrowtail Distr(\longrightarrow)$	$\pi{\downarrow} \in IS, |\pi|, \Delta(\pi)$
	H	TTHR	$D : Paths^\star_{abs} \times \mathbb{R}_{\geq 0} \rightarrowtail Distr(\longrightarrow)$	$abs(\pi)$ with $\pi{\downarrow} \in IS, \Delta(\pi)$
T	P	TPR	$D : IS \times \mathbb{R}_{\geq 0} \rightarrowtail Distr(\longrightarrow)$	$\pi{\downarrow} \in IS, \Delta(\pi, |\pi|) - \Delta(\pi, |\pi| - 1)$
	H	THR $(\mathcal{GM})$	$D : Paths^\star \rightarrowtail Distr(\longrightarrow)$	π with $\pi{\downarrow} \in IS$

Furthermore, the optimal scheduler for computing time-bounded reachability probabilities is *total time-dependent positional deterministic* (TTPD) [50]. More classes of randomised schedulers are depicted in Table 2. The deterministic version of each class can be obtained under the assumption that $Distr(\longrightarrow)$ is Dirac.

Example 3. We define a scheduler over the IMC in Figure 1, which always chooses action α in state s_0 with probability 1. In addition, it selects α and β in state s_6 with probability p and $1 - p$, respectively, provided that a path in the set $A(T_1, T_5) = \{s_0 \xrightarrow{0,\alpha} s_1 \xrightarrow{t_1,\perp} s_5 \xrightarrow{t_5,\perp} s_6 : t_1 < T_1 \wedge t_5 < T_5\}$ has been observed. Otherwise, action β (in state s_6) is almost surely picked. Assume that $p = 0.5$, $T_1 = 1$ and $T_5 = 3$, then the scheduler is in the THR $(\mathcal{GM})$ class. It becomes deterministic (THD) by setting $p = 1$ or $p = 0$. By taking $p = 1$, $T_1 = \infty$ and $T_5 = \infty$, $A(T_1, T_5)$ becomes time-abstract and the scheduler is then time-abstract history-dependent deterministic (TAHD). On the other hand when $A(T_1, T_5)$ is replaced by the set $B = \{\pi \in Paths^\star : \pi{\downarrow} = s_6 \wedge \Delta(\pi, |\pi|) \leq 4\}$, the scheduler is total time-dependent and positional deterministic (TTPD) or randomised (TTPR), depending on the value of p. ∎

2.4 Probability Measures

The model induced by an IMC after the nondeterministic choices are resolved by a scheduler is pure stochastic and then can be analysed. To that end the unique probability measure [40, 50] for probability space $(Paths^\omega, \mathfrak{F}_{Paths^\omega})$ is proposed. Given a state s, a general measurable scheduler D and a set Π of infinite paths, then $\Pr_{s,D}(\Pi)$ denotes the probability of visiting all paths in Π under scheduler D starting from state s. We omit the details due to lack of space.

Zenoness. Due to the presence of immediate state changes, an IMC might exhibit *Zeno behaviour*, where infinitely many interactive transitions are taken in finite time. This is an unrealistic phenomenon characterised by paths π, where $\Delta(\pi, n)$ for $n \to \infty$ does not diverge to ∞. In other words, the time spent in the system may stop increasing if the system follows path π. Accordingly, an IMC $\mathcal{I}$ with initial state s_0 is non-Zeno, if for all schedulers D, $\Pr_{s_0,D}(\{\pi \in \mathit{Paths}^{\omega} \mid \lim_{n \to \infty} \Delta(\pi, n) < \infty\}) = 0$. As the probability of a Zeno path in a finite CTMC is zero [5], IMC $\mathcal{I}$ is non-Zeno, if and only if no strongly connected component with states $T \subseteq IS$ is reachable from s_0. In the remainder of this paper we restrict to models without zenoness.

2.5   Composition

Compositionality is one of the key properties of IMCs. Complex models consisting of various interacting IMCs can be aggregated in a stepwise manner. This allows e. g. to model each subsystem separately and obtain a model of the whole system by applying the following parallel composition.

Definition 5 (Parallel Composition). *Let $\mathcal{I}_1 = (S_1, \mathit{Act}_1, \longrightarrow_1, \dashrightarrow_1, s_{0,1})$ and $\mathcal{I}_2 = (S_2, \mathit{Act}_2, \longrightarrow_2, \dashrightarrow_2, s_{0,2})$ be IMCs. The parallel composition of $\mathcal{I}_1$ and $\mathcal{I}_2$ wrt. synchronisation set $\mathit{Syn} \subseteq (\mathit{Act}_1 \cap \mathit{Act}_2) \setminus \{\tau\}$ of actions is defined by:*

$$\mathcal{I}_1 \| \mathcal{I}_2 = (S_1 \times S_2, \mathit{Act}_1 \cup \mathit{Act}_2, \longrightarrow, \dashrightarrow, (s_{0,1}, s_{0,2}))$$

where $\longrightarrow$ and $\dashrightarrow$ are defined as the smallest relations satisfying

1. $s_1 \xrightarrow{\alpha}_1 s_1'$ *and* $s_2 \xrightarrow{\alpha}_2 s_2'$ *and* $\alpha \in \mathit{Syn}$, $\alpha \neq \tau$ *implies* $(s_1, s_2) \xrightarrow{\alpha} (s_1', s_2')$
2. $s_1 \xrightarrow{\alpha}_1 s_1'$ *and* $\alpha \notin \mathit{Syn}$ *implies* $(s_1, s_2) \xrightarrow{\alpha} (s_1', s_2)$ *for any* $s_2 \in S_2$
3. $s_2 \xrightarrow{\alpha}_2 s_2'$ *and* $\alpha \notin \mathit{Syn}$ *implies* $(s_1, s_2) \xrightarrow{\alpha} (s_1, s_2')$ *for any* $s_1 \in S_1$
4. $s_1 \dashrightarrow^{\lambda}_1 s_1'$ *implies* $(s_1, s_2) \dashrightarrow^{\lambda} (s_1', s_2)$ *for any* $s_2 \in S_2$
5. $s_2 \dashrightarrow^{\lambda}_2 s_2'$ *implies* $(s_1, s_2) \dashrightarrow^{\lambda} (s_1, s_2')$ *for any* $s_1 \in S_1$.

The two IMCs have to synchronise on actions in *Syn*, i. e. any action $\alpha \in \mathit{Syn}$ needs to be performed by both IMCs at the same time, except if α is an internal action (first condition). The second and third conditions state that any other action can be performed autonomously by any of the two IMCs. According to the last two conditions, Markovian transitions are interleaved independently. This is justified by the memoryless property of the annotated exponential distributions.

Given a set of IMCs B which need to be synchronised, the computational effort of the composition process is crucially dependent on the order in which these IMCs are aggregated. Crouzen and Hermanns [20] suggested an algorithm based on heuristics to determine a composition order which induces low computing costs. In a first step, the algorithm determines candidate subsets of B up to a certain size. For each subset a metric is calculated which estimates how good the composition of the IMCs in this subset is in keeping the cost of the overall composition low. The IMCs in the subset with the maximal metric are then composed and minimised, as described in Section 4. This process iterates until only one IMC remains in B.

The composition of two or more IMCs involves two steps: After synchronisation on a set of actions, those actions which require no further synchronisation are hidden.

Definition 6 (Hiding). *The* hiding *IMC* $\mathcal{I} = (S, Act, \rightarrow, \dashrightarrow, s_0)$ *wrt. the set A of actions is the IMC* $\mathcal{I} \backslash A = (S, Act \backslash A, \rightarrow', \dashrightarrow, s_0)$ *where* $\rightarrow'$ *is the smallest relation defined by*

1. $s \xrightarrow{\alpha} s'$ and $\alpha \notin A$ implies $s \xrightarrow{\alpha}' s'$
2. $s \xrightarrow{\alpha} s'$ and $\alpha \in A$ implies $s \xrightarrow{\tau}' s'$

Through hiding, interactive transitions annotated with actions in A are transformed into τ-transitions. Further, we distinguish between two classes of IMCs:

- *closed* IMCs, where all interactive transitions are hidden, such that the IMC is not subject to any further synchronisation, and
- *open* IMCs, which still have visible interactive transitions, and can interact with other IMCs.

As we will see next, closed IMCs are closely related to CTMDPs.

2.6   IMCs versus CTMDPs

The modelling of a system usually involves the composition of various communicating subsystems. Therefore, open IMCs are used to describe those subsystems. Once all open IMCs are composed to a single closed IMC, it is subject to analysis. Note that a CTMDP combines the two transition relations of an IMC in one transition rate matrix. We recapitulate a transformation from an IMC to a CTMDP [40, 38, 37] which preserves important properties of the original IMC, and thus can be used to apply CTMDP analysis techniques [16, 6] on the transformed model.

IMC vs CTMDP. In general, closed IMCs are a generalisation of CTMDPs in which interactive and Markovian transitions are loosely coupled. Therefore, every CTMDP can be converted into an equivalent IMC in a straightforward way. The equivalent IMC is contained in a restricted subclass called *strictly alternating IMCs* that behaves exactly like CTMDPs. Note that in a strictly alternating IMC, Markovian and interactive transitions are exhibited in a strict alternation. The idea of the transformation from an IMC to a CTMDP [40] is to convert a given IMC to a strictly alternating IMC which is essentially a CTMDP.

Given an IMC $\mathcal{I}$, the following steps [38] are applied: (1) obtain an *alternating IMC* by transformation of hybrid states into interactive states, (2) turn all successors of any Markovian state into interactive states to obtain a *Markov*

Alternating IMC, (3) transform any immediate successor of all interactive states into Markovian states to obtain an *Interactive Alternating IMC*. By employing these transformation steps, an arbitrary IMC turns into a strictly alternating IMC. The strictly alternating IMC can then be transformed into a corresponding CTMDP in a straightforward way. Here we explain each step by an example.

Alternating IMC. In the first step, IMC $\mathcal{I}$ is transformed into an alternating IMC which does not contain any hybrid state. Owing to closeness of the IMC and imposing Assumption 1 interactive transitions take precedence over Markovian transitions. Hence all emanating Markovian transitions of a hybrid state can be safely eliminated.

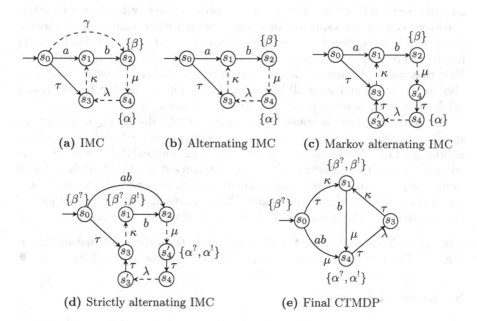

(a) IMC   **(b)** Alternating IMC   **(c)** Markov alternating IMC

(d) Strictly alternating IMC   **(e)** Final CTMDP

Fig. 2. Step by step transformation of an IMC into a CTMDP

Markov Alternating IMC. The aim of the second step is to make sure that predecessors and successors of any Markov state are interactive. In this step, a fresh interactive state with internal action τ is inserted in between two consecutive Markovian states. Due to immmediate firing of the τ transition, the timing behaviour of the IMC is preserved.

Example 4. The state-labelled IMC (see Section 3) in Figure 2a is closed and subject to analysis. The result of the first two steps of the transformation, namely the alternating IMC and the Markov alternating IMC, are illustrated in Figures 2b and 2c, respectively. ∎

Strictly Alternating IMC. After this step we make sure that there is no sequence of interactive transitions, therefore each interactive state is preceded and
succeeded by Markovian states. As discussed earlier, a sequence of consecutive
interactive transitions occur in zero time and thus can be seen as a single transition labelled by a word of taken actions. Note that the sequence always ends in a
Markovian state. There are interactive states in between that have only outgoing
and incoming interactive transitions, which are eliminated from the state space.
We call those states *vanishing*, and all others *persistent*.

The above transformation is not enough to reconstruct all information from
the original model. In order to preserve the semantic structure of the model
after eliminating vanishing states, their state labels (atomic propositions) must
be adopted by persistent states. In this regard, state labels are decorated with
an extra *may* and/or *must* tag. In other words, if starting from an interactive
persistent state s, all sequences of interactive transitions ending in Markovian
states visit label α, then s will be labelled by $\alpha^!$ (s *must* satisfy α). On the other
hand, if there exists such a sequence, s will be labelled by $\alpha^?$ (s *may* satisfy α).
Note that *must* labelling implies *may* labelling, as a label that must occur, may
also occur. At the end since all labelling information is inherited by interactive
persistent states, labels of other states will be removed.

An alternating IMC is transformed into a strictly alternating one after the
specified Markov and interactive alternating steps are applied. Since in a strictly
alternating IMC, Markovian and interactive transitions exhibit in a strict alternation, the strictly alternating IMC can be interpreted as a CTMDP. It has been
proven [38, 40] that the above transformation steps preserve the uniformity and
timed reachability of the original model. The transformation is a crucial part of
the evaluation of STATEMATE models as will be discussed in Section 6.2.

Example 5. The result of the transformation into the strictly alternating IMC is
shown in Figure 2d and the transformed CTMDP is illustrated in Figure 2e.   ∎

3 Model Checking

Consider we are confronted with a IMC originated from some high level-formalism
and a performability requirement. How can one describe this performability property and then compute the set of satisfying states in the IMC? First of all we need
a logic representing the desired property. Then the basic computational procedure
of the satisfaction set is a simple recursive descent of the logical formulae.

In this section we provide an overview of the current model checking capabilities of IMCs to provide an answer to the preceded question. We first introduce a
logic which is used to specify a wide range of properties and thereafter describe
algorithms to check those properties for IMCs.

3.1 Continuous Stochastic Logic

This section describes Continuous Stochastic Logic [5] (CSL), which is suitable
to express a broad range of performance and dependability measures. CSL is an

extension of Probabilistic Computation Tree Logic (PCTL) [30, 9] to continuous-time Markov models. This section reviews CSL and its related model checking algorithms as introduced in [59, 50] and enriches it with expected reachability and long-run average operators as described in [27]. CSL works on *state-labelled* IMCs.

Definition 7 (State-Labelled IMC). *A state-labelled IMC is a tuple* $\mathcal{I} = (S, Act, \rightarrow, \dashrightarrow, s_0, L)$ *where* $L : S \rightarrowtail 2^{AP}$ *is a* state labelling function *with* AP *as a set of* atomic propositions. *All other elements are as in Definition 3.*

Hence, given an IMC $\mathcal{I}$ and a finite set of *atomic propositions* AP, a *state labelling function* $L : S \rightarrowtail 2^{AP}$ decorates each state with a set of atomic propositions which do hold in that state.

Syntax of CSL. Let $\mathfrak{I}$ be the set of all nonempty nonnegative real intervals with real bounds, then Continuous Stochastic Logic (CSL) for IMCs is defined as follows.

Definition 8 (CSL Syntax). *Let* $a \in AP$, $p \in [0,1]$, $t \in \mathbb{R}_{\geq 0}$, $I \in \mathfrak{I}$ *an interval and* $\trianglelefteq \in \{<, \leq, \geq, >\}$, *CSL state and path formulae are described by*

$$\Phi ::= a \mid \neg\Phi \mid \Phi \wedge \Phi \mid \mathcal{P}_{\trianglelefteq p}(\phi) \mid \mathcal{E}_{\trianglelefteq t}(\Phi) \mid \mathcal{L}_{\trianglelefteq p}(\Phi)$$
$$\phi ::= \mathcal{X}^I \Phi \mid \Phi \mathcal{U} \Phi \mid \Phi \mathcal{U}^I \Phi$$

Except for the last two operators of the state formulae this logic corresponds to the CSL logic defined in [59]. Note that $\mathcal{P}_{\trianglelefteq p}(\phi)$ denotes the probability of the set of paths that satisfy ϕ. The formula $\mathcal{E}_{\trianglelefteq t}(\Phi)$ describes the expected time to reach some state satisfying Φ and $\mathcal{L}_{\trianglelefteq p}(\Phi)$ denotes the average time spent in states satisfying Φ in the long-run.

Given an infinite path $\pi \in Paths^{\omega}$, π satisfies $\mathcal{X}^I \Phi$ if the first transition of π occurs within time interval I and leads to a state that satisfies Φ. Similarly, the bounded until formula $\Phi \mathcal{U}^I \Psi$ is satisfied by π if π visits states that satisfy formula Φ until it reaches a state that satisfies formula Ψ within the time interval I. In contrast to the bounded until, an unbounded until formula does not constrain the time at which π may visit a state which satisfies Ψ. This corresponds to the time interval $[0, \infty)$.

Semantics of CSL. To define the semantics of CSL we first introduce some important notations. We denote with $\gamma(\pi, n)$ the time interval during which a given path π stays in its n-th state. More formally, it equals $[\Delta(\pi, n), \Delta(\pi, n+1)]$ if $\Delta(\pi, n) < \Delta(\pi, n+1)$, and $\{\Delta(\pi, n)\}$ otherwise. Let $V_{\Phi} : Paths \rightarrow \mathbb{R}_{\geq 0}^{\infty}$ be the random variable which defines the elapsed time before visiting some state $s \models \Phi$ for the first time. In other words, for an infinite path $\pi = s_0 \xrightarrow{\sigma_0, t_0} s_1 \xrightarrow{\sigma_1, t_1} \cdots$ we have $V_{\Phi}(\pi) = \min\{t \in \mathbb{R}_{\geq 0} \mid s \in \pi@t \wedge s \models \Phi\}$. Furthermore, let $\mathbf{I}_{\Phi}$ be the characteristic function of Φ, such that $\mathbf{I}_{\Phi}(s) = 1$ if $s \models \Phi$ and otherwise 0. The fraction of time spent in states satisfying Φ on an infinite path π is given by the random variable $A_\Phi(\pi) = \lim_{t\to\infty} \frac{1}{t} \int_0^t I_\Phi(\pi @ u) du$ [2, 46]. The formal semantics of CSL formulae is then defined as follows.

Definition 9 (CSL Semantics). *Let* $\mathcal{I} = (S, Act, \rightarrow, \dashrightarrow, AP, L, \nu)$ *be a state-labelled IMC,* $s \in S$, $a \in AP$, $p \in [0,1]$, $t \in \mathbb{R}_{\geq 0}$, $I \in \mathfrak{I}$, $\trianglelefteq \in \{<, \leq, \geq, >\}$, *and* $\pi \in Paths^\omega$. *We define the satisfaction relation* $\vDash$ *for state formulae:* $s \vDash a$ *iff* $a \in L(s)$, $s \vDash \neg\Phi$ *iff* $s \nvDash \Phi$, $s \vDash \Phi \wedge \Psi$ *iff* $s \vDash \Phi \wedge s \vDash \Psi$, *and*

$$s \vDash \mathcal{P}_{\trianglelefteq p}(\phi) \qquad \textit{iff } \forall D \in \mathcal{GM}. \, \mathrm{Pr}_{s,D}(\{\pi \in Paths^\omega \mid \pi \vDash \phi\}) \trianglelefteq p$$

$$s \vDash \mathcal{E}_{\trianglelefteq t}(\Phi) \qquad \textit{iff } \forall D \in \mathcal{GM}. \int_{Paths^\omega} V_\Phi(\pi) \, \mathrm{Pr}_{s,D}(\,d\pi) \trianglelefteq t$$

$$s \vDash \mathcal{L}_{\trianglelefteq p}(\Phi) \qquad \textit{iff } \forall D \in \mathcal{GM}. \int_{Paths^\omega} A_\Phi(\pi) \mathrm{Pr}_{s,D}(\,d\pi) \trianglelefteq p$$

For path formulae:

$$\pi \vDash \mathcal{X}^I \Phi \qquad \textit{iff } \pi[1] \vDash \Phi \wedge \Delta(\pi, 1) \in I$$

$$\pi \vDash \Phi \, \mathcal{U}^I \, \Psi \quad \textit{iff } \exists n \in \mathbb{N}_0.\gamma(\pi, n) \cap I \neq \emptyset \wedge \pi[n] \vDash \Psi \wedge \forall k = 0 \ldots n-1.\pi[k] \vDash \Phi$$

$$\pi \vDash \Phi \, \mathcal{U} \, \Psi \quad \textit{iff } \exists n \in \mathbb{N}_0.\pi[n] \vDash \Psi \wedge \forall k = 0 \ldots n-1.\pi[k] \vDash \Phi$$

Example 6. Consider a system with the two atomic propositions *up* and *down*. We are interested in the availability of the system and want to know if we are in an *up* state at least 90 percent of the time. This CSL property is described with the long-run average operator $\mathcal{L}_{\geq 0.9}(up)$. It is satisfied, if we are in the set of *up* states with more than 90% in the long-run. We denote the states that satisfy this property with the atomic proposition *available*.

Besides the availability of the system, we are also interested in its safety. Therefore, we want to validate that the probability to reach a *down* state via *up* state is at most 0.01 during the first 5 time units . This condition is expressed by the CSL formula $\mathcal{P}_{\leq 0.01}(up \, \mathcal{U}^{[0,5]} \, down)$. We denote all states that satisfy this property with the atomic proposition *safe*.

With these propositions, one can e.g. investigate if the average time to reach some *available* and *safe* state is at most 10 time units. This is be determined by the CSL formula $\mathcal{E}_{\leq 10}(available \wedge safe)$. ∎

3.2 Probability Bounds

Model checking a CSL formula Φ over an IMC $\mathcal{I}$ entails the computation of all sub-formulas Ψ of Φ by determining the satisfaction sets $Sat(\Psi) = \{s \in S \mid s \vDash \Psi\}$. Just like for other branching-time logics, we recursively compute those sets by starting with the inner most formula, represented by an atomic proposition. In general, we have $Sat(a) = \{s \in S \mid a \in L(s)\}$ for an atomic proposition $a \in AP$, $Sat(\neg\Psi) = S \setminus Sat(\Psi)$ for negation formulae, and $Sat(\Psi_1 \wedge \Psi_2) = Sat(\Psi_1) \cap Sat(\Psi_2)$ for the conjunction of formulae.

Probability Bounds. The proper calculation of $Sat(\mathcal{P}_{\unlhd p}(\phi))$, however, requires deeper considerations. $Sat(\mathcal{P}_{\unlhd p}(\phi))$ is defined as:

$$\{s \in S \mid \forall D \in \mathcal{GM}. \, \mathrm{Pr}_{s,D}(\{\pi \in Paths^{\omega} \mid \pi \vDash \phi\}) \unlhd p\}.$$

In a nutshell, determining this set requires the calculation of the maximum or minimum (depending on $\unlhd$) probability measures induced by all ϕ-satisfying paths starting from state s, where the maximum or minimum are to be taken over all measurable schedulers. Let $p_{\max}^{\mathcal{I}}(s, \phi)$ and $p_{\min}^{\mathcal{I}}(s, \phi)$ be those values respectively. In the following, we show how to compute them for different types of path formulae ϕ. We only consider the maximum, since the minimum can be handled analogously.

Next Formula. Assume that $\phi = \mathcal{X}^{I} \Phi$ and $Sat(\Phi)$ have been already computed. Let $a = \inf I$ and $b = \sup I$. If $s \in MS$ is a Markovian state, then nondeterminism does not occur, and the computation can be done as for CTMCs [5], i.e. $p_{\max}^{\mathcal{I}}(s, \mathcal{X}^{I}\Phi) = \sum_{s' \in Sat(\Phi)} \mathbf{P}(s, s')(e^{-E(s)a} - e^{-E(s)b})$. For $s \in IS$, we determine the possibility to move directly from s to a Φ-satisfying state. Hence $p_{\max}^{\mathcal{I}}(s, \mathcal{X}^{I}\Phi) = 1$ if $\exists s' \in S, \alpha \in Act.s \xrightarrow{\alpha} s' \wedge s' \vDash \Phi \wedge 0 \in I$, and it is zero otherwise.

Unbounded Until Formula. The evaluation of a given unbounded until formula in an IMC can be reduced to the computation of *unbounded reachability*, which in turn can be reduced to the computation of reachability in a time-abstract model. It utilises the same technique that is used for the model checking of an unbounded until formula in CTMCs [5]. Let $\mathcal{I}$ be an IMC and $\phi = \Phi \, \mathcal{U} \, \Psi$ be an unbounded until formula. We assume that $Sat(\Phi)$ and $Sat(\Psi)$ have already been computed. At first, we reduce the problem to the computation of unbounded reachability in the IMC $\mathcal{I}_{\neg \Phi}$, which is built by turning all states $Sat(\neg \Phi)$ in $\mathcal{I}$ into absorbing states. This is achieved by replacing all outgoing transitions of these states with a single Markovian self loop with an arbitrary rate, so that once a path has entered an absorbing state it cannot leave it anymore. The reasoning behind this transformation is that as soon as a path reaches some state in $Sat(\neg \Phi) \setminus Sat(\Psi)$, regardless of which states will be visited in future, it does not satisfy ϕ. Consequently, making these states absorbing does not affect the evaluation of an unbounded until formula. More formally, let $\Diamond G$ be the set of paths that eventually reach some goal states $G \subseteq S$, then $\forall s \in S. \, p_{\max}^{\mathcal{I}}(s, \Phi \, \mathcal{U} \, \Psi) = p_{\max}^{\mathcal{I}_{\neg \Phi}}(s, \Diamond Sat(\Psi))$.

In a second step, the unbounded reachability problem in $\mathcal{I}_{\neg \Phi}$ can be transformed into the computation of unbounded reachability in a time-abstract model. We can use a time-abstract model, since the sojourn time in Markovian states is not of importance in the evaluation of unbounded reachability. In other words, it does not matter at which point in time a transition from a Markovian state s to its successor s' occurs. It is sufficient to know the probability $\mathbf{P}(s, s')$ of eventually reaching s' from s. Therefore, it suffices to compute the unbounded reachability in a discrete model in which all interactive transitions of $\mathcal{I}_{\neg\Phi}$ are mimicked and all Markovian transitions are replaced with the corresponding discrete branching probabilities. The discrete model is called the embedded Markov Decision Process induced from $\mathcal{I}_{\neg\Phi}$ and denoted as $emb(\mathcal{I}_{\neg\Phi})$. Formally speaking, the unbounded reachability property in $\mathcal{I}_{\neg\Phi}$ is preserved by the transformation in its embedded MDP, or $\forall s \in S.\ p_{\max}^{\mathcal{I}_{\neg\Phi}}(s, \Diamond Sat(\Psi)) = p_{\max}^{emb(\mathcal{I}_{\neg\Phi})}(s, \Diamond Sat(\Psi))$. In the final step, we can compute the unbounded reachability property in $emb(\mathcal{I}_{\neg\Phi})$ by using, for example, the algorithms described in [7, Chapter 10].

Time-Bounded Until Formula. The computation of a time-bounded until formula is more complicated and requires some innovation. As above, the problem can be transformed into the computation of reachability in a first step. Let $\mathcal{I}$ be an IMC, $\phi = \Phi\,\mathcal{U}^I\,\Psi$ with $I \in \mathfrak{I}$ be a CSL formula, and $\Diamond^I G$ denote the set of paths that reach goal states $G \subseteq S$ within interval I. We assume that $Sat(\Phi)$ and $Sat(\Psi)$ has already been computed. Similarly to the unbounded until, all states in $Sat(\Psi)$ are considered to be goal states and all states in $Sat(\neg\Phi)$ are made absorbing. The analysis of time-bounded until analysis is then replaced by the analysis of time-bounded reachability, utilising the following theorem.

Theorem 1 (Bounded Until [50]). *Let $\mathcal{I} = (S, Act, \rightarrow, \dashrightarrow, s_0)$ be an IMC as before, and $\phi = \Phi\,\mathcal{U}^I\,\Psi$ with $I \in \mathfrak{I}$ be a CSL path formula and $G = Sat(\Psi)$. We construct $\mathcal{I}_{\neg\Phi}$ from $\mathcal{I}$ by making all states in $Sat(\neg\Phi)$ absorbing. Then $\forall s \in S.\ p_{\max}^{\mathcal{I}}(s, \Phi\,\mathcal{U}^I\,\Psi) = p_{\max}^{\mathcal{I}_{\neg\Phi}}(s, \Diamond^I G)$.*

The computation of time-bounded reachability is explained in the following section.

3.3   Time-Bounded Reachability

This section presents the algorithm introduced in [59, 50] which approximates the probabilities of a time-bounded reachability analysis in IMCs. The algorithm is based on a discretisation technique with a predefined approximation error. Given IMC $\mathcal{I}$, interval $I \in \mathfrak{I}$, a set of goal states $G \subseteq S$ and $s \in S$, the technique provides a fixpoint characterisation for the computation of $p_{\max}^{\mathcal{I}}(s, \Diamond^I G)$ (and similarly for $p_{\min}^{\mathcal{I}}(s, \Diamond^I G)$). The characterisation implies that TTPD schedulers are sufficient for this purpose, i.e. $p_{\max}^{\mathcal{I}}(s, \Diamond^I G) = \sup_{D \in \text{TTPD}} \Pr_{s,D}(\Diamond^I G)$. In other words, it suffices to find the optimal scheduler among all TTPD schedulers, which maximises time-bounded reachability. Note that similar results exist for the minimum.

Example 7. Consider the IMC in Figure 3 and assume we want to compute the maximum reachability probability from the initial state s_0 to the goal state s_5 within 3 time units. Thanks to the simple structure of the IMC, the fixpoint characterisation gives us the closed form of the maximum reachability as well as the optimal TTPD schedule. The optimal decision in state s_1 depends on the time when it is visited. Hence, the scheduler takes action α if the time is less than $3 - \ln(3)$ time units, and action β otherwise. ∎

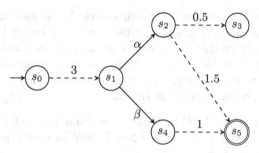

Fig. 3. An exemplary IMC

The fixpoint characterisation yields an integral equation system which is in general not tractable [5]. To circumvent this problem, the fixpoint characterisation is approximated by a discretisation technique. The time horizon is divided into equally-sized subintervals with length δ, where δ is assumed to be small enough such that at most one Markovian transition fires with a high probability. Under this assumption we can transform the IMC into its induced interactive probabilistic chain [19], the discrete version of IMCs.

Definition 10 (Interactive Probabilistic Chain). *An* interactive proba-bilistic chain *(IPC) is a tuple* $\mathcal{D} = (S, Act, \longrightarrow, \dashrightarrow_d, s_0)$, *where* S, Act, $\longrightarrow$ *and* s_0 *are as in Definition 3 and* $\dashrightarrow_d \subseteq S \times Distr(S)$ *is the set of probabilistic transitions.*

A probabilistic transition specifies the probability with which a state evolves to its successors after one time step. The notion of probabilistic transitions re-sembles the one-step transition matrix in DTMCs. The concepts of closed and open models can be transferred to IPCs. Additionally, since we do not con-sider continuous time, paths in an IPC can be seen as time-abstract paths in an IMC, implicitly still counting discretisation steps, and thus discrete time. The most general scheduler classes for IPCs are time-abstract history-dependent ran-domised (TAHR) schedulers.

Discretisation from IMC to IPC. Below we describe the discretisation technique that transforms an IMC into an IPC. Afterwards, we explain how reachability computation in an IMC can be approximated by an analysis on the corresponding IPC with a proven error bound.

Definition 11 (Discretisation [50]). *Given an IMC* $\mathcal{I} = (S, Act, \longrightarrow, \dashrightarrow, s_0)$ *and a discretisation constant* δ, $\mathcal{I}_\delta = (S, Act, \longrightarrow, \dashrightarrow_\delta, s_0)$ *is the induced IPC from* $\mathcal{I}$ *with respect to discretisation constant* δ, *where* $\dashrightarrow_\delta = \{(s, \mu^s) \mid s \in MS\}$ *and*

$$\mu^s(s') = \begin{cases} (1 - e^{-E(s)\delta})\mathbf{P}(s, s') & s' \neq s \\ (1 - e^{-E(s)\delta})\mathbf{P}(s, s') + e^{-E(s)\delta} & s' = s \end{cases}$$

This discretisation approximates the original model by assuming that at most one Markovian transition fires in each time-interval of length δ. Accordingly, μ^s specifies the probability that either one or no Markovian transition occurs from state s within each discretisation step. Using the fixpoint characterisation above, it is now possible to relate the probabilities of a reachability analysis in an IMC $\mathcal{I}$ to reachability probabilities in its IPC $\mathcal{I}_\delta$.

Example 8. Consider the IMC in Figure 1 and assume that all actions are internal. Given discretisation constant $\delta > 0$, Figure 4a shows the induced IPC of the original model w.r.t. δ.   ∎

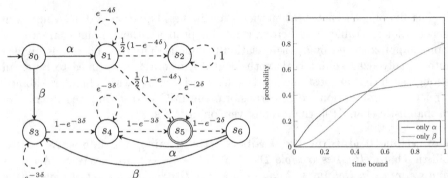

(a) The induced IPC of the original model. δ is an arbitrary positive discretisation constant.

(b) Approximate maximum time-bounded reachability computed by discretisation

Fig. 4. Time-bounded reachability for the IMC depicted in Figure 1

Theorem 2 (Discretisation error [50]). *Let $\mathcal{I} = (S, Act, \rightarrow, \dashrightarrow, s_0)$ be an IMC, $G \subseteq S$ and an interval I with rational bounds such that $a = \inf I, b = \sup I$ with $0 \le a < b$ and $\lambda = \max_{s \in MS} E(s)$. Let $\delta > 0$ be such that $a = k_a \delta, b = k_b \delta$ for some $k_a, k_b \in \mathbb{N}$. Then, for all $s \in S$ it holds that*

$$p_{\max}^{\mathcal{I}_\delta}(s, \Diamond^{(k_a, k_b]} G) - k_a \frac{(\lambda \delta)^2}{2} \le p_{\max}^{\mathcal{I}}(s, \Diamond^I G) \le p_{\max}^{\mathcal{I}_\delta}(s, \Diamond^{(k_a, k_b]} G)$$

$$+ k_b \frac{(\lambda \delta)^2}{2} + \lambda \delta.$$

Theorem 2 states that the time-bounded reachability property in an IMC $\mathcal{I}$ can be arbitrarily closely approximated by evaluating the same property in the induced IPC $\mathcal{I}_\delta$. The error bound decreases linearly with smaller discretisation steps δ. It has been recently improved in [33].

The remaining problem is to compute the maximum (or minimum) probability to reach G in an IPC within step bound $k \in \mathbb{N}$. Let $\Diamond^{[0,k]} G$ be the set of infinite paths in an IPC that reach a state in G within k steps, and let $p_{\max}^{\mathcal{D}}(s, \Diamond^{[0,k]} G)$ denote the maximum probability of those paths that start from state s and are subject to scheduler $\mathcal{D}$. Then, we have $p_{\max}^{\mathcal{D}}(s, \Diamond^{[0,k]} G) = \sup_{D \in TA} \Pr_{s,D}(\Diamond^{[0,k]} G)$. This expression can be solved by using an adaptation of the well-known value iteration scheme for MDPs to IPCs [59].

The algorithm unfolds the IPC backwards in an iterative manner, starting from the goal states. Each iteration intertwines the analysis of Markovian states and the analysis of interactive states. The main idea is that a path from interactive states to G is split into two parts:(1) reaching Markovian states from interactive states in zero time and (2) reaching goal states from Markovian states in interval $[0, j]$, where j is the step count of the iteration. The computation of the former can be reduced to an unbounded reachability analysis in the MDP induced by interactive states and rewards on Markovian states. For the latter, the algorithm operates on the previously computed reachability probabilities from all Markovian states up to step count j. We can generalise this recipe to step interval-bounded reachability [59].

Example 9. We want to compute the maximum reachability probability from the initial state s_0 to state s_5 of the IMC shown in Figure 1. Consider the induced IPC shown in Figure 4a which discretises the IMC. The maximum step-bounded reachability of the IPC is illustrated in Figure 4b. The optimal decision in state s_0 depends on the time bound. When the time bound is small the optimal action in state s_0 is α, whereas for larger time bounds taking action β yields the maximum reachability. The discretisation constant $\delta = 1.27e - 7$ is chosen on the basis of Theorem 2 to guarantee that the error bound is at most 1e-6. Hence, the computation is completed after 8e+6 iterations. ∎

3.4   Time-Bounded Reachability in Open IMCs

IMCs feature compositional behaviour which allows them to communicate with their environment. As discussed in Section 2, the class of IMCs which can interact with other IMCs, in particular via parallel composition, is called *open*. Lately, model checking of open IMCs has been studied, where the IMC is considered to be placed in an unknown environment that may delay or influence its behaviour via synchronisation [15]. The approach is restricted to a subclass of IMCs that are non-Zeno and do not contain states that have both internal and external actions enabled at the same time. Let IMC $\mathcal{I}$ satisfy these restrictions and be subject to an environment E, which can be seen as the composition of several other IMCs and has the same external actions as $\mathcal{I}$. IMC $\mathcal{I}$ is then turned into a two-player *controller-environment game*, in which the controller controls $\mathcal{I}$ and the environment controls E. In each state of $\mathcal{I}$ the controller selects one of the enabled internal transitions, if there are some. Otherwise, the environment either chooses an external action and synchronises $\mathcal{I}$ and E, or it chooses an internal action. Given a set of goal states G and time bound b, the controller tries to maximise the probability to reach the target set G within b time units. The environment tries to prevent the controller from reaching its goal by either delaying synchronisation steps or forcing the controller to take non-optimal paths. In this setup, the time-bounded reachability can be computed by the approximation scheme laid out in [59], which we have discussed above.

3.5   Expected Time

This section presents an algorithm to obtain the minimum and maximum expected time to reach a given set of goal states in an IMC, introduced in [27]: We describe the expected time objective with a fixpoint characterisation, and its transformation into a *stochastic shortest path* (SSP) problem. This SSP problem can then be used to solve the expected time CSL formula. Note that we only consider well-defined IMCs without Zeno paths.

Expected Time Objective. Let's assume that we already computed $Sat(\Phi)$, and denote this set as our set of goal states G. We want to compute the minimum expected time to reach a state in G from a given state $s \in S$. Thus, we have to consider all possible paths π induced by a given scheduler D. We define the random variable $V_G : Paths \rightarrow \mathbb{R}_{\geq 0}$ as the elapsed time before visiting a state in G . For an infinite path $\pi = s_0 \xrightarrow{\sigma_0, t_0} s_1 \xrightarrow{\sigma_1, t_1} \ldots$ let $V_G(\pi) = \min\{t \in \mathbb{R}_{\geq 0} | G \cap \pi@t \neq \varnothing\}$ with $\min(\varnothing) = \infty$ [27]. Then the minimal expected time to reach G from $s \in S$ is given by:

$$\mathsf{eT}^{\min}(s, \Diamond G) = \inf_D \mathbb{E}_{s,D}(V_G) = \inf_D \int_{Paths} V_G(\pi) \Pr_{s,D}(\,\mathrm{d}\pi). \tag{1}$$

Formula (1) expresses that we have to find a scheduler D which minimises the time until reaching a state in G . We therefore need to consider all paths induced by scheduler D. Note that, by definition of V_G, it is sufficient to consider the time before entering a goal state. Hence, we can transform all goal states into absorbing Markovian states without affecting the expected time reachability. This may result in a much smaller state space, since we can neglect those states that become unreachable from the initial state.

Theorem 3 ([27]). *The function* $\mathsf{eT}^{\min}$ *is a fixpoint of the Bellman operator*

$$v(s) = \begin{cases} \dfrac{1}{E(s)} + \displaystyle\sum_{s' \in S} \mathbf{P}(s, s') \cdot v(s') & \textit{if } s \in MS \setminus G \\[2mm] \min_{s \xrightarrow{\alpha} s'} v(s') & \textit{if } s \in IS \setminus G \\[2mm] 0 & \textit{if } s \in G. \end{cases}$$

Theorem 3 encodes expression (1) in a Bellman equation, in which we aim to find optimal values $v(s)$ for all states $s \in S$. If we are already in a goal state, we have by definition that $V_G(\pi) = 0$ with $\pi = s_0 \xrightarrow{\sigma_0, t_0} \ldots$ and $s_0 \in G$. If $s \in IS$ and s has only one outgoing interactive transition, then the expected time is the same as the one of its successor state. In case there is a nondeterministic choice between interactive transitions in s, the next transition is determined by the scheduler. Since we look for the infimum over all schedulers D, we choose the action which induces the lowest expected time in the successor state. If $s \in MS$, we add the sojourn time in state s to the time to reach a state in G over all paths starting in s induced by scheduler D. In other words, we add the sojourn time in state s to the expected sojourn time of each successor state s' weighted with the probability to reach s'.

As a result of Theorem 3, the nondeterminism in $\mathsf{eT}^{\min}(s, \Diamond G)$ can be resolved by using a stationary deterministic scheduler [27]. This implies that the scheduler chooses an action that results in the minimum expected time for each interactive state with a nondeterministic choice. To yield an effective algorithm as well as to show the correctness of Theorem 3, we transform the expected time computation into a non-negative stochastic shortest path (SSP) problem for MDPs. A SSP problem derives the minimum expected cost to reach a set of goal states in a MDP.

Definition 12 (SSP Problem). *A non-negative stochastic shortest path problem (SSP problem) is a tuple* $\mathsf{ssp} = (S, Act, \mathbf{P}, s_0, G, c, g)$, *where* $(S, Act, \mathbf{P}, s_0)$ *is an MDP,* $G \subseteq S$ *is a set of* goal states, $c : S \setminus G \times Act \to \mathbb{R}_{\geq 0}$ *is a cost function and* $g : G \to \mathbb{R}_{\geq 0}$ *is a terminal cost function.*

Given a smallest index $k \in \mathbb{N}$ of a path π with $\pi[k] = s_k \in G$, the accumulated costs to reach G on π is given by $\sum_{i=0}^{k-1} c(s_i) + g(s_k)$. The transformation of an IMC into an SSP problem is realized with the following definition.

Definition 13 (SSP for Minimum Expected Time Reachability). *The SSP of IMC* $\mathcal{I} = (S, Act, \to, \dashrightarrow, s_0)$ *for the expected time reachability of* $G \subseteq S$ *is* $\mathsf{ssp}_{\mathsf{eT}^{\min}}(\mathcal{I}) = (S, Act \cup \{\bot\}, \mathbf{P}, s_0, G, c, g)$ *where* $g(s) = 0$ *for all* $s \in G$ *and for all* $s, s' \in S$ *and* $\sigma \in Act \cup \{\bot\}$:

$$\mathbf{P}(s, \sigma, s') = \begin{cases} \frac{\mathbf{R}(s,s')}{E(s)} & \textit{if } s \in MS \wedge \sigma = \bot \\ 1 & \textit{if } s \in IS \wedge s \xrightarrow{\sigma} s' \\ 0 & \textit{otherwise, and} \end{cases}$$

$$c(s, \sigma) = \begin{cases} \frac{1}{E(s)} & \textit{if } s \in MS \setminus G \wedge \sigma = \bot \\ 0 & \textit{otherwise.} \end{cases}$$

The Markovian states are equipped with costs, since they are the states in which time advances. The cost to traverse a Markovian state along path π is determined by the sojourn time. Observe that the Bellman equation in Theorem 3 coincides with the definition of the SSP problem. The uniqueness of the minimum expected cost of an SSP problem [3, 8] implies that $\mathsf{eT}^{\min}(s, \Diamond G)$ is the unique fixpoint of $v(s)$ [27].

Example 10. Consider IMC $\mathcal{I}$ depicted in Figure 1 with $G = \{s_5\}$ being the set of goal states and s_0 being the initial state. We want to obtain $\mathsf{eT}^{\min}(s_0, \Diamond G)$. In a first step, we make goal state s_5 absorbing. Afterwards, we transform the resulting IMC into the SSP problem depicted in Figure 5. From this SSP problem we can derive the following LP problem, where x_i represents values for s_i:

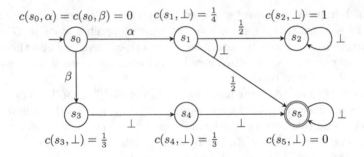

$$c(s_0, \alpha) = c(s_0, \beta) = 0 \quad c(s_1, \perp) = \tfrac{1}{4} \quad c(s_2, \perp) = 1$$

$$c(s_3, \perp) = \tfrac{1}{3} \quad\quad c(s_4, \perp) = \tfrac{1}{3} \quad\quad c(s_5, \perp) = 0$$

Fig. 5. Resulting $\mathsf{ssp}_{\mathsf{eTmin}}$ of the IMC depicted in Figure 1

Maximise $x_0 + x_1 + x_3 + x_4$ subject to:

$$x_0 \leq x_1 \qquad x_1 \leq \frac{1}{4} + \frac{1}{2}x_2 + \frac{1}{2}x_5 \qquad x_3 \leq \frac{1}{3} + x_4 \qquad x_5 = 0$$

$$x_0 \leq x_2 \qquad x_2 \leq 1 + x_2 \qquad\qquad x_4 \leq \frac{1}{3} + x_5$$

By solving these equations we obtain $x_0 = \frac{2}{3}, x_1 = \infty, x_2 = \infty, x_3 = \frac{2}{3}, x_4 = \frac{1}{3}$, which yields $\mathsf{eT}^{\min}(s_0, \Diamond G) = \frac{2}{3}$. ∎

An analogous approach can be applied to obtain the maximum expected time. In this case, we search for the supremum over all schedulers, and thus, we resolve nondeterministic choices in such a way that the scheduler chooses the actions that maximises the expected time.

3.6   Long-Run Average

In this section we present an algorithm to compute the long-run average (LRA) time spent in a set of goal states, as introduced in [27]. We describe the long-run average objective and a three step procedure to obtain the long-run average and, thus, compute the LRA CSL formula. Again, we only consider well-defined IMCs without Zeno paths.

Long-Run Average Objective. We assume that $Sat(\Phi)$ has already been computed with the technique explained before, and we denote this set as our set of goal states G. Random variable $A_{G,t}(\pi) = \frac{1}{t} \int_0^t \mathbf{1}_G(\pi @ u) du$ defines the fraction of time that is spent in G on an infinite path π in $\mathcal{I}$ up to time bound $t \in \mathbb{R}_{\geq 0}$ [2]. Note that $\mathbf{1}_G(s) = 1$ if and only if $s \in G$ and otherwise 0. For the computation of the long-run average we consider the limit $t \to \infty$ for random variable $A_{G,t}$, denoted by A_G. The expectation of A_G under scheduler D and initial state s then yields the long-run average time spent in G, where the minimum long-run average time spent in G starting from s is defined by:

$$\mathrm{LRA}^{\min}(s, G) = \inf_D \mathrm{LRA}^D(s, G) = \inf_D \mathbb{E}_{s,D}(A_G). \tag{2}$$

In contrast to the computation of the expected time and time-bounded reachability, we may assume w.l.o.g. that $G \subseteq MS$, since the long-run average time spent in any interactive state is always 0 (see Section 2). In the remainder of this section we give the basic intuition of how to compute the minimum long-run average. The general idea is given by the following three-step procedure:

1. Determine the maximal end components $\{\mathcal{I}_1, \ldots, \mathcal{I}_k\}$ of IMC $\mathcal{I}$.
2. Determine $\text{LRA}^{\min}(G)$ for each maximal end component $\mathcal{I}_j$.
3. Reduce the computation of $\text{LRA}^{\min}(s_0, G)$ in IMC $\mathcal{I}$ to an SSP problem.

The first step can be performed by a graph-based algorithm [1, 17], whereas the latter two can be expressed as LP problems.

Definition 14 (End Component). *An* end component *of IMC $\mathcal{I}$ is a sub-IMC defined by the tuple (S', A) where $S' \subseteq S$ and $A \subseteq Act$ such that:*
- *for all Markovian states $s \in S'$ with $s \xrightarrow{\lambda} s'$ it follows that $s' \in S'$, and*
- *for all interactive states $s \in S'$ and for all $\alpha \in A$ with $s \xrightarrow{\alpha} s'$ it follows that $s' \in S'$, where at least one action is enabled in $s \in S'$.*

Further, the underlying graph of (S', A) must be a strongly connected component.

Note that a *maximal end component* (MEC) is an end component which is not contained in any larger end component.

Long-Run Average in MECs. For the second step we show that for *unichain* IMCs the computation of $\text{LRA}^{\min}(s, G)$ can be reduced to the determination of long-run ratio objectives in MDPs. An IMC is unichain if and only if under any stationary deterministic scheduler it yields a strongly connected graph structure. Note that an MEC is a unichain IMC. At first, we define long-run ratio objectives for MDPs, and then show how to transform them to LRA objectives in unichain IMCs.

Let $\mathcal{M} = (S, Act, \mathbf{P}, s_0)$ be an MDP and $c_1, c_2 : S \times Act_\perp \to \mathbb{R}_{\geq 0}$ be cost functions. The operational interpretation is that cost $c_1(s, \alpha)$ is incurred when α is selected in state s, similarly for c_2. The long-run ratio between the accumulated costs c_1 and c_2 along an infinite path π in MDP $\mathcal{M}$ is defined as:

$$\mathcal{R}(\pi) = \lim_{n \to \infty} \frac{\sum_{i=0}^{n-1} c_1(s_i, \alpha_i)}{\sum_{j=0}^{n-1} c_2(s_j, \alpha_j)}.$$

Example 11. Consider the infinite path $\pi = (s_0 \xrightarrow{2} s_1 \xrightarrow{3} s_2 \xrightarrow{1} s_3 \xrightarrow{4} s_0)^\omega$ where $c_2(s_i, \cdot)$ denotes the transition labels and $c_1(s_0, \cdot) = 2$ and $c_1(s_i, \cdot) = 0$ for $1 \leq i \leq 3$. Table 3 depicts the computation of the long-run ratio until $n = 6$. By setting the limit $n \to \infty$ we obtain a fixpoint with $\mathcal{R}(\pi) = \frac{1}{5}$. ∎

The minimum long-run ratio objective for state s of MDP $\mathcal{M}$ is then defined by:

$$R^{\min}(s) = \inf_D \mathbb{E}_{s,D}(\mathcal{R}) = \inf_D \sum_{\pi \in Paths_{\text{abs}}} \mathcal{R}(\pi) \cdot \text{Pr}_{s,D}^{\text{abs}}(\pi).$$

Table 3. Example computation for the long-run ratio

n	1	2	3	4	5	6
$\mathcal{R}(\pi)$	$\frac{2}{2}=1$	$\frac{2}{2+3}=\frac{2}{5}$	$\frac{2}{2+3+1}=\frac{1}{3}$	$\frac{2}{2+3+1+4}=\frac{1}{5}$	$\frac{2+2}{2+3+1+4+2}=\frac{1}{3}$	$\frac{2+2}{2+3+1+4+2+3}=\frac{4}{15}$

Paths$_{\text{abs}}$ denotes the time-abstract paths of the MDP $\mathcal{M}$ and $\text{Pr}^{\text{abs}}_{s,D}$ represents the probability measure on the sets of paths starting from s induced by scheduler D in $\mathcal{M}$. $R^{\min}(s)$ can be obtained by solving a linear programming problem [1]. With real variable k representing $R^{\min}$ and x_s representing each $s \in S$ we have:

Maximise k subject to:

$$x_s \leq c_1(s,\alpha) - k \cdot c_2(s,\alpha) + \sum_{s' \in S} \mathbf{P}(s,\alpha,s') \cdot x_{s'} \quad \text{for each } s \in S, \alpha \in Act.$$

This system of inequations can be solved by linear programming algorithms, e.g. with the simplex method [54].

Example 12. We take path π from Example 11 and assume that it is the only path in an MDP $\mathcal{M}$. Deriving the system of linear inequations with variables k, x_{s_i} for $0 \leq i \leq 3$ then yields:

Maximise k subject to:

$$x_{s_0} \leq 2 - 2 \cdot k + x_{s_1} \qquad x_{s_2} \leq -1 \cdot k + x_{s_3}$$
$$x_{s_1} \leq -3 \cdot k + x_{s_2} \qquad x_{s_3} \leq -4 \cdot k + x_{s_0}$$

By solving the inequation system we obtain $k = \frac{1}{5}$, which is the minimum long-run ratio on $\mathcal{M}$. Note that this value equals the product of the long-run ratio as obtained in Example 11 and the probability that path π is chosen, which is 1 in our case. ∎

This result can now be transferred to a unichain IMC by transforming it into an MDP with two cost functions.

Definition 15. *Let* $\mathcal{I} = (S, Act, \rightarrow, \dashrightarrow, s_0)$ *be an IMC and* $G \subseteq S$ *a set of goal states. We define the MDP* $\text{mdp}(\mathcal{I}) = (S, Act_\perp, \mathbf{P}, s_0)$ *with cost functions* c_1 *and* c_2, *where* $\mathbf{P}$ *is defined as in Definition 13 and*

$$c_1(s,\sigma) = \begin{cases} \frac{1}{E(s)} & \text{if } s \in MS \cap G \wedge \sigma = \perp \\ 0 & \text{otherwise,} \end{cases} \qquad c_2(s,\sigma) = \begin{cases} \frac{1}{E(s)} & \text{if } s \in MS \wedge \sigma = \perp \\ 0 & \text{otherwise.} \end{cases}$$

Observe that cost function c_2 keeps track of the average sojourn time in all states $s \in S$ whereas c_1 only does so for states $s \in G$.

For a unichain IMC $\mathcal{I}$, $LRA^{\min}(s, G)$ equals the long-run ratio $R^{\min}(s)$ in the transformed MDP $\text{mdp}(\mathcal{I})$ [27]. Further, in a unichain IMC we have that $\text{LRA}^{\min}(s, G)$ and $\text{LRA}^{\min}(s', G)$ are the same for any two states s and s'. Therefore, we will omit the state and write $\text{LRA}^{\min}(G)$ when considering unichain IMCs.

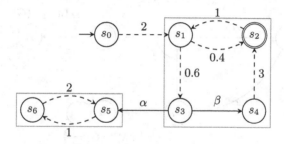

Fig. 6. IMC with two maximal end components

Reducing LRA *Objectives to an SSP Problem.* Let $\mathcal{I}$ be an IMC with initial state s_0 and maximal end components $\{\mathcal{I}_1, \ldots, \mathcal{I}_k\}$ for $k > 0$, where $\mathcal{I}_j$ has state space S_j. Using this decomposition of $\mathcal{I}$ into maximal end components, we obtain the following result:

Theorem 4 ([27]). [1] *For IMC* $\mathcal{I} = (S, Act, \rightarrow, \dashrightarrow, s_0)$ *with MECs* $\{\mathcal{I}_1, \ldots, \mathcal{I}_k\}$ *with state spaces* $S_1, \ldots, S_k \subseteq S$, *and set of goal states* $G \subseteq S$:

$$\text{LRA}^{\min}(s_0, G) = \inf_D \sum_{j=1}^{k} \text{LRA}_j^{\min}(G) \cdot \Pr^D(s_0 \models \Diamond\Box S_j),$$

where $\Pr^D(s_0 \models \Diamond\Box S_j)$ *is the probability to eventually reach and continuously stay in* S_j *from* s_0 *under policy* D *and* $\text{LRA}_j^{\min}(G)$ *is the* LRA *of* $G \cap S_j$ *in unichain MA* $\mathcal{I}_j$.

Intuitively we have to find a scheduler that minimises the product of the probability to eventually reach and stay in a MEC and the minimum LRA, over all possible combinations of MECs. We illustrate this procedure more clearly in the following example.

Example 13. Consider the IMC in Figure 6 with $G = \{s_2\}$. It consists of the two maximal end components MEC$_1$ with $S_1 = \{s_1, s_2, s_3, s_4\}$ and $Act(s_3) = \{\beta\}$, and MEC$_2$ with $S_2 = \{s_5, s_6\}$. Note that only MEC$_1$ contains a goal state. Hence, the long-run average for MEC$_2$ is automatically 0, wheras for MEC$_1$ it is greater than 0. Since we are looking for the minimum long-run average of s_2 and are starting in s_0, we choose action α in s_3 so that we end up in MEC$_2$. According to Theorem 4 we have to look for the scheduler that minimises the LRA in such a way that we eventually always stay in the desired MEC. With the choice of α we can neglect the LRA for MEC$_1$, since we will almost surely leave MEC$_1$, and thus obtain $\text{LRA}^{\min}(s_0, G) = 0$. ∎

The computation of the minimum LRA for IMCs is now reducible to a non-negative SSP problem. In IMC $\mathcal{I}$ we replace each maximal end component $\mathcal{I}_j$

---

[1] This theorem corrects a small flaw in the theorem for IMCs in [27].

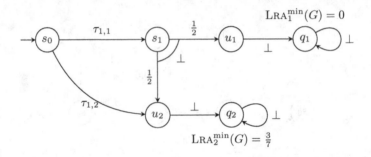

Fig. 7. Resulting SSP for $\text{LRA}^{\min}$ of the IMC depicted in Figure 1

with two fresh states q_j and u_j. Intuitively, q_j represents the MEC $\mathcal{I}_j$ and u_j represents a decision state that has a transition to q_j and contains all outgoing interactive transitions of S_j. Let U denote the set of u_j states and Q the set of q_j states. For simplification, we assume w.l.o.g. that all actions are unique and replace actions of a state $s_i \in S$ by $\tau_{i,j}$ where $j \in \{1 \ldots n_i\}$ with $n_i \in \mathbb{N}$ defined as the number of nondeterministic choices in state s_i.

Definition 16 (SSP for Long-Run Average). *Let $\mathcal{I}$, S, $G \subseteq S$, $\mathcal{I}_j$ and S_j be as before. The SSP induced by $\mathcal{I}$ for the long-run average fraction of time spent in G is the tuple* $\text{ssp}_{\text{LRA}^{\min}}(\mathcal{I}) = \left(S \setminus \bigcup_{i=1}^{k} S_i \cup U \cup Q, Act \cup \{\bot\}, \mathbf{P}', s_0, U, c, g \right)$, *where $g(q_j) = \text{LRA}_j^{\min}(G)$ for $q_j \in Q$ and $c(s, \sigma) = 0$ for all s and $\sigma \in Act_\bot$. $\mathbf{P}'$ is defined as follows: Let $S' = S \setminus \bigcup_{i=1}^{k} S_i$. $\mathbf{P}'$ equals $\mathbf{P}$ for all $s, s' \in S'$ and for the new states in U:*

$$\mathbf{P}'(u_i, \tau_{k,l}, s') = \mathbf{P}(s_k, \tau_{k,l}, s') \quad \text{if } s' \in S' \wedge s_k \in S_i \wedge l \in \{1 \ldots n_k\} \quad \text{and}$$
$$\mathbf{P}'(u_i, \tau_{k,l}, u_j) = \mathbf{P}(s_k, \tau_{k,l}, s') \quad \text{if } s_k \in S_i \wedge s' \in S_j \wedge l \in \{1 \ldots n_k\}$$

Finally, we have: $\mathbf{P}'(q_i, \bot, q_j) = 1 = \mathbf{P}'(u_i, \bot, q_i)$ *and* $\mathbf{P}'(s, \sigma, u_i) = P(s, \sigma, S_i)$.

Here, $P(s, \sigma, S_i)$ is a shorthand for $\sum_{s' \in S'} \mathbf{P}(s, \sigma, s')$. An example of the SSP transformation of IMC $\mathcal{I}$ from Figure 1 is given in Figure 7.

Example 14. Consider the IMC in Figure 1 with two maximal end components MEC_1 with $S_1 = \{s_2\}$ and MEC_2 with $S_2 = \{s_3, s_4, s_5, s_6\}$. For each MEC we introduce new states u_i and q_i, which substitute the states of MEC_i. Further, we substitute α with $\tau_{1,1}$ and β with $\tau_{1,2}$. Note that both MECs are bottom strongly connected components, which means that, under all schedulers of IMC $\mathcal{I}$, we cannot leave the MEC after entering it. Therefore, decision state u_i has only one outgoing transition to the corresponding q_i state. After the transformation to the IMC in Figure 6, the decision state of MEC_1 has a nondeterministic choice between β, to stay in MEC_1, and α, to leave it. ∎

Note that an analogous approach can be applied to obtain the maximum LRA. The main difference is that, in this case, we look for the supremum over all schedulers. In the second and the third step we now resolve the nondeterminsitic choices according to maximise the LRA.

4   Abstraction

In the previous chapter we introduced a number of IMC properties and presented algorithms for their computation. For each presented algorithm the runtime is crucially depends on the size of the considered IMC. On average, the complexity of most algorithms grows polynomially in the size of the state space, but in the worst case it grows exponentially resulting in extremely long computation times for complex models. Abstraction provides the means to reduce the state space of investigated IMCs and thereby to reduce the complexity of the verification of certain properties. In this section, we will first define the most important behavioural equivalences, namely strong and weak bisimulation, and outline efficient algorithms to compute bisimulation quotients. We remark that bisimulation can be decided in polynomial time but most coarser behavioural equivalences like trace and testing equivalences are PSPACE-complete [41].

4.1   Behavioural Equivalences

Behavioural equivalences relate states which are indistinguishable for an external observer of the system. In the following we will present the concepts of strong and weak bisimulation. As for non-probabilistic systems, these behavioural equivalences relate states that can mimic each other's behaviour. Weak bisimulation relaxes strong bisimulation by allowing that interactive transitions with visible actions may be interleaved with transitions annotated with the internal action τ. In the context of model checking, the most important application of behavioural equivalences is to provide the means for 'quotienting' a system with respect to the behavioural equivalence to reduce its state space.

Definition 17 (Strong Bisimulation). *Let* $\mathcal{I} = (S, Act, \rightarrow, \dashrightarrow, s_0)$ *be an IMC. An equivalence relation* $R \subseteq S \times S$ *is a strong bisimulation on* $\mathcal{I}$*, iff for all* $(s, t) \in R$*,* $a \in Act$ *and* $C \in S/R$ *we have that:*

- $s \xrightarrow{a} s'$ *for some* $s' \in C$ *iff* $t \xrightarrow{a} t'$ *for some* $t' \in C$
- $\mathbf{R}(s, C) = \mathbf{R}(t, C)$ *whenever* $s \xrightarrow{\tau}\!\!\!\!\!\not\;$.

Here, $\mathbf{R}(s, C)$ is a shorthand for $\sum_{s' \in C} \mathbf{R}(s, s')$, as defined in Section 2. The first condition expresses the classical bisimulation condition, requiring that for related states $s\,R\,t$ every interactive transition $s \xrightarrow{a} s'$ can be mimicked by an interactive transition $t \xrightarrow{a} t'$ such that the target states are again related, i.e. $s'\,R\,t'$. The second condition expresses that related states $s\,R\,t$ need to agree on the cumulative rates of moving from s, respectively t, to any equivalence class C; it thereby corresponds to conditions for lumpability of Markov chains and probabilistic bisimulation of DTMCs [44]. This condition is only required for states which can perform Markovian transitions. Due to the maximal progress assumption, these can only be states that have no internal action τ enabled, denoted by $\xrightarrow{\tau}\!\!\!\!\!\not\;\;$. Bisimulation relations on IMCs are closed under union which allows to define the largest bisimulation $\sim$ by the union on all bisimulations of the considered IMC. As shown in [35, Theorem. 4.3.1] both parallel composition and hiding are defined with respect to $\sim$. Moreover, time-bounded and unbounded reachability properties are preserved by bisimilar states [51, Theorem 4]. This allows us to reason over IMCs in a compositional manner.

Strong bisimulation is rigid in the sense that it requires the mimicking of interactive transitions for visible and internal actions τ. To achieve a higher degree of abstraction, we relate states that cannot be distinguished by an external observer by considering the visible actions only. For interactive transitions we apply the same machinery as for LTS. We denote by $\xrightarrow{\tau^*}$ the transitive reflexive closure of interactive transitions labelled with the internal action τ. Weak interactive transitions are then given by $\xRightarrow{a} = \xrightarrow{\tau^*} \circ \xrightarrow{a} \circ \xrightarrow{\tau^*}$. On the other hand, this does not work for Markovian transitions, since sequencing of Markovian transitions leads to the formation of the more general phase-type distributions. Thus, Markovian transitions need to be mimicked in the same way as in strong bisimulation.

Definition 18 (Weak Bisimulation). *Let $\mathcal{I} = (S, Act, \rightarrow, \dashrightarrow, s_0)$ be an IMC. An equivalence relation $R \subseteq S \times S$ is a* weak bisimulation *on $\mathcal{I}$, iff for all $(s, t) \in R$, $a \in Act$ and $C \in S/R$ we have that:*

- *$s \xRightarrow{a} s'$ for some $s' \in C$ if and only if $t \xRightarrow{a} t'$ for some $t' \in C$*
- *$\mathbf{R}(s', C) = \mathbf{R}(t', C)$ for some $t \xrightarrow{\tau^*} t'$ and $t' \xrightarrow{\tau}\!\!\!\!\!\not\;\;$ whenever $s \xrightarrow{\tau^*} s'$ and $s' \xrightarrow{\tau}\!\!\!\!\!\not\;\;$.*

Weak bisimulation is closed under union and we denote the largest weak bisimulation by $\approx$. Similarly to strong bisimulation, parallel composition and hiding are compatible with $\approx$ [35, Theorem. 4.4.1]. Moreover, weak bisimilarity preserves maximal time-bounded reachability properties [36, Theorem 10]. Strong and weak bisimulation are suitable to compare systems and to reduce their state space by deriving strong bisimilar (resp. weak bisimilar) IMCs with smaller state spaces constructed from the equivalence classes of the strong bisimilarity (resp. weak bisimilarity) and with the interactive and Markovian transitions defined in the natural way [36, Definition 10]. We remark that, in order to reason over the refinement of IMCs, there are appropriate notions of strong and weak simulations available, for which parallel composition and hiding are precongruences [36].

Example 15. Consider the IMC in Figure 8. We have $s_3 \approx s_4$ and $s_3 \sim s_4$ since these states can mimic each other's behaviour. It follows that $s_1 \approx s_2$ and $s_1 \sim s_2$; the accumulated transition rate into the bisimulation class $C_0 = \{s_3, s_4\}$ is the same for both states. Because s_0 reaches s_1 and s_2 via internal τ transitions we have $s_0 \approx s_1 \approx s_2$, but s_0 is not strongly bisimilar to s_1 and s_2. ∎

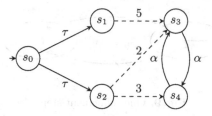

Fig. 8. An interactive Markov chain with 5 states

4.2 Algorithmic Computation of the Strong Bisimulation Quotient

Given an IMC $\mathcal{I}$, we want to determine its counterpart $\mathcal{I}'$ in which strongly bisimilar states are collapsed into one state so that $\mathcal{I} \sim \mathcal{I}'$. The result $\mathcal{I}'$ of this collapsing process is called the bisimulation quotient. In the following, we present an algorithm based on partition refinement techniques [35]. The core idea of the algorithm is to partition the states in S, and refine the resulting partition step by step. The refinement procedure of one particular partition consists of two stages that validate the two conditions of Definition 14 with respect to a so-called *splitter*. A splitter is a tuple formed by a set of states C and an action a. Given a partition P of S, in each set of P we group all those states together that can reach C via an a-transition, and those that cannot. This process is illustrated in Figure 9. More formally, given a splitter (C, a) we refine the partition according to the first condition in Definition 14 by applying

$$Refine(P, a, C) := \left(\bigcup_{X \in P} \left(\bigcup_{\nu \in \{true, false\}} \left\{ \{s \in X \mid \gamma(s, a, C) = \nu\} \right\} \right) \right) \setminus \{\emptyset\}.$$

The function $\gamma : S \times \text{Act} \times S^* \to \{true, false\}$ applied to (s, a, C) returns true if there is an a-transition from s to at least one state in C. Thus, the function *Refine* splits each set of the partition by grouping those states together that can reach states in C by at least one interactive a-transition and those that cannot. Similarly, the second condition for strongly bisimilar states is validated by refining the partition with

$$M\_Refine(P, C) := \left(\bigcup_{X \in P} \left(\bigcup_{\nu \in \mathbb{R}^+} \left\{ \{s \in X \mid R(s, C) = \nu\} \right\} \right) \right) \setminus \{\emptyset\}.$$

In other words, *M_Refine* splits each set of the partition P into classes so that all states in one class reach the set C with the same accumulated rate. Note that the result might be a set of singletons in case that all values of $R(s, C)$ are different. These two functions highlight the naming of the tuple (a, C) as splitter: It splits sets of states according to the two conditions of strongly bisimilar states and thereby refines the initial partition step by step.

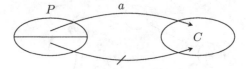

Fig. 9. One refinement step

Equipped with these two functions we construct the algorithm. The algorithm starts with a splitter (a, C) build from an arbitrary action $a \in Act$ an the set of states C either comprising all states which can perform a τ step or all states which cannot perform τ. The original partition consists of one set which contains all states. We then apply *Refine* and *M_Refine* iteratively by choosing a different splitter in each step. After each iteration we possibly obtain a finer partition and, thus, we need to add newly formed classes to the set of splitters. A final observation speeds this procedure up: States with outgoing internal τ-transition will never take Markovian transitions due to the maximal progress assumption. We therefore do not need to apply *M_Refine* on these states. Furthermore, they cannot be strongly bisimilar to states without outgoing τ-transition. For these reasons we evaluate these two classes of states separately from the very start.

Algorithm 3 (Computation of the Strong Bisimulation Quotient)

STRONG-BISIM-QUOTIEN(S, R)

```
1    S_Part←  {{ P' ∈ S|P' ⟶τ↛ }}\{∅};
2    U_Part←  {{ P' ∈ S|P' ⟶τ }}\{∅};
3    Spl← Act× (S_Part ∪ U_Part);
4    while Spl not empty
5         do
6              Choose (a, C) in Spl;
7              Old:=S_Part ∪ U_Part;
8              S_Part←Refine(S_Part,a,C);
9              U_Part← Refine(U_Part,a,C);
10             S_Part← M_Refine(S_Part,C);
11             New←(S_Part ∪ U_Part)-Old;
12             Spl← (Spl-{(a,C)}) ∪ (Act × New);
13   return S_Part ∪ U_Part
```

The algorithm can be implemented with a time complexity of $\mathcal{O}((m_I + m_M)\log n)$, where m_I is the number of interactive transitions, m_M the number of Markovian transitions and n the number of states [35].

4.3   Algorithmic Computation of the Weak Bisimulation Quotient

The computation of the weak bisimulation quotient of an IMC is more involved and we will only briefly outline the ideas formalised in [35], which uses an adaptation of the partition refinement technique described above. In contrast to strong bisimulation, weak bisimulation does not mimic internal τ actions to achieve a higher degree of abstraction. We therefore have to identify those states that have no outgoing interactive transition annotated with a τ action, also called *stable* *states*, and those that have at least one, called *unstable states*. We then partition the whole state space by grouping those states together that can reach stable states by taking only internal τ transitions, and those that cannot. We call the first class C_1 and the latter C_2 . This step requires the a priori computation of the transitive reflexive closure $\xrightarrow{\tau^*}$ of internal τ actions. The algorithm is then similar to the computation of the strong bisimulation quotient: We initialise the set of splitters and refine the classes C_1 and C_2 separately in a stepwise fashion. The class C_2 is refined with function *Refine* and C_1 with *Refine* and *M_RefineS*, where *Refine* is as before and *M_RefineS* an adaptation of *M_Refine*. The algorithm then computes the weak bisimulation quotient in $\mathcal{O}((m_I' + m_M)n)$ time where m_I' is the number of interactive transitions after transitive closure of internal transitions [35].

4.4   Bisimulation Quotient of Acyclic IMCs

In case that the considered IMC is acyclic, the minimum strong bisimulation can be determined at a much lower time-complexity of $\mathcal{O}(m)$ as suggested in [21], where m is the total number of transitions.

Definition 19 (Acyclic IMC). *An IMC P is acyclic, if it does not contain any plausible path π with $k \in \mathbb{N}_{\geq 0}$ and $\pi[k] = s$ such that $\exists n.k < n \leq |\pi|$ with $\pi[n] = s$ for all $s \in S$. A path is plausible, if it does not contain any Markovian transition such that the maximum progress assumption is violated.*

Since S is finite and P is acyclic, there is at least one state which cannot be left by a plausible path. The idea of the following algorithm is to order the states according to their longest distance to such an absorbing state. To do so we define the notion of ranks for IMCs.

Definition 20 (Rank Function). *The rank function $R: S \to \mathbb{N}$ is defined by $R(s) = \max\{|\pi| \mid \pi \in Paths^P(s)\}$.*

Here, $Paths^P(s)$ denotes the set of all plausible paths – which are finite – such that $R(s) < \infty$ for all $s \in S$. The observation which sets the groundwork for the algorithm below is that any two strongly bisimilar states have the same rank. Vice-versa, two states with the same rank are strongly bisimular if and only if they fulfil the two conditions of strongly bisimilar states. Note that transitions go only from states with a higher rank to states with a lower rank. This observation is exploited in the following algorithm. First, check states with rank 1 (states with rank 0 are by default strongly bisimilar) for strong bisimilarity. This computation requires only states with rank 0. We apply the same procedure iteratively to all states with rank 2, then to all states with rank 3 and so forth. In each iteration states with the same rank are analysed by looking at their transitions to states at the next lower rank. The time-complexity $\mathcal{O}(m)$ is defined by the depth-first search which is required for the rank computation [21]. A similar algorithm can be adopted for the computation of the weak bisimulation quotient.

5   Extensions

The remarkable progress in the theoretic developments of IMCs in the last decade has also triggered research on related concepts. In this section we review inhomogeneous IMCs and Markov automata as specific extensions of IMCs.

5.1   Inhomogeneous IMCs

Inhomogeneous IMCs extend IMCs by allowing the annotations of Markovian transitions with functions rather than real values. So far we assumed that the rates of the Markovian transitions were static and independent of time, i.e. the dynamics of IMCs were assumed to be time-homogeneous. However, in many real-world applications the progress of time crucially influences the system's dynamics. Hardware components tend to degrade over time due to oxidation and deterioration, so that their failure rate is a monotonically increasing function over time rather than a static value. Another example is the power extraction rate of a battery, which depends on the remaining amount of stored energy. Those natural phenomena can be accurately captured with the help of time-inhomogeneous IMCs (I^2MCs) as introduced in [29]. Technically, the Markovian transitions $\xrightarrow{\lambda}$ of IMCs (labelled with a positive real number λ) are generalized to transitions of the form $s \xrightarrow{r_{s,s'}(t)} s'$ of I^2MCs (labelled with a continuous functions $r_{s,s'} : \mathbb{R}_{\geq 0} \to \mathbb{R}_{\geq 0}$). The rate to execute the transition $s \xrightarrow{r_{s,s'}(t)} s'$ from s to s' at time t, is determined by $r_{s,s'}(t)$. The state transition probabilities respecting the respective race condition if multiple transitions leave a single state can be formulated analogously [29].

A process algebra and congruence results for weak and strong bisimulation for I^2MCs can be found in [29]. On the other hand, the model checking of inhomogeneous systems is quite intricate since one needs to account for the time-dependent dynamics and it still needs to be investigated further.

5.2   Markov Automata

Markov automata (MA) constitute a compositional behavioural model for continuous time stochastic and nondeterministic systems [23]. Markov automata are on the one hand rooted in interactive Markov chains by extending the expressiveness of IMCs with instantaneous random switching. They are on the other hand an orthogonal composition of probabilistic automata and continuous time Markov chains. Formally, $\mathcal{M} = (S, Act, \to, \dashrightarrow, s_0)$ is a Markov automaton, where all components of $\mathcal{M}$, but $\to$, are as Definition 3 and the set of interactive transitions $\to$ is a subset of $S \times Act \times Distr(S)$. The definition of interactive transitions characterises the ability of random switching in Markov automata. Consequently, IMCs are special cases of Markov automata, where all distributions which prevail in the set of interactive transitions $\to$ are Dirac. Markov automata are expressive enough to capture the complete semantics of generalised stochastic Petri nets and of stochastic activity networks [22]. Due to these attractive semantic and compositionality features, there is a growing interest in tool and technique support for modelling and analysis with MA [57, 26, 32].

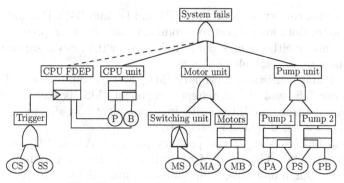

Fig. 10. The cardiac assist system DFT

6   Case Studies

IMCs have shown their practical relevance in diverse areas where memoryless continuous delay prevails. They serve as the semantics of several modelling and engineering formalisms such as generalised stochastic Petri nets [50], and Architectural Analysis and Design Language (AADL) [13]. Furthermore, they have proven their practical importance in various applications like Globally Asynchronous Locally Synchronous (GALS) designs [19], supervisory control [48, 49], satellite design [24], water-treatment facilities [34], and train control systems [10].

In the following, we will demonstrate how IMCs provide a precise formal semantics and enable compositional design and verification by examples about the industrial specification formalisms of dynamic fault trees [12] and STATEMATE [10].

6.1   Dynamic Fault Trees with Input/Output IMCs

Fault trees (FT) constitute a prominent formalism in reliability analysis to model how the failure propagation of a system's components induces a failure of the whole system [13, 12]. Its intuitive graphical syntax is often used for reliability analysis in industry. The leaves of a FT represent component failures called *basic events*, and all non-leaves indicate how component failures propagate through the system, modelled by so called *gates*. The root node represents the system failure, called the *top-level event*.

Static fault trees allow the use of the logic AND, OR and VOTING gates. Dynamic Fault Trees (DFT) extend them with a number of dynamic gates, to model common patterns in reliability engineering: functional dependencies can be specified via the FDEP gate; spare management via the SPARE gate ; and sequencing via the PAND gate. Further, each basic event is equipped with a probability distribution showing how the failure behaviour evolves over time.

Example 16 (Cardiac assist system). Figure 10 depicts a DFT representing a cardiac assist system (CAS) [11] consisting of three types of subsystems: the CPU, the motor unit, and the pump unit. If either one of these subsystem fails then the entire CAS fails, as modelled by the top-level OR gate.

The CPU unit consists of a primary (P) and backup (B) CPU, indicated by the SPARE gate. Both are subject to a common cause failure represented by the CPU FDEP gate: if either a crossbar switch (CS) or the system supervisor (SS) fails, both become unavailable.

The motor unit consists of a primary (MA) and backup motor (MB). If the primary motor fails and the switching component (MS) is still available, the backup motor is turned on. If the switching component fails afterwards, this can be ignored, as modelled by the PAND gate.

The pump unit consists of two primary pumps (PA and PB) which share a backup pump (PS). Thus, one of the primary pumps can be replaced after failing, and the pump unit fails, if all pumps are unavailable, represented by the AND gate.   ■

Given a DFT, one is typically interested in calculating the reliability of the whole system over time. An efficient way to do so is the transformation of a DFT into an Input/Output Interactive Markov Chains (I/O-IMC). I/O-IMCs [12] extend IMCs by integrating features from input/output automata. Interactive transitions are partitioned into *input actions* and *output actions*. Input actions can only be taken, if another I/O-IMC executes a matching output action. This refinement enables one to define which component triggers a synchronization and which merely reacts. This can be readily exploited to model the components' dependencies with respect to failure propagation in DFTs.

The process chain from a DFT to its semantical IMC is depicted in Figure 11. The behaviour of each leaf and gate is encoded as an I/O-IMC. The interaction between components is modelled by means of input and matching output actions. Composition and abstraction techniques explained in Section 4 can then be used to aggregate a DFT into one IMC representation for the whole system. Finally, the IMC can be analysed either directly or after the transformation to a CTMDP.

6.2   Compositional Performability Evaluation for STATEMATE

STATEMATE [31] is a statechart-based tool set used by engineers in several avionic and automotive companies like AIRBUS and BMW [10]. In this section we explain an approach proposed in [10], which enables performability evaluation of STATEMATE models. It applies various construction, optimisation and analysis techniques including compositional modelling using IMCs. In fact, the use of IMCs plays a crucial role in the model construction part of the methodology. We first recapitulate an example taken from [10] to show the applicability of STATEMATE and then explain the function of IMCs in the methodology.

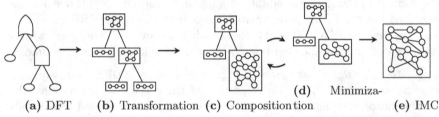

(a) DFT   (b) Transformation (c) Composition (d)   Minimiza-tion   (e) IMC

Fig. 11. Graphical overview of the compositional aggregation of DFT models

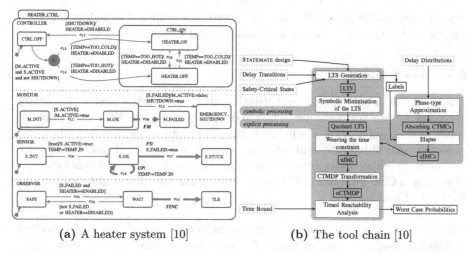

(a) A heater system [10]   (b) The tool chain [10]

Fig. 12. An example of STATEMATE design and the tool chain for quantitative evaluation of STATEMATE

Example 17. Figure 12a shows a STATEMATE *design* that represents the functional behaviour of a heating system. It consists of a CONTROLLER, a MONITOR, a SENSOR, and an OBSERVER. The SENSOR repeatedly measures and stores the temperature. However, it is an unreliable component, which means that it might become inactive due to a defect and, therefore, does not update the temperature value. As soon as the MONITOR detects a sensor failure, it shuts the system down in order to avoid severe damage. The CONTROLLER constantly checks the current temperature and turns the heater on/off when the temperature is too cold/too hot. The OBSERVER's role is to observe whether a *safety-critical* state has been reached. In this example, the high level state TLE of the OBSERVER, which specifies a situation where the sensor has failed while the heater is still on, is safety-critical. There are also *delay transitions* denoted by bold arrows. They specify the event that is triggered as a delay passes. Here they signal a component failure; for example FM and FC indicate failures of the monitor and the sensor respectively.   ■

The question that is naturally raised in such models is whether the risk of reaching some safety-critical state within a certain time bound is below a certain threshold. Such a question can be answered by using the tool chain (Figure 12b) devised in [10]. The input of the tool chain consists of several parts including the STATEMATE model, specification of safety-critical states and distributions of delay transitions. Based on these inputs, the tool chain computes the *worst-case probability* to reach some safety-critical state within the provided time bound. The given delay distribution is specified as a uniform IMC (uIMC), which is then composed with the model generated from input STATEMATE design. The result is later transformed into a uniform CTMDP (uCTMDP) by applying the steps in Section 2.6. Finally, a worst case time-bounded reachability analysis [6] is performed on the uCTMDP. As a case study, the proposed technique has been successfully employed in the area of train control systems [10].

6.3   Tool Support

IMCA [27] is a tool for the quantitative analysis of IMCs and was recently extended to Markov automata. In particular, it supports the verification of IMCs against unbounded reachability, time- and interval-bounded reachability, expected-time objectives, and long-run average objectives. Hence, it supports the model-checking algorithms presented in this paper, whereas it is not capable of parsing a CSL formula. IMCA computes the maximum and minimum values for a set of goal states.

CADP.   [18] supports construction, minimisation and analysis of extended Markovian models including IMCs. It compiles and generates the state space from a specification. The compositional verification engine of CADP then composes a network of communicating IMCs. The tool set also enables minimisation modulo strong and branching bisimulation. Furthermore, it supports the steady-state and transient analysis of the final model via numerical verification techniques or simulation. The analysis is, however, restricted to models that exhibit only spurious nondeterminism.

SCOOP. [57] is a tool that symbolically optimises process-algebraic specifications of probabilistic processes including IMCs. Optimisations such as dead-variable reduction and confluence reduction are applied automatically by SCOOP. The optimised state spaces are ready to be analysed by, for instance, CADP or IMCA. Moreover, SCOOP and IMCA constitute a tool chain called MaMa [28], which supports construction, minimisation and analysis of IMCs, among other models (e. g. MA).

MRMC. [42] is a model checker for discrete-time and continuous-time Markov reward models. It supports the verification of PCTL and CSL as well as their reward extensions CSRL and PCTRL. There is also a CTMDP extension available[2] which provides recent analysis techniques based on [16].

7   Conclusion

This paper presents an overview about IMCs, a fruitful combination of CTMCs and LTSs which facilitates the modeling of and reasoning about probabilistic systems. A great strength of IMCs is their compositional semantics which established them as a prominent formalism for a wide range of applications. We presented the theoretical framework of IMCs and introduced related concepts such as composition and schedulers. The main reason for the application of IMCs in

---

[2] http://depend.cs.uni-sb.de/tools/ctmdp/

system models is the plethora of available analysis techniques. Given a certain model, one typically wants to examine qualitative aspects such as reachability of certain system states but also quantitative aspects such as time-dependent probabilities and long-run behaviour. We provided an overview of state-of-the-art algorithms to answer these questions. A key to efficient computations is the reduction of the state space of the underlying model by exploitation of behavioural equivalences. In this context we introduced the notion of strong and weak bisimulation and presented algorithms to derive bisimulation quotients. Equipped with these techniques, IMCs have been successfully applied to a number of real-world problems and integrated into various models, especially as semantical model. We presented IMCs as the semantics of two industrial specification formalisms: Dynamic fault trees and STATEMATE.

Despite extensive progress on theoretical results and numerous tools and application developments over the last decade, IMCs still form a highly active field of research. First progress towards the measurability and analysis of IMCs has been made in [40] and future works might exploit IMCs' close entanglement with CTMDPs. Two extensions of IMCs, time-inhomogeneous IMCs and Markov automata, have been introduced. The former exhibits inhomogeneity in its embedded CTMC, while the latter adds the capability of random switching to interactive transitions. Tool support for IMCs is already available, for instance by CADP and IMCA. The latter has recently been integrated with SCOOP as a fully-fledged tool chain, called MAMA, which supports modelling, reduction and analysis of Markov automata, and IMCs as a special case of Markov automata. However, there is still room for improvements in time-bounded computations and possible extensions to Markov reward analysis. Given the advantages of IMCs, especially their expressiveness and compositional semantics, one can investigate opportunities to widen their application range to new concepts and formalisms.

References

[1] de Alfaro, L.: Formal Verification of Probabilistic Systems. Ph.D. thesis, Stanford University (1997)

[2] de Alfaro, L.: How to specify and verify the long-run average behavior of probabilistic systems. In: Proceedings of the 13th Annual IEEE Symposium on Logic in Computer Science (LICS), pp. 454–465. IEEE (1998)

[3] de Alfaro, L.: Computing minimum and maximum reachability times in probabilistic systems. In: Baeten, J.C.M., Mauw, S. (eds.) CONCUR 1999. LNCS, vol. 1664, pp. 66–81. Springer, Heidelberg (1999)

[4] Ash, R., Doléans-Dade, C.: Probability & Measure Theory. Academic Press (2000)

[5] Baier, C., Haverkort, B.R., Hermanns, H., Katoen, J.P.: Model-checking algorithms for continuous-time Markov chains. IEEE Transactions on Software Engineering 29(6), 524–541 (2003)

[6] Baier, C., Hermanns, H., Katoen, J.P., Haverkort, B.R.: Efficient computation of time-bounded reachability probabilities in uniform continuous-time Markov decision processes. Theoretical Computer Science 345(1), 2–26 (2005)

[7] Baier, C., Katoen, J.P.: Principles of model checking, vol. 950. MIT Press (2008)

[8] Bertsekas, D.P., Tsitsiklis, J.N.: An analysis of stochastic shortest path problems. Mathematics of Operations Research 16(3), 580–595 (1991)

[9] Bianco, A., de Alfaro, L.: Model checking of probabalistic and nondeterministic systems. In: Thiagarajan, P.S. (ed.) FSTTCS 1995. LNCS, vol. 1026, pp. 499–513. Springer, Heidelberg (1995)

[10] Böde, E., Herbstritt, M., Hermanns, H., Johr, S., Peikenkamp, T., Pulungan, R., Rakow, J., Wimmer, R., Becker, B.: Compositional dependability evaluation for STATEMATE. IEEE Transactions on Software Engineering 35(2), 274–292 (2009)

[11] Boudali, H., Dugan, J.B.: A Bayesian network reliability modeling and analysis framework. IEEE Transactions on Reliability 55, 86–97 (2005)

[12] Boudali, H., Crouzen, P., Stoelinga, M.: Dynamic fault tree analysis using input/output interactive Markov chains. In: Proceedings of the 37th Annual IEEE/IFIP International Conference on Dependable Systems and Networks (DSN), pp. 708–717 (2007)

[13] Bozzano, M., Cimatti, A., Katoen, J.P., Nguyen, V.Y., Noll, T., Roveri, M.: Safety, dependability and performance analysis of extended AADL models. The Computer Journal 54(5), 754–775 (2011)

[14] Bravetti, M., Hermanns, H., Katoen, J.P.: YMCA: Why Markov chain algebra? Electronic Notes in Theoretical Computer Science (ENTCS) 162, 107–112 (2006)

[15] Brázdil, T., Hermanns, H., Krcál, J., Kretínský, J., Rehák, V.: Verification of open interactive Markov chains. In: IARCS Annual Conference on Foundations of Software Technology and Theoretical Computer Science (FSTTCS), pp. 474–485 (2012)

[16] Buchholz, P., Hahn, E.M., Hermanns, H., Zhang, L.: Model checking algorithms for CTMDPs. In: Gopalakrishnan, G., Qadeer, S. (eds.) CAV 2011. LNCS, vol. 6806, pp. 225–242. Springer, Heidelberg (2011)

[17] Chatterjee, K., Henzinger, M.: Faster and dynamic algorithms for maximal end-component decomposition and related graph problems in probabilistic verification. In: Proceedings of the Twenty-second Annual ACM-SIAM Symposium on Discrete Algorithms (SODA), pp. 1318–1336. SIAM (2011)

[18] Coste, N., Garavel, H., Hermanns, H., Lang, F., Mateescu, R., Serwe, W.: Ten Years of Performance Evaluation for Concurrent Systems Using CADP. In: Margaria, T., Steffen, B. (eds.) ISoLA 2010, Part II. LNCS, vol. 6416, pp. 128–142. Springer, Heidelberg (2010)

[19] Coste, N., Hermanns, H., Lantreibecq, E., Serwe, W.: Towards performance prediction of compositional models in industrial GALS designs. In: Bouajjani, A., Maler, O. (eds.) CAV 2009. LNCS, vol. 5643, pp. 204–218. Springer, Heidelberg (2009)

[20] Crouzen, P., Hermanns, H.: Aggregation ordering for massively compositional models. In: Proceedings of the10th International Conference onApplication of Concurrency to System Design (ACSD), pp. 171–180. IEEE (June 2010)

[21] Crouzen, P., Hermanns, H., Zhang, L.: On the minimisation of acyclic models. In: van Breugel, F., Chechik, M. (eds.) CONCUR 2008. LNCS, vol. 5201, pp. 295–309. Springer, Heidelberg (2008)

[22] Eisentraut, C., Hermanns, H., Zhang, L.: Concurrency and composition in a stochastic world. In: Gastin, P., Laroussinie, F. (eds.) CONCUR 2010. LNCS, vol. 6269, pp. 21–39. Springer, Heidelberg (2010)

[23] Eisentraut, C., Hermanns, H., Zhang, L.: On probabilistic automata in continuous time. In: Proceedings of the 25th Annual IEEE Symposium on Logic in Computer Science (LICS), pp. 342–351. IEEE (2010)

[24] Esteve, M.A., Katoen, J.P., Nguyen, V.Y., Postma, B., Yushtein, Y.: Formal correctness, safety, dependability, and performance analysis of a satellite. In: Proceedings of the 34th International Conference on Software Engineering (ICSE), pp. 1022–1031. IEEE (2012)

[25] Giacalone, A., Jou, C., Smolka, S.A.: Algebraic reasoning for probabilistic concurrent systems. In: Proceedings of the IFIP TC2 Working Conference on Programming Concepts and Methods, pp. 443–458 (1990)

[26] Guck, D.: Quantitative Analysis of Markov Automata. Master's thesis, RWTH Aachen University (2012)

[27] Guck, D., Han, T., Katoen, J.-P., Neuhäußer, M.R.: Quantitative timed analysis of interactive Markov chains. In: Goodloe, A.E., Person, S. (eds.) NFM 2012. LNCS, vol. 7226, pp. 8–23. Springer, Heidelberg (2012)

[28] Guck, D., Hatefi, H., Hermanns, H., Katoen, J.-P., Timmer, M.: Modelling, reduction and analysis of Markov automata. In: Joshi, K., Siegle, M., Stoelinga, M., D'Argenio, P.R. (eds.) QEST 2013. LNCS, vol. 8054, pp. 55–71. Springer, Heidelberg (2013)

[29] Han, T., Katoen, J.-P., Mereacre, A.: Compositional modeling and minimization of time-inhomogeneous Markov chains. In: Egerstedt, M., Mishra, B. (eds.) HSCC 2008. LNCS, vol. 4981, pp. 244–258. Springer, Heidelberg (2008)

[30] Hansson, H., Jonsson, B.: A logic for reasoning about time and reliability. Formal Aspects of Computing 6(5), 512–535 (1994)

[31] Harel, D., Lachover, H., Naamad, A., Pnueli, A., Politi, M., Sherman, R., Shtull-Trauring, A., Trakhtenbrot, M.B.: STATEMATE: A working environment for the development of complex reactive systems. IEEE Transactions on Software Engineering 16(4), 403–414 (1990)

[32] Hatefi, H., Hermanns, H.: Model checking algorithms for Markov automata. Electronic Communications of the ECEASST 53 (2012)

[33] Hatefi, H., Hermanns, H.: Improving time bounded reachability computations in interactive markov chains. In: FSEN, pp. 250–266 (2013)

[34] Haverkort, B.R., Kuntz, M., Remke, A., Roolvink, S., Stoelinga, M.: Evaluating repair strategies for a water-treatment facility using Arcade. In: Proceedings of the 40th Annual IEEE/IFIP International Conference on Dependable Systems and Networks (DSN), pp. 419–424 (2010)

[35] Hermanns, H.: Interactive Markov Chains. Springer, Berlin (2002)

[36] Hermanns, H., Katoen, J.-P.: The how and why of interactive Markov chains. In: de Boer, F.S., Bonsangue, M.M., Hallerstede, S., Leuschel, M. (eds.) FMCO 2009. LNCS, vol. 6286, pp. 311–338. Springer, Heidelberg (2010)

[37] Hermanns, H., Johr, S.: Uniformity by construction in the analysis of nondeterministic stochastic systems. In: Proceedings of the 37th Annual IEEE/IFIP International Conference on Dependable Systems and Networks (DSN), pp. 718–728 (2007)

[38] Hermanns, H., Johr, S.: May we reach it? Or must we? In what time? With what probability? In: MMB, pp. 125–140. VDE Verlag (2008)

[39] Hermanns, H., Katoen, J.P., Meyer-Kayser, J., Siegle, M.: ETMCC: Model checking performability properties of Markov chains. In: Proceedings of the 33rd International Conference on Dependable Systems and Networks (DSN). IEEE Computer Society (2003)

[40] Johr, S.: Model checking compositional Markov systems. Ph.D. thesis, Saarland University (2008)

[41] Kanellakis, P.C., Smolka, S.A.: CCS expressions, finite state processes, and three problems of equivalence. Information and Computation 86(1), 43–68 (1990)

[42] Katoen, J.P., Zapreev, I.S., Hahn, E.M., Hermanns, H., Jansen, D.N.: The Ins and Outs of the probabilistic model checker MRMC. In: Proceedings of the 6th International Conference on the Quantitative Evaluation of Systems (QEST), pp. 167–176. IEEE (2009)

[43] Kwiatkowska, M., Norman, G., Parker, D.: PRISM: Probabilistic symbolic model checker. In: Field, T., Harrison, P.G., Bradley, J., Harder, U. (eds.) TOOLS 2002. LNCS, vol. 2324, pp. 200–204. Springer, Heidelberg (2002)

[44] Larsen, K.G., Skou, A.: Bisimulation through probabilistic testing (preliminary report). In: Proceedings of the 16th ACM SIGPLAN-SIGACT Symposium on Principles of Programming Languages (POPL), pp. 344–352. ACM (1989)

[45] Larsen, K.G., Skou, A.: Bisimulation through probabilistic testing. Information and Computation 94, 1–28 (1991)

[46] López, G.G.I., Hermanns, H., Katoen, J.-P.: Beyond memoryless distributions: Model checking semi-Markov chains. In: de Luca, L., Gilmore, S. (eds.) PROBMIV 2001, PAPM-PROBMIV 2001, and PAPM 2001. LNCS, vol. 2165, pp. 57–70. Springer, Heidelberg (2001)

[47] López, N., Núñez, M.: An overview of probabilistic process algebras and their equivalences. In: Baier, C., Haverkort, B.R., Hermanns, H., Katoen, J.-P., Siegle, M. (eds.) Validation of Stochastic Systems. LNCS, vol. 2925, pp. 89–123. Springer, Heidelberg (2004)

[48] Markovski, J.: Towards supervisory control of interactive Markov chains: Controllability. In: 11th International Conference on Application of Concurrency to System Design (ACSD), pp. 108–117 (2011)

[49] Markovski, J.: Towards supervisory control of interactive Markov chains: Plant minimization. In: 9th IEEE International Conference on Control and Automation (ICCA), pp. 1195–1200 (2011)

[50] Neuhäußer, M.R.: Model checking nondeterministic and randomly timed systems. Ph.D. thesis, RWTH Aachen University (2010)

[51] Neuhäußer, M.R., Katoen, J.-P.: Bisimulation and logical preservation for continuous-time Markov decision processes. In: Caires, L., Vasconcelos, V.T. (eds.) CONCUR 2007. LNCS, vol. 4703, pp. 412–427. Springer, Heidelberg (2007)

[52] Pulungan, R.: Reduction of Acyclic Phase-Type Representations. Ph.D. thesis, Universität des Saarlandes, Saarbruecken, Germany (2009)

[53] Puterman, M.L.: Markov decision processes: discrete stochastic dynamic programming, vol. 414. John Wiley & Sons (2009)

[54] Schrijver, A.: Theory of linear and integer programming. John Wiley & Sons (1998)

[55] Segala, R., Lynch, N.: Probabilistic simulations for probabilistic processes. Nordic Journal of Computing 2, 250–273 (1995)

[56] Sen, K., Viswanathan, M., Agha, G.: VESTA: A statistical model-checker and analyzer for probabilistic systems. In: Proceedings of the 2nd International Conference on the Quantitative Evaluation of Systems (QEST), pp. 251–252. IEEE (2005)

[57] Timmer, M., Katoen, J.-P., van de Pol, J., Stoelinga, M.I.A.: Efficient modelling and generation of markov automata. In: Koutny, M., Ulidowski, I. (eds.) CONCUR 2012. LNCS, vol. 7454, pp. 364–379. Springer, Heidelberg (2012)

[58] Younes, H.L.S.: Ymer: A statistical model checker. In: Etessami, K., Rajamani, S.K. (eds.) CAV 2005. LNCS, vol. 3576, pp. 429–433. Springer, Heidelberg (2005)

[59] Zhang, L., Neuhäußer, M.R.: Model checking interactive markov chains. In: Esparza, J., Majumdar, R. (eds.) TACAS 2010. LNCS, vol. 6015, pp. 53–68. Springer, Heidelberg (2010)

A Theory for the Semantics of Stochastic and Non-deterministic Continuous Systems[*]

Carlos E. Budde[1,2], Pedro R. D'Argenio[1,2],
Pedro Sánchez Terraf[1,2], and Nicolás Wolovick[1]

[1] FaMAF, Universidad Nacional de Córdoba, Argentina
[2] CONICET, Córdoba, Argentina
{cbudde,dargenio,sterraf,nicolasw}@famaf.unc.edu.ar

Abstract. The description of complex systems involving physical or biological components usually requires to model complex continuous behavior induced by variables such as time, distance, speed, temperature, alkalinity of a solution, etc. Often, such variables can be quantified probabilistically to better understand the behavior of the complex systems. For example, the arrival time of events may be considered a Poisson process or the weight of an individual may be assumed to be distributed according to a log-normal distribution. However, it is also common that the uncertainty on how these variables behave makes us prefer to leave out the choice of a particular probability and rather model it as a purely non-deterministic decision, as it is the case when a system is intended to be deployed in a variety of very different computer or network architectures. Therefore, the semantics of these systems needs to be represented by a variant of probabilistic automata that involves continuous domains on the state space and the transition relation.

In this paper, we provide a survey on the theory of such kind of models. We present the theory of the so-called labeled Markov processes (LMP) and its extension with internal non-determinism (NLMP). We show that in these complex domains, the bisimulation relation can be understood in different manners. We show the relation between the different bisimulations and try to understand their expressiveness through examples. We also study variants of Hennessy-Milner logic that provides logical characterizations of some of these bisimulations.

1 Introduction

The interplay of probabilistic and non-deterministic choices in systems that live in a continuous state space is becoming more common. For example, they arise naturally on software applications for mobile devices. This type of system has discrete state (memory hierarchy) as well as continuous state (position, orientation, acceleration, battery voltage, etc.). These continuous quantities are disrupted by the environment, and such disruption may be stochastically quantifiable. Besides, many algorithms make internal decisions sampling according to discrete probabilities. Moreover, they operate in meshes of devices where the relative speeds of execution among them are not known in

[*] Supported by ANPCyT project PICT-2012-1823, SeCyT-UNC projects 05/B284 and 05/B497 and program 05/BP02, and EU 7FP grant agreement 295261 (MEALS).

A. Remke and M. Stoelinga (Eds.): ROCKS Autumn School 2012, LNCS 8453, pp. 67–86, 2014.
© Springer-Verlag Berlin Heidelberg 2014

advance, therefore there is no information on how these devices interleave their operations in the time-line. Observations of discrete values like enabled or disabled buttons, and observations of continuous values like displayed roll angle in a cell phone, are part of these systems.

Examples of this kind which need to interact with physical or biological components abound and they exceed the modeling capabilities of Markov processes with continuous-state spaces or continuous time evolution (or both): they also need the consideration of non-determinism. Many formal frameworks have been defined to study them from a process theory or process algebra perspective [5,6,8,9,14–16,20,21,44,45, etc.]. A prominent and extensive work on this area is the one that builds on top of the so-called *labeled Markov processes (LMP)* [14, 20, 21, 39]. This is due to its solid and well understood mathematical foundations. An LMP allows for many transition probability functions (or Markov kernels) leaving each state (instead of only one as in usual Markov processes). Each transition probability function is a measure ranging on a (possibly continuous) measurable space, and the different transition probability functions can be singled out through labels. Thus this model *does not* consider *internal* non-determinism. From the modeling point of view, this is a significant drawback for this theory since internal non-determinism immediately arises in the analysis of systems, e.g., because of abstracting internal activity (such as weak bisimulation [37]) or because of state abstraction techniques (such as in model checking [2, 12]).

Many variants of continuous Markov processes that include internal non-determinism have been defined [5,6,8,9,15,16,45], including a continuous probabilistic variant of the (strong) bisimulation. Contrarily to LMPs, these models lack the sufficient structure to ensure that bisimilar models share the same observable behavior. Although [8,9] deal with the same unstructured type of model, they lift the burden of checking measurability to the semantic tools (such as bisimulation or schedulers). In particular, this results in the definition of a bisimulation as a relation between measures rather than states.

Contrarily to [8, 9] we preferred to follow the approach of Desharnais, Panangaden, et al. and extend LMPs with internal non-determinism using the power of the mathematics provided by measure theory. This led us to develop a theory of *non-deterministic labeled Markov processes (NLMP)* [7,10,17,18,48]. An NLMP has a non-deterministic transition function T_a for each label a that, given a state, it returns a measurable *set* of probability measures, rather than only one probability measure as in LMPs.

In this paper, we provide a survey to the theory of labeled Markov processes (LMP) and its extension with internal non-determinism (NLMP). Moreover, we introduce a structured version of NLMPs (SNLMP) where action labels are also endowed with a measurable structure.

The natural notion of identity on measurable spaces is given by the σ-algebra: two points can be considered indistinguishable if they cannot be separated by the σ-algebra (i.e. there is no measurable set that contains one point but not the other). As a consequence, it is expected that bisimulation respects this principle in a setting where states are endowed with a σ-algebra. However, Danos et al. [14] showed that this is not the case in the LMP model and that there are bisimulation relations that may distinguish *more* than what the underlying σ-algebra can distinguish. That is, states that cannot be separated by any measurable set (and hence always equated in the σ-algebra) may not be related by some bisimulation relation. This is awkward since measurable sets are the smallest distinguishable objects in a σ-algebra. Therefore, this raises the question of whether this problem extends to the bisimulation equivalence. To overcome this, [14] defines the so-called *event bisimulation* (in opposition to the previously existing *state* bisimulation—name which we will use from now on). The same situation arises on NLMPs. Moreover two candidates for state bisimulation appear if internal non-determinism is considered [17, 18, 48]. We recall here the definitions of the bisimulations on the different settings. In addition, we report the relations between the different bisimulations and try to understand their expressiveness through examples.

Behavioral equivalences like bisimulation have been characterized using logics with modalities, notably the Hennessy-Milner logic [33] (see also [29]). Similarly, there are Hennessy-Milner-like logics to completely characterize the different event bisimulation equivalences in LMPs [14], NLMPs [17, 18, 48], and SNLMPs [7] which we also present in this survey and show how they relate among themselves and the other bisimulations.

The next section summarizes some preliminaries on measure theory required to understand the paper. The rest of the text is structured according to each of the models. Section 3 reviews the theory of LMPs which deals with label-deterministic models. Section 4 reviews the theory of NLMPs which deals with internal non-determinism. Section 5 introduces SNLMPs which are a restriction of NLMPs where non-determinism in general is requested to be structured in a measurable space by endowing the set of labels with a σ-algebra. The paper concludes in Section 6 by reviewing some additional results that strongly relates to these models. We remark that the proofs of all results presented in Section 3 can be found in [14], the proofs of Section 4 can be found in [17] or, with more detail, in [48], and the proofs of Section 5 can be found in [7]. The results reported in Section 5 have only appeared as part of the theses [7, 48].

2   Preliminaries on Measure Theory

In this section, we recall some fundamental notions of measure theory that will be useful throughout the paper.

Given a set S and a collection Σ of subsets of S, we call Σ a *σ-algebra* iff $S \in \Sigma$ and Σ is closed under complement and denumerable union. By $\sigma(\mathcal{G})$ we denote the *σ-algebra generated by* the family $\mathcal{G} \subseteq 2^S$, i.e., the minimal σ-algebra containing $\mathcal{G}$. Each element of $\mathcal{G}$ is called a *generator* and $\mathcal{G}$ is called the *generator set*. We call the pair (S, Σ) a *measurable space*. A *measurable set* is a set $Q \in \Sigma$. Let (L, Λ) and (S, Σ) be measurable spaces. A *measurable rectangle* is a set $A \times B$ with $A \in \Lambda$ and $B \in \Sigma$. The *product σ-algebra* on $L \times S$ is the smallest σ-algebra containing all measurable rectangles, and is denoted by $\Lambda \otimes \Sigma$.

A function $\mu : \Sigma \to [0, 1]$ is a *probability measure* if (i) it is σ-additive, i.e. $\mu(\bigcup_{i \in \mathbb{N}} Q_i) = \sum_{i \in \mathbb{N}} \mu(Q_i)$ for all countable family of pairwise disjoint measurable sets $\{Q_i \mid i \in \mathbb{N}\} \subseteq \Sigma$, and (ii) $\mu(S) = 1$. By δ_a we denote the Dirac probability measure concentrated in $\{a\}$. Let $\Delta(S)$ denote the set of all probability measures over the measurable space (S, Σ). We let $\mu, \mu', \mu_1, \ldots$ range over $\Delta(S)$. Let (S_1, Σ_1) and (S_2, Σ_2) be two measurable spaces. A function $f : S_1 \to S_2$ is said to be *measurable* if for all $Q_2 \in \Sigma_2$,

$f^{-1}(Q_2) \in \Sigma_1$, i.e., its inverse image maps measurable sets to measurable sets. In this case we denote $f : (S_1, \Sigma_1) \to (S_2, \Sigma_2)$ and say that f is Σ_1-Σ_2 *measurable*.

Along the article we will often set our examples on *Borel σ-algebras*. A σ-algebra is *Borel* if it is generated by the set of all open sets in a topology. Particularly, the Borel σ-algebra on the real line is $\mathcal{B}(\mathbb{R}) = \sigma(\{(a, b) \mid a, b \in \mathbb{R} \text{ and } a < b\})$. Similarly, $\mathcal{B}([0, 1])$ is the Borel σ-algebra on the interval $[0, 1]$ generated by the open sets in the interval $[0, 1]$.

There is a standard construction by Giry [28] to endow $\Delta(S)$ with a σ-algebra[1] as follows: $\Delta(\Sigma)$ is defined as the σ-algebra generated by the sets of probability measures $\Delta^B(Q) \doteq \{\mu \mid \mu(Q) \in B\}$, with $Q \in \Sigma$ and $B \in \mathcal{B}([0, 1])$. If $p \in [0, 1]$, we will write $\Delta^{\geq p}(Q), \Delta^{>p}(Q), \Delta^{<p}(Q)$, etc. for $\Delta^B(Q)$ with $B = [p, 1], (p, 1], [0, p)$, respectively. It is known that the set $\{\Delta^{\geq p}(Q) \mid p \in (\mathbb{Q} \cap [0, 1]), Q \in \Sigma\}$ generates $\Delta(\Sigma)$. We let $\xi, \zeta, \xi', \zeta', \xi_1, \zeta_1, \ldots$ range over $\Delta(\Sigma)$.

3   Labeled Markov Processes

Labeled Markov processes (LMP) were developed in [20, 21] by Desharnais et al. An LMP has a labeled set of *actions* where an action represents the interaction with the environment. Thus, an LMP is a reactive model in which there are different transition probabilities for each action. In this model, uncertainty is (only) considered to be probabilistic; therefore, the LMP model can be regarded as a generalization of deterministic processes. The thesis of Josée Desharnais [20] and the book of Prakash Panangaden [39] contain a thorough study on LMPs.

Definition 1. *A labeled Markov process is a triple $(S, \Sigma, \{\tau_a \mid a \in L\})$ where Σ is a σ-algebra on the set of states S, and for each label $a \in L$, the transition probability function $\tau_a : S \to \Delta(S) \cup \{\mathbf{0}\}$ is a measurable function. Here, we let $\mathbf{0} : \Sigma \to [0, 1]$ be the* null measure *such that $\mathbf{0}(Q) = 0$ for all $Q \in \Sigma$.*

The value $\tau_a(s)(Q)$ represents the probability of making a transition to a state in Q provided that the system is in state s and action a has been accepted. Therefore, the transition probability is actually a *conditional probability*: the probability of Q is conditioned to the fact that the system is in state s and that it reacts to action a. Originally, [20, 39] allow $\tau_a(s)$ to be a *subprobability* measure (i.e., $\tau_a(s)(S) \leq 1$) where the value $1 - \tau_a(s)(S)$ represents the probability of refusing a. As we are going to review other models we prefer to deal only with full probability measures and let $\tau_a(s) = \mathbf{0}$ indicate that action a is refused at the state s with probability 1.

Example 2. Consider a computer system that measures a movement of a particle in the real line $\mathbb{R}$. The particle moves according to the dynamics of a Brownian motion, but the system can only measure the position of the particle at discrete time. We also want to distinguish whether the particle has passed a particular threshold $h \in \mathbb{R}$.

---

[1] The application $S \mapsto \Delta(S)$ gives rise to an endofunctor Δ of the category of measurable spaces and measurable maps. The base space of $\Delta(S, \Sigma)$ is $\Delta(S)$. By an innocuous abuse of notation, we call $\Delta(\Sigma)$ the σ-algebra of this measurable space; hence $\Delta(S, \Sigma) = (\Delta(S), \Delta(\Sigma))$.

The movement of the particle is described by a Wiener process which states that if the particle is observed at position r, the new position after a delay of t time units is a random variable with distribution $N(r, t)$, i.e., a normal distribution with mean r and variance t. More precisely, this distribution is defined by

$$\mu_r^t([l, u]) = \frac{1}{\sqrt{2\pi t}} \int_l^u e^{\frac{(x-r)^2}{2t}} \, dx.$$

To construct the LMP modeling the system, we let $\mathbb{R}$ be the set of states with the usual Borel σ-algebra $\mathcal{B}(\mathbb{R})$, where each state indicates the current position of the particle. The LMP has two types of probability transitions: one that represents the measure of the position of the particle after $n \in \mathbb{N}$ time units, and the other that indicates if the system is below or above the threshold through the labels low and $high$ respectively. Thus, for each $n \in \mathbb{N}$ and state $r \in \mathbb{R}$, $\tau_n(r) = \mu_r^n$ represents the probability at position r to read that the particle has jumped to a new position in a given interval after n (discrete) time units have elapsed. Besides, $\tau_{low}(r) = $ **if** $r < h$ **then** δ_r **else 0** is used to indicate whether the system is below or above the threshold; if r is below the threshold $\tau_{low}(r)$ is a self-loop with probability 1, otherwise it refuses action low. Similarly, $\tau_{high}(r) = $ **if** $r < h$ **then 0 else** δ_r indicates whether the system is above the threshold.

The system is then modeled by the LMP $(\mathbb{R}, \mathcal{B}(\mathbb{R}), \{\tau_a \mid a \in \mathbb{N} \cup \{low, high\}\})$.   □

Probabilistic bisimulation was introduced by Larsen and Skou [35] in a discrete setting very much like the LMP, only that distributions run on discrete sets. This notion has been adapted by Desharnais et al. in [20, 21] to the continuous case of LMPs. The idea behind the bisimulation equivalence is that from two equivalent states, an a-transition should lead with equal probability to any measurable aggregate of equivalence classes (properly speaking, to any measurable set that results of an arbitrary union of equivalence classes).

Given a relation $R \subseteq S \times S$, a set $Q \subseteq S$ is R-closed if $R(Q) \subseteq Q$. Notice that if R is symmetric, Q is R-closed if and only if for all $s, t \in S$ such that $s \, R \, t$, $s \in Q \Leftrightarrow t \in Q$. Using this definition, a symmetric relation R can be lifted to an equivalence relation in $\Delta(S)$ as follows: $\mu \, R \, \mu'$ iff for every R-closed $Q \in \Sigma$, $\mu(Q) = \mu'(Q)$.

Using this idea Desharnais et al. defined the notion of bisimulation that was called *state* bisimulation in [14] to stress the fact that the relation is defined directly on states.

Definition 3. *$R \subseteq S \times S$ is a* state bisimulation *on the LMP $(S, \Sigma, \{\tau_a \mid a \in L\})$ if it is symmetric and for all $s, t \in S$, $a \in L$, $s \, R \, t$ implies that $\tau_a(s) \, R \, \tau_a(t)$.*

We say that two states s and t are state bisimilar *(or state bisimulation equivalent), denoted by $s \sim_s t$, if there is a state bisimulation R such that $s \, R \, t$.*

Relation $\sim_s$ can be proved to be a state bisimulation and also an equivalence relation [20, 21].

The definition of state bisimulation is point-wise and not event-wise as one should expect in a measure-theoretic realm, since R has no measurability restrictions. Indeed, as shown in [14], a state bisimulation can distinguish more states than what the underlying σ-algebra can distinguish. Suppose the set of states $\{1, 2, 3, 4\}$ with the σ-algebra $\{\varnothing, \{1, 2\}, \{3, 4\}, \{1, 2, 3, 4\}\}$. No matter what the transition function is, the identity relation is a state bisimulation. However, the identity relation distinguishes states that cannot be distinguished through measurable sets (i.e., events) on the σ-algebra. Take for instance states 1 and 2: they are not related by the identity relation but they cannot be distinguished through transition probability functions because the transition probability function has to be measurable with respect to the σ-algebra.

The question is whether this problem extends to the state bisimulation equivalence $\sim_s$. To understand the problem, [14] introduced a measure-theory aware notion of behavioral equivalence.

Definition 4. *An* event bisimulation *on an LMP* $(S, \Sigma, \{\tau_a \mid a \in L\})$ *is a sub-σ-algebra* Ξ *of Σ s.t.* $(S, \Xi, \{\tau_a \mid a \in L\})$ *is an LMP.*

We extend the notion of event bisimulation to relations. We say that a relation R is an event bisimulation if there is an event bisimulation Ξ such that $R = \mathcal{R}(\Xi)$, where $\mathcal{R}(\Xi) \doteq \{(s, t) \in S \times S \mid \forall Q \in \Xi : s \in Q \Leftrightarrow t \in Q\}$. More generally, we say that two states $s, t \in S$ are *event bisimilar*, denoted by $s \sim_e t$, if there is an event bisimulation Ξ such that $s \, \mathcal{R}(\Xi) \, t$. The fact that $\sim_e$ is an equivalence relation is an immediate corollary of Theorem 5.

The article [14] shows that R is a state bisimulation iff $\Sigma(R)$, defined by $\Sigma(R) = \{Q \in \Sigma \mid Q \text{ is } R\text{-closed}\}$, is an event bisimulation. This is an important result that leads to prove that the largest state bisimulation $\sim_s$ is also an event bisimulation. That is $\sim_s \subseteq \sim_e$.

Another way to understand the semantics of a process is through a modal logic. The semantics of a process is defined by the set of properties that it satisfies. Particularly, a formula in a Hennessy-Milner-like logic defines a possible observation of the execution of the system [29, 33].

Besides, it would be useful if the semantics from the point of view of the logic agrees with that defined by the bisimulation. Thus, if two states are not bisimilar there must be an observation (a formula) that distinguishes them.

In [20, 21] a variant of the Hennessy-Milner logic for LMPs is introduced. The logic, that we call $\mathcal{L}^0$, is given by the following productions:

$$\varphi \equiv \top \mid \varphi_1 \wedge \varphi_2 \mid \langle a \rangle_q \varphi$$

where $a \in L$ and $q \in \mathbb{Q} \cap [0, 1]$.

Formulas in $\mathcal{L}^0$ are interpreted as sets of states. Thus, a formula φ is satisfied by a state s if and only if $s \in [\![\varphi]\!]$.

$$[\![\top]\!] := S \qquad [\![\varphi_1 \wedge \varphi_2]\!] := [\![\varphi_1]\!] \cap [\![\varphi_2]\!] \qquad [\![\langle a \rangle_q \varphi]\!] := \{s \in S : \tau_a(s, [\![\varphi]\!]) \geq q\}$$

Notice, in particular, that $\langle a \rangle_q \varphi$ is satisfied by a state s if there is an a-labeled transition from s reaching a set of states that satisfy φ with probability at least q. Therefore, for the semantics to be well defined, $[\![\varphi]\!]$ should be measurable for any formula φ. To show this, first notice that all operations involved on the definition of the semantics preserve measurability (in particular τ_a is a measurable function). Then, by structural induction on the formula φ, it is straightforward to conclude that $[\![\varphi]\!]$ is measurable. Let $[\![\mathcal{L}^0]\!] := \{[\![\varphi]\!] : \varphi \in \mathcal{L}^0\}$. By the previous observation, $[\![\mathcal{L}^0]\!] \subseteq \Sigma$.

It has been proved in [20, 21] that if the set of states is an analytic space[2] and the set of labels L is countable, then $\mathcal{L}^0$ characterizes state bisimulation; that is, for any two given states $s, t \in S$, $s \sim_s t$ if and only if $s \, \mathcal{R}(\mathcal{L}^0) \, t$ (i.e., for all $\varphi \in \mathcal{L}^0$, $s \in [\![\varphi]\!] \Leftrightarrow t \in [\![\varphi]\!]$). Besides, [14] showed that $\mathcal{L}^0$ completely characterizes the event bisimulation in general. More precisely, they proved that $\sigma([\![\mathcal{L}^0]\!])$, the σ-algebra generated by $[\![\mathcal{L}^0]\!]$, is the smallest σ-algebra that is also an event bisimulation.

Summarizing the results, we have:

Theorem 5. *For every LMP* $(S, \Sigma, \{\tau_a \mid a \in L\})$, $\sim_s \subseteq \sim_e = \mathcal{R}(\mathcal{L}^0)$. *Moreover, if* (S, Σ) *is an analytic Borel space and* L *is countable then* $\sim_s = \sim_e = \mathcal{R}(\mathcal{L}^0)$.

It was shown in [42] that this result does not generalize to arbitrary measurable spaces.

As we mentioned, an LMP is an inherently deterministic model in the sense that the label determines a unique transition probability function. Thus, this model *does not* consider *internal* non-determinism. From the process algebra point of view, this is a significant drawback of this theory since internal non-determinism immediately arises in the modeling and analysis of systems. For example, internal non-determinism arises by abstracting internal activity (to later use weak bisimulation [37]) or by using state abstraction techniques (such as in model checking [12]). This can be seen more clearly in the next example.

Example 6. This time we consider two computer systems, each measuring a different particle moving in the real line. One of them moves half as fast as the other, and for this reason, the system that monitors this new particle, also samples half as fast. These two systems can be modeled in a single LMP where the state space is defined by $\mathbb{R} \times \{1, 2\}$ and each state (r, k) indicates that the particle k is in position r. As in Ex. 2, we also consider a threshold h.

The complete LMP is defined by: $(\mathbb{R} \times \{1, 2\}, \mathcal{B}(\mathbb{R} \times \{1, 2\}), \{\tau_a \mid a \in \mathbb{N} \cup \{low, high\}\})$, where $\tau_{low}(r, k) = \textbf{if } r < h \textbf{ then } \delta_{(r,k)} \textbf{ else } 0$, $\tau_{high}(r, k) = \textbf{if } r < h \textbf{ then } 0 \textbf{ else } \delta_{(r,k)}$, and for all $n \in \mathbb{N}$, $\tau_n(r, 1) = \mu_r^n \circ f_1^{-1}$, $\tau_{2n}(r, 2) = \mu_r^n \circ f_2^{-1}$, and $\tau_{2n-1}(r, 2) = \mathbf{0}$, with $f_k(x) = (x, k)$ for all $x \in \mathbb{R}$.

Notice that the probability of the first particle going beyond the threshold is the same as the probability of the second particle going beyond the threshold if both are at the same position (i.e., in states $(r, 1)$ and $(r, 2)$ respectively) and the time to reach the threshold is not important. This is easy to see since each τ_n transition of the fast particle ($k = 1$) can be matched with a τ_{2n} transition of the slow particle ($k = 2$) and vice versa. Nevertheless, it is also clear that $(r, 1)$ and $(r, 2)$ are not (event nor state) bisimilar. In particular, if $q = \mu_{h-d}^n([h, \infty))$, $(h - d, 1) \in [\![\langle n \rangle_q \langle high \rangle_1 \top]\!]$ but $(h - d, 2) \notin [\![\langle n \rangle_q \langle high \rangle_1 \top]\!]$.

The distinction occurs only because the label $n \in \mathbb{N}$ (the elapsed time) is observable. Abstracting from time would require hiding this class of label, and any reasonable hiding operation will immediately end up with an object not expressible in terms of an LMP. $\qquad\square$

---

[2] A topological space is *Polish* if it is separable and completely metrizable. Examples of Polish spaces are the Euclidean spaces $\mathbb{R}^n$ and all countable discrete spaces. Polish spaces are closed under countable product, and hence $A^{\mathbb{N}}$ (with A a countable discrete space) is Polish. Finally, an *analytic* space is the continuous image of a Polish space.

4   Non-deterministic Labeled Markov Processes

Non-deterministic Labeled Markov Processes (NLMP) were introduced in [17, 18] as a generalization of LMPs that enable the modeling of internal non-determinism. That is, in an NLMP, two different but equally labeled transition probabilities are allowed to leave the same state.

There have been several attempts to define non-deterministic continuous probabilistic transition systems and all of them are straightforward extensions of (simpler) discrete versions. There are two fundamental differences in the NLMP model. The first one is that the *non-deterministic transition function T_a* now maps states to *measurable sets of probability measures* rather than arbitrary sets as previous approaches do. This is motivated by the fact that the non-determinism has to be resolved using schedulers. If we allowed the target set of states to be an arbitrary subset (as in [6, 9, 15]), the system as a whole could suffer from non-measurability issues, which would mean that it could not be quantified. (Rigorously speaking, labels should also be provided with a σ-algebra in order to define schedulers, but we omit it in this first approach.) The second difference is inspired by the definition of LMP: we ask that, for each label $a \in L$, T_a is a measurable function. One of the reasons for this restriction is to have well defined modal operators of a probabilistic Hennessy-Milner logic, like in the LMP case.

Definition 7. *A non-deterministic labeled Markov process (NLMP for short) is a structure $(S, \Sigma, \{T_a \mid a \in L\})$ where Σ is a σ-algebra on the set of states S, and for each label $a \in L$, $T_a : S \to \Delta(\Sigma)$ is measurable.*

For the requirement that T_a is measurable, we need to endow $\Delta(\Sigma)$ with a σ-algebra. This is a key construction for the development of the theory of NLMPs.

Definition 8. *$H(\Delta(\Sigma))$ is the minimal σ-algebra containing all sets $H_\xi \doteq \{\zeta \in \Delta(\Sigma) \mid \zeta \cap \xi \neq \emptyset\}$ with $\xi \in \Delta(\Sigma)$.*

This construction is similar to that of the Effros-Borel spaces [34] and resembles the so-called hit-and-miss topologies [38]. Note that the generator set H_ξ contains all measurable sets that "hit" the measurable set ξ. Also observe that $T_a^{-1}(H_\xi)$ is the set of all states that "hit" the set of measures ξ through label a (i.e., $T_a^{-1}(H_\xi) = \{s \mid T_a(s) \cap \xi \neq \emptyset\}$). This forms the basis to existentially quantify over the non-determinism, and it is fundamental for the behavioral equivalences and the logic.

The next two examples (inspired by an example in [8]) show why T_a is required to map into measurable sets and to be measurable. For these examples we let the state space and σ-algebra be the real unit interval with the standard Borel σ-algebra.

Example 9. Let $\mathcal{V} = \{\delta_q \mid q \in V\}$, where V is a non-measurable set in $[0, 1]$ (that is, $V \subseteq [0, 1]$ and $V \notin \Sigma$). It can be shown that $\mathcal{V}$ is not measurable in $\Delta(\Sigma)$. Let $T_a(s) = \mathcal{V}$ for all $s \in [0, 1]$. The resolution of the internal non-determinism by means of so-called schedulers (also adversaries or policies) [41, 46], would require to assign probabilities to all possible choices. This amounts to measure the non-measurable set $T_a(s)$. This is why we require that T_a maps into measurable sets. □

Example 10. Let $T_a(s) = \{\mu\}$ for a fixed measure μ, and let $T_b(s) = $ **if** $(s \in V)$ **then** $\{\delta_1\}$ **else** $\emptyset$, for every $s \in [0, 1]$, with V being a non-measurable set. Notice that both

$T_a(s)$ and $T_b(s)$ are measurable sets for every $s \in [0, 1]$. Assume that there is a scheduler that chooses to first do a and then b starting at some state s. The probability under such scheduler of reaching state 1 after preforming both transitions cannot be measured since it requires to apply μ to the set $T_b^{-1}(H_{\Delta(s)}) = V$ which is not measurable. Besides, we will later need that sets $T_a^{-1}(H_\xi)$ are measurable so that the semantics of the logic $\mathcal{L}^1$ maps into measurable sets.   □

Notice that an LMP can be regarded as an NLMP without internal non-determinism, that is, an NLMP in which $T_a(s)$ is either a singleton or the empty set for all $a \in L$ and $s \in S$. In fact, an LMP can be encoded as an NLMP by taking $T_a(s) = \{\tau_a(s)\} \setminus \{0\}$. For this, it is necessary that singletons $\{\mu\}$ are measurable in $\Delta(\Sigma)$ for the NLMP to be well defined. (In general, it suffices that Σ is countably generated to ensure that singleton sets are measurable [17].) Moreover, it is also necessary that function T_a is measurable, which is actually the case. Indeed, it is not difficult to verify that T_a is measurable iff τ_a is measurable [48].

Example 11. Taking the previous definition, the LMP of the two particles moving on the real line of Ex. 6, can be translated into the NLMP $(\mathbb{R} \times \{1, 2\}, \mathcal{B}(\mathbb{R} \times \{1, 2\}), \{T_a \mid a \in \mathbb{N} \cup \{low, high\}\})$, where $T_{low}(r, k) = $ **if** $(r < h)$ **then** $\{\delta_{(r,k)}\}$ **else** $\varnothing$, $T_{high}(r, k) = $ **if** $(r < h)$ **then** $\varnothing$ **else** $\{\delta_{(r,k)}\}$, and for all $n \in \mathbb{N}$, $T_n(r, 1) = \{\mu_r^n \circ f_1^{-1}\}$, $T_{2n}(r, 2) = \{\mu_r^n \circ f_2^{-1}\}$, and $T_{2n-1}(r, 2) = \varnothing$, with $f_k(x) = (x, k)$ for all $x \in \mathbb{R}$.

Notice that if we abstract the time just like in process algebra, then we obtain the NLMP $(\mathbb{R} \times \{1, 2\}, \mathcal{B}(\mathbb{R} \times \{1, 2\}), \{T_a \mid a \in \{\epsilon, low, high\}\})$, with T_{low} and T_{high} as before and $T_\epsilon(r, k) = \{\mu_r^n \circ f_k^{-1} \mid n \in \mathbb{N}\}$. Clearly this last set is measurable in $\Delta(\mathcal{B}(\mathbb{R} \times \{1, 2\}))$ since it is countable (singleton sets are measurable in $\Delta(\mathcal{B}(\mathbb{R} \times \{1, 2\}))$, see [17]). Besides, it can be proved that T_ϵ is measurable which shows that this abstraction defines a proper NLMP.   □

The original definition of bisimulation given by Larsen and Skou [35] has been generalized to a continuous setting in, e.g., [5,6,15,16,45]. These definitions closely resemble the definition of Larsen and Skou, the only difference being that two measures are considered equivalent if they agree in every measurable union of equivalence classes induced by the relation. In our setting, this definition can be instantiated as follows

Definition 12. *A relation R is a* state bisimulation *on an NLMP $(S, \Sigma, \{T_a \mid a \in L\})$ if it is symmetric and for all $a \in L$, s R t implies that for all $\mu \in T_a(s)$, there is $\mu' \in T_a(t)$ s.t. μ R μ'. We say that $s, t \in S$ are* state bisimilar, *denoted by $s \sim_s t$, if there is a state bisimulation R such that s R t.*

The relation $\sim_s$ is the largest state bisimulation and it is also an equivalence relation [17,48]. The proof of this follows the standard strategy of the classic bisimulation (see [37]). Apart from the probabilistic treatment, it only differs in that the composition $R \circ R'$ is granted to be state bisimulation if R and R' are *reflexive* state bisimulations. (If one of R or R' is not reflexive, $R \circ R'$ may not be a state bisimulation.) Besides, it is easy to show that a state bisimulation on an LMP is also a state bisimulation on the encoding NLMP and vice versa.

The next example revisits Example 6. Using the timed abstracted version of Example 11, it shows that it is possible to prove that the two particles behave the same (modulo state bisimulation).

Example 13. Take the time-abstracted NLMP of Ex. 11, and let $R = \{((r, 1), (r, 2)) \mid r \in \mathbb{R}\}$. Notice that any measurable R-closed set has the form $Q_B = B \times \{1, 2\}$ for some $B \in \mathcal{B}(\mathbb{R})$. Hence $\mu_r^n \circ f_1^{-1}(Q_B) = \mu_r^n(B) = \mu_r^n \circ f_2^{-1}(Q_B)$ and therefore $(\mu_r^n \circ f_1^{-1}) R (\mu_r^n \circ f_2^{-1})$. Since $\mu \in T_\epsilon(r, 1)$ implies that $\mu = \mu_r^n \circ f_1^{-1}$ for some n, then there is some $\mu' \in T_\epsilon(r, 2)$ such that $\mu R \mu'$ (in fact, $\mu' = \mu_r^n \circ f_2^{-1}$). From here (and the cases of T_{low} and T_{high}, which we omit) it follows that R is a state bisimulation.   □

In the case of an LMP $(S, \Sigma, \{\tau_a \mid a \in L\})$, an event bisimulation is a sub-σ-algebra $\Xi \subseteq \Sigma$ such that all transition probability functions are Ξ-$\Delta(\Xi)$ measurable. We state our generalization following the same idea.

Definition 14. *An* event bisimulation *on an NLMP* $(S, \Sigma, \{T_a \mid a \in L\})$ *is a sub-σ-algebra Ξ of Σ s.t. $T_a : (S, \Xi) \to (\Delta(\Sigma), H(\Delta(\Xi)))$ is measurable for each $a \in L$.*

Note that T_a is the same function from S to $\Delta(\Sigma)$ only that, for Ξ to be an event bisimulation, it should be measurable from Ξ to $H(\Delta(\Xi))$. Here, $H(\Delta(\Xi))$ is the sub-σ-algebra of $H(\Delta(\Sigma))$ generated by $\{H_\xi \mid \xi \in \Delta(\Xi)\}$.

Just like for LMPs, the notion of event bisimulation can be extended to relations: R is an event bisimulation if there is an event bisimulation Ξ s.t. $R = \mathcal{R}(\Xi)$. More generally, we say that two states $s, t \in S$ are *event bisimilar*, denoted by $s \sim_e t$, if there is an event bisimulation Ξ such that $s \mathcal{R}(\Xi) t$. The fact that $\sim_e$ is an equivalence relation is an immediate corollary of Theorem 16 given below. We remark that an event bisimulation on an LMP is also an event bisimulation on the encoding NLMP and vice versa.

For NLMPs, we introduce a third kind of bisimulation that we call *hit bisimulation*[3]. Rather than looking point-wise at probability measures as state bisimulations do, the definition of the hit bisimulation follows the idea of Def. 8 and verifies that both $T_a(s)$ and $T_a(t)$ *hit* the same measurable sets of probability measures which measure only R-closed sets.

Definition 15. *A relation $R \subseteq S \times S$ is a* hit bisimulation *on the NLMP $(S, \Sigma, \{T_a \mid a \in L\})$ if it is symmetric and for all $a \in L$, $s R t$ implies that, for all $\xi \in \Delta(\Sigma(R))$, $T_a(s) \cap \xi \neq \emptyset \Leftrightarrow T_a(t) \cap \xi \neq \emptyset$. We say that $s, t \in S$ are* hit bisimilar, *denoted by $s \sim_h t$, if there is a hit bisimulation R such that $s R t$.*

The relation $\sim_h$ is the largest hit bisimulation and an equivalence relation. Hit bisimulations relate to the event bisimulations in NLMPs very much like the the state bisimulations relate to the event bisimulations in LMPs. In particular, R is a hit bisimulation if and only if $\Sigma(R)$ is an event bisimulation. This is indeed an important result that is central to eventually prove that $\sim_h$ is also an event bisimulation and hence $\sim_h \subseteq \sim_e$. The fact that $\sim_h$ is an equivalence relation is actually a consequence of the fact that it is also an event bisimulation (every event bisimulation is an equivalence relation directly from its definition). The details of all these results appeared in [17, 48].

A state bisimulation R is also a hit bisimulation. The proof of this relies on the fact that if $\xi \in \Delta(\Sigma(R)), \mu \in \xi$ and $\mu R \mu'$, then $\mu' \in \xi$. The rest of the proof is straightforward from the definitions. An immediate consequence is that $\sim_s \subseteq \sim_h$. As we will see later, the inclusion is proper.

---

[3] In our original works [17, 18, 48], we called the state and hit bisimulations, "traditional" and "state" respectively. We are changing the names here as we find them more appropriate.

Nevertheless both notions of bisimulation agree on NLMPs that are *image denumerable*. That is, a hit bisimulation R is also a state bisimulation on any NLMP satisfying that for all $a \in L, s \in S, T_a(s)$ is denumerable. As a consequence of this, a state bisimulation on an LMP is a hit bisimulation on the translated NLMP and vice versa, since the translated NLMP is deterministic and hence image denumerable ($|T_a(s)| \leq 1$ for all $a \in L$ and $s \in S$).

Like for LMPs, we can also provide a Hennessy-Milner-like logic for NLMPs that characterizes event bisimulation in general and all the bisimulations under some conditions. As we will see in Ex. 18, $\mathcal{L}^0$ is not sufficiently expressive to characterize event bisimulation in NLMPs. Therefore, we need a richer logic. The logic we present below was introduced in [17,18] and is related to the logic of Parma and Segala [40]. The main difference is that we consider two kinds of formulas: one that is interpreted on states, and another that is interpreted on measures. The syntax is as follows,

$$\varphi \equiv \top \mid \varphi_1 \wedge \varphi_2 \mid \langle a \rangle \psi$$
$$\psi \equiv \bigvee_{i \in I} \psi_i \mid \neg \psi \mid [\varphi]_{\geq q}$$

where $a \in L$, I is a denumerable index set, and $q \in \mathbb{Q} \cap [0,1]$. We denote by $\mathcal{L}^1$ the set of all formulas generated by the first production and by $\mathcal{L}^1_{\Delta}$ the set of all formulas generated by the second production.

The semantics is defined with respect to an NLMP (S, Σ, T). Formulas in $\mathcal{L}^1$ are interpreted as sets of states, and formulas in $\mathcal{L}^1_{\Delta}$ are interpreted as sets of measures on the state space as follows,

$$[\![\top]\!] = S \qquad\qquad [\![\bigvee_{i \in I} \psi_i]\!] = \bigcup_i [\![\psi_i]\!]$$
$$[\![\varphi_1 \wedge \varphi_2]\!] = [\![\varphi_1]\!] \cap [\![\varphi_2]\!] \qquad\qquad [\![\neg\psi]\!] = [\![\psi]\!]^c$$
$$[\![\langle a \rangle \psi]\!] = T_a^{-1}(H_{[\![\psi]\!]}) \qquad\qquad [\![[\varphi]_{\geq q}]\!] = \Delta^{\geq q}([\![\varphi]\!])$$

In particular, notice that $\langle a \rangle \psi$ is satisfied at a state s whenever there is some measure $\mu \in T_a(s)$ that satisfies ψ, and that $[\varphi]_{\geq q}$ is satisfied by a measure μ whenever $\mu([\![\varphi]\!]) \geq q$. As in the case of LMPs, the sets $[\![\varphi]\!]$ and $[\![\psi]\!]$ are measurable in Σ and $\Delta(\Sigma)$, respectively. For the rest of the section, fix $[\![\mathcal{L}^1]\!] = \{[\![\varphi]\!] \mid \varphi \in \mathcal{L}^1\}$.

Note that some other operators can be encoded as syntactic sugar. For instance, we can define $[\varphi]_{>r} \equiv \bigvee_{q \in \mathbb{Q} \cap [0,1] \wedge q > r} [\varphi]_{\geq q}$ for any real $r \in [0,1]$, and $[\varphi]_{\leq r} \equiv \neg[\varphi]_{>r}$.

It can be shown that $\mathcal{L}^1$ characterizes event bisimulation for NLMPs. Following the lines of the proof for the logical characterization of event bisimilarity for LMPs, it can be proved that $\sigma([\![\mathcal{L}^1]\!])$ is the smallest σ-algebra that is also an event bisimulation. A mild generalization of the concept of event bisimulation, namely families of sets being *stable*[4] plays a role in the proof; it is immediate from the definition that stable σ-algebras are exactly the event bisimulations. It is shown in [17, Sect. 5] that $[\![\mathcal{L}^1]\!]$ is the smallest stable family of subsets that is closed under finite intersections. A key lemma that appears in [47] ensures that $\sigma(C)$ is stable whenever C is, and the result follows:

---

[4] The family $C \subseteq \Sigma$ is *stable* for an NLMP (S, Σ, T) if for all $a \in L$ and $\xi \in \Delta(C), T_a^{-1}(H_\xi) \in C$. This notion of stability was further generalized by Doberkat [25] to the concept of *congruence* for stochastic systems.

Theorem 16. *The logic $\mathcal{L}^1$ completely characterizes event bisimulation. In other words,* $\mathcal{R}(\mathcal{L}^1) = \sim_e$.

A consequence of this theorem together with the previously discussed relations between the different bisimulations, is that both state and hit bisimulation are sound for $\mathcal{L}^1$, i.e., they preserve the validity of formulas.

Theorem 17. $\sim_s \subseteq \sim_h \subseteq \sim_e = \mathcal{R}(\mathcal{L}^1)$.

As we will review in Examples 20 and 21, these inclusions are proper in general. Nevertheless, for image-finite NLMPs over analytic spaces it can be proved that the same logic is complete for state bisimilarity, and hence all notions are the same. (An NLMP is image finite if $T_a(s)$ is finite for all $a \in L$ and $s \in S$.) In fact, the sub-logic of $\mathcal{L}^1$ defined by

$$\varphi \equiv \top \mid \varphi_1 \wedge \varphi_2 \mid \langle a \rangle \textstyle\bigwedge_{i=1}^{n} [\varphi_i]_{\bowtie_i q_i} \tag{1}$$

where $\bowtie_i \in \{>, <\}$ and $q_i \in \mathbb{Q} \cap [0, 1]$, is complete for $\sim_s$ under those restriction. Let $\mathcal{L}^{1-}$ be the set of all formulas generated by (1).

It should be noted that the expression $\langle a \rangle \bigwedge_{i=1}^{n} [\varphi_i]_{\bowtie_i q_i}$ may not be expressed as a conjunction of formulas $\langle a \rangle [\varphi_i]_{\bowtie_i q_i}$ because the probabilistic bounds must be satisfied by the *same* non-deterministic transition. The next example from [10] illustrates this fact.

Example 18. Take the discrete NLMPs depicted in Fig. 1. States s and t are not bisimilar since given a measure $\mu \in T_a(s)$, there is no $\mu' \in T_a(t)$ such that $\mu(Q) = \mu'(Q)$ for all $Q \in \{\{x\}, \{y\}, \{z\}\}$ (which are the only relevant possible R-closed sets). A logic having a modality that can only describe one behavior after a label will not be able to distinguish between s and t. For example, $[\![\langle a \rangle [\varphi]_{>q}]\!] = \{w \mid T_a(w) \cap \Delta^{>q}([\![\varphi]\!]) \neq \varnothing\}$ will always have s and t together, or none of them. Observe that negation, denumerable conjunction or disjunction, do not add any distinguishing power (on an image finite setting).

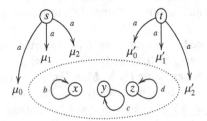

		μ_0	μ_1	μ_2	μ_0'	μ_1'	μ_2'
$\{x\}$		$\frac{1}{3}$	$\frac{2}{3}$	0	$\frac{2}{3}$	$\frac{1}{3}$	0
$\{y\}$		0	$\frac{1}{3}$	$\frac{2}{3}$	0	$\frac{2}{3}$	$\frac{1}{3}$
$\{z\}$		$\frac{2}{3}$	0	$\frac{1}{3}$	$\frac{1}{3}$	0	$\frac{2}{3}$

Fig. 1. s and t are not bisimilar

Notice, however, that the $\mathcal{L}^{1-}$ formula $\langle a \rangle ([\langle b \rangle \top]_{<\frac{2}{3}} \wedge [\langle c \rangle \top]_{>\frac{1}{3}})$ is satisfied by s but not by t. □

The essential need for this new modal operator also shows that our σ-algebra $H(\Delta(\Sigma))$ in Def. 8 cannot be simplified to $\sigma(\{H_{\Delta^B(Q)} : B \in \mathcal{B}([0, 1]), Q \in \Sigma\})$. States s and t in the example above should be observationally distinguished from each other. Formally, this amounts to saying that there must be some label a and some measurable $\Theta \in H(\Delta(\Sigma))$ such that $T_a^{-1}(\Theta)$ separates s from t. Therefore, the same must be true for some generator Θ, but this does not hold for the family $\{H_{\Delta^B(Q)} : B \in \mathcal{B}([0,1]), Q \in \Sigma\}$.

The proof that our logic is complete for state bisimilarity follows from [17, Lemma 5.8] that states that given an NLMP (S, Σ, T) with (S, Σ) analytic, if we have a countable logic $\mathfrak{L}$ with $[\![\mathfrak{L}]\!] \subseteq \Sigma$ satisfying some local criteria, then that logic must characterize $\sim_s$ completely. (Logic $\mathfrak{L}$ is countable if the number of formulas in $\mathfrak{L}$ is countable.) In our case, $\mathcal{L}^1$ is not countable but the sub-logic $\mathcal{L}^{1-}$ given by (1) satisfies all the requirements and hence $\sim_s = \mathcal{R}(\mathcal{L}^{1-})$. Since $\mathcal{R}(\mathcal{L}^1) \subseteq \mathcal{R}(\mathcal{L}^{1-})$, by Theorem 17 we have the following result.

Theorem 19. *Let (S, Σ, T) be an image finite NLMP with (S, Σ) being analytic. For all $s, t \in S$,*

$$s \sim_s t \;\Leftrightarrow\; s \sim_h t \;\Leftrightarrow\; s \sim_e t \;\Leftrightarrow\; s\, \mathcal{R}(\mathcal{L}^1)\, t$$

There are two delimiting results on possible generalizations of this theorem. First, the hypothesis of an analytic state space cannot be dropped completely (even for deterministic processes). This was seen in [42], where it is shown that state bisimilarity for LMPs is not characterized by $\mathcal{L}^0$ in general. Secondly, a generalization of the same arguments to image-countable processes is not feasible since there is no countable logic having formulas with measurable extensions that characterize state bisimilarity on such processes [43].

In fact, as we have already anticipated, the inclusions in Theorem 17 are proper in the general case. In the following, we construct counterexamples over standard Borel spaces witnessing that all our notions of bisimilarity are different in the case of uncountable non-determinism.

Moreover, it suffices to consider a non-probabilistic variant of NLMP, in which transitions only map into a set of Dirac measures. These structures look very much like LTSs, with the additional requirement that the set of states is endowed with a σ-algebra that the transition should respect. More formally, let (S, Σ) be a standard Borel space and $\delta(Q) = \{\delta_s \mid s \in Q\}$ for each $Q \in \Sigma$. An NLMP $\mathbf{S} = (S, \Sigma, \{T_a \mid a \in L\})$ is called *non-probabilistic* if for all $a \in L$ and $s \in S$, $T_a(s) \subseteq \delta(S)$.

Example 20. We will first construct a non-probabilistic NLMP witnessing the fact that state bisimilarity is strictly finer than the other notions. Consider the standard Borel space $(S_1, \Sigma_1) = ([0,1] \cup [2,3] \cup \{s, t, x\}, \mathcal{B}([0,1] \cup [2,3] \cup \{s, t, x\}))$ where $s, t, x \in \mathbb{R} \setminus [0,3]$ are different. Let V be a non-Borel subset of $[2.5, 3]$. Clearly, $[0,1]$ is equinumerous with $[2,3] \setminus V$; pick a bijection f between them. Now, let $L_1 = \{a\} \cup [0,1]$ be the set of labels and let $\mathbf{S_1} = (S_1, \Sigma_1, \{T_a \mid a \in L_1\})$ be non-probabilistic such that

$$T_a(s) = \{\delta_d \mid d \in [2,3]\} \qquad T_r(r) = T_r(f(r)) = \{\delta_x\} \qquad \text{if } r \in [0,1]$$
$$T_a(t) = \{\delta_d \mid d \in [0,1]\} \qquad T_c(y) = \varnothing \qquad\qquad \text{otherwise.}$$

Now, take $\mathcal{F}$ to be $\{\{s, t\}, \{r, f(r)\}_{r \in [0,1]}\}$ and $R = \mathcal{R}(\sigma(\mathcal{F}))$. It is not hard to prove that $\mathbf{S_1}$ is a non-probabilistic NLMP, $\sigma(\mathcal{F})$ is an event bisimulation and R is a hit bisimulation that relate s and t. Also, it can be seen that s and t are not state bisimilar. But this shows that $\sim_s$ differs from $\sim_e$ and $\sim_h$. $\qquad\square$

Example 21. By modifying slightly S_1 we can show that the largest event bisimulation $\sim_e$ is not contained in $\sim_h$. Take V to be the interval $(2.5, 3]$ and let $(S_2, \Sigma_2) = (S_1, \Sigma_1)$. We complete the construction of a non-probabilistic NLMP S_2 by picking any bijection f between $[0, 1]$ and $[2, 2.5]$. The transition is defined just like for S_1 only that using the the new f. We also use family $\mathcal{F}$ but defined with the new f. The same arguments for S_1 go through here, showing that $s \sim_e t$ but $s \not\sim_h t$.   □

Some observations on the counterexamples are in order. First, counterexample S_1 relies on the fact that hit bisimulation cannot distinguish a non-measurable set V while state bisimulation can. From our point of view, such distinction should *not* be possible since V has no measure. Second, counterexample S_2 makes a difference on the measurable set V that the event bisimulation cannot distinguish. In our opinion, such distinction *should* be possible since some scheduler may lead to such set of states with certain probability. Note that in this example, states in V do not allow the system to reach state x from s, while x can always be reached from t. In particular, if the scheduler chooses uniformly the branching on the a-transition in both cases, the system starting from s will deadlock with probability 0.5 immediately while no deadlock is possible after a when starting from t. In this sense, we argue that hit bisimulation is the most appropriate definition.

Somehow, this is disappointing since logic $\mathcal{L}^1$ has a natural definition but, as it completely characterizes event bisimulation, it will not be able to test the presence of states like those in V in S_2. This is due to the fact that the modality $\langle a \rangle\_$ can only test one transition at a time and, together with the other operators, any $\mathcal{L}^1$ formula can only test countably many transitions at a time. Notice that a state in the set V can only be distinguished through a formula testing that no action (in the *uncountable* set $[0, 1]$) can be performed.

Therefore, both examples call for adding structure to the set of labels on the NLMP. In the first case, endowing the set of labels with a σ-algebra exclude the "bad behaved" NLMPs like S_1 from the set of definable objects. In the second case, this will allow to define a richer logic that can test measurable sets of labels in a single formula. Regardless of these situations, a σ-algebra on the labels is also necessary for the definition of schedulers and probabilistic trace semantics [48].

5   Structured Non-deterministic Labeled Markov Processes

In view of the previous observations we developed a variant of NLMPs that requires that the set of labels has a measurable space associated. Since one of the aims of introducing structure on labels is to be able to define schedulers that resolve the (continuous) non-determinism of the model, we need to adapt the transition probability function to the new setting so that the different measurability aspects interact properly.

First, notice that a transition label is intended to represent the occurrence of a single action. Therefore, we will assume that, if L is the set of labels and Λ its associated σ-algebra, all singleton subsets of L are measurable in Λ.

Recall that a scheduler is a function that, given a particular execution history of the system, randomly selects a transition from those enabled at the last state of the execution. That is, given the fact that the execution finishes at state s in a given NLMP, the scheduler has to randomly chose first a label a and then a measure in $T_a(s)$. More precisely, a scheduler will have to randomly choose an element in $\theta = \{(a, \mu) \mid \mu \in T_a(s)\}$. So, we actually need θ to be measurable in $\Lambda \otimes \Delta(\Sigma)$.

Therefore, a *structured* NLMP has a single transition function $T : S \to \Lambda \otimes \Delta(\Sigma)$ that assigns to each state a measurable set of pairs of label-probability measure on states. As in the case of LMP and NLMP, we will also need that the transition function is a measurable mapping. Hence, we need to endow $\Lambda \otimes \Delta(\Sigma)$ with a σ-algebra. We proceed in a similar way to Def. 8.

Definition 22. $H(\Lambda \otimes \Delta(\Sigma))$ *is the smallest σ-algebra containing all sets* $H_{\lambda \times \xi} = \{\theta \in \Lambda \otimes \Delta(\Sigma) \mid \theta \cap (\lambda \times \xi) \neq \emptyset\}$, *with* $\lambda \in \Lambda$ *and* $\xi \in \Delta(\Sigma)$.

Here we follow a slightly different approach to that of Def. 8 by taking only hit sets induced by rectangles rather than arbitrary measurable sets in the product σ-algebra.

Now, we can formally define the structured version of NLMPs:

Definition 23. *A structured non-deterministic labeled Markov process (SNLMP for short) is a structure* $(S, \Sigma, L, \Lambda, T)$ *where* Σ *is a σ-algebra on the set of states S, Λ is a σ-algebra on the set of labels L so that* $\{a\} \in \Lambda$ *for all $a \in L$, and* $T : S \to \Lambda \otimes \Delta(\Sigma)$ *is measurable.*

An SNLMP can be straightforwardly encoded as an NLMP by taking $T_a(\cdot) = T(\cdot)|_a$, where $\theta|_a \doteq \{\mu \in \Delta(S) \mid (a, \mu) \in \theta\}$ is the a-section of θ, known to be measurable if θ is measurable. Also, it is not difficult to see that, in our setting, the section seen as a function $(\cdot)|_a$ is a measurable function. This ensures the required properties of T_a. As it can be expected, NLMPs can not be encoded as SNLMPs in general. This is confirmed in the following example.

Example 24. Consider the NLMP S_1 of Ex. 20. To translate it into an SNLMP, take $T(d) = \{(a, \mu) \mid \mu \in T_a(d)\}$ for all $d \in S_1$. Notice that

$$T(s) = \{a\} \times \{\delta_d \mid d \in [2, 3]\} \qquad T(r) = T(f(r)) = \{(r, \delta_x)\} \quad \text{if } r \in [0, 1]$$
$$T(t) = \{a\} \times \{\delta_d \mid d \in [0, 1]\} \qquad T(y) = \emptyset \qquad \text{otherwise.}$$

Though clearly $T(d)$ is a measurable set for any $d \in S_1$, T is not a measurable function. In effect, $T^{-1}(H_{[0,1] \times \Delta(S)}) = \{d \mid T(d) \cap ([0, 1] \times \Delta(S)) \neq \emptyset\} = [0, 1] \cup ([2, 3] \setminus V)$ which is not measurable, since V was chosen to be a non-Borel subset of $[2.5, 3]$. $\qquad \square$

Example 25. Notice, however, that S_2 in Ex. 21 can be encoded as an SNLMP provided function f^{-1} is measurable. This is immediate after observing that

$$T^{-1}(H_{\lambda \times \xi}) = \{s \mid a \in \lambda \wedge \{\delta_d \mid d \in [2, 3]\} \cap \xi \neq \emptyset\}$$
$$\cup \{t \mid a \in \lambda \wedge \{\delta_d \mid d \in [0, 1]\} \cap \xi \neq \emptyset\}$$
$$\cup \{d \mid d \in \lambda \cup f(\lambda) \wedge \delta_x \in \xi\} \qquad \square$$

All bisimulations introduced for NLMPs have their counterpart in SNLMPs. In fact, state bisimulation and hit bisimulation are defined exactly in the same way as for NLMPs by taking $T_a(\cdot) = T(\cdot)|_a$. For the event bisimulation, we also have to consider the fact that, in addition to states, labels are also observed through events.

Definition 26. *An* event bisimulation *on an SNLMP* $(S, \Sigma, L, \Lambda, T)$ *is a sub-σ-algebra Ξ of Σ s.t. $T : (S, \Xi) \to (\Lambda \otimes \Delta(\Sigma), H(\Lambda \otimes \Delta(\Xi)))$ is measurable.*

Just like for LMPs and NLMPs, the notion of event bisimulation can be extended to relations and the largest event bisimulation relation $\sim_e$ can be analogously defined.

This way of defining event bisimulation raises the question on why not redefining also the hit bisimulation so that it considers the new hit sets containing pairs of labels and probability measures as in Def. 22. It turns out that this variant does not alter the definition of hit bisimulation as we state in the following.

Theorem 27. *Consider the SNLMP $(S, \Sigma, L, \Lambda, T)$ and let $R \subseteq S \times S$ be a symmetric relation over states. The following characterizations for R are equivalent:*

(1) if $s\ R\ t$ then $T(s)|_a \cap \xi \neq \varnothing \Leftrightarrow T(t)|_a \cap \xi \neq \varnothing$, for all $a \in L$ and $\xi \in \Delta(\Sigma(R))$;
(2) if $s\ R\ t$ then $T(s) \cap (\lambda \times \xi) \neq \varnothing \Leftrightarrow T(t) \cap (\lambda \times \xi) \neq \varnothing$, for all $\lambda \in \Lambda$ and $\xi \in \Delta(\Sigma(R))$;
(3) if $s\ R\ t$ then $T(s) \cap \theta \neq \varnothing \Leftrightarrow T(t) \cap \theta \neq \varnothing$, for all $\theta \in \Lambda \otimes \Delta(\Sigma(R))$.

Clearly (3) implies (2) which implies (1). The proof that (1) implies (3) relies on the fact that $(((\{a\} \times \Delta(S)) \cap \theta)|_a \in \Delta(\Sigma(R))$. Notice that (1) is in fact the same definition of hit bisimulation as given in Def. 15 interpreting T_a as $T(\cdot)|_a$.

All results presented for the different bisimulations on NLMPs repeat on SNLMPs. In particular, it also holds that R is a hit bisimulation if and only if $\Sigma(R)$ is an event bisimulation (with the new definition of event bisimulation). Details can be found in [7].

Schedulers aside, the other reason to define SNLMPs was motivated by Ex. 21 in which the logic $\mathcal{L}^1$ failed to distinguish states s and t in $\mathbf{S}_2$. As we saw in Ex. 25, $\mathbf{S}_2$ is also an SNLMP. Therefore we would like to define a new logic that can distinguish s and t. To understand the difference, notice that t can perform and a-transition and reach a state where no transition labeled with $r \in [0, 1]$ can be performed with probability 1. This behavior could be described by a formula like $\langle a \rangle \neg [\langle [0, 1] \rangle [\top]_{\geq 1}]_{\geq 1}$.

Indeed, the logic $\mathcal{L}^2$ is the same logic as $\mathcal{L}^1$ where the modal construct $\langle a \rangle \psi$ has been replaced by $\langle \lambda \rangle \psi$ with $\lambda \in \Lambda$. The semantics of this new operator is given by

$$[\![\langle \lambda \rangle \psi]\!] \doteq T^{-1}\left(H_{\lambda \times [\![\psi]\!]}\right).$$

The semantics for the rest of the operations of $\mathcal{L}^2$ are defined just like for $\mathcal{L}^1$. Again, $[\![\psi]\!]$ is measurable for all $\varphi \in \mathcal{L}^2$.

Because singletons are measurable in Λ, $\langle \{a\} \rangle \psi \in \mathcal{L}^2$ provided $a \in L$ and $\psi \in \mathcal{L}^2$ (we will use $\langle a \rangle \psi$ as a shorthand). Therefore $\mathcal{L}^2$ is at least as expressive as $\mathcal{L}^1$. Moreover, it is strictly more expressive since $t \in [\![\langle a \rangle \neg [\langle [0, 1] \rangle [\top]_{\geq 1}]_{\geq 1}]\!]$ but $s \notin [\![\langle a \rangle \neg [\langle [0, 1] \rangle [\top]_{\geq 1}]_{\geq 1}]\!]$ and hence $\mathcal{L}^2$ can distinguish states s and t in $\mathbf{S}_2$.

It can be shown that $\mathcal{L}^2$ characterizes the event bisimulation for SNLMPs. The proof follows the same strategy as that of Theorem 16.

Theorem 28. *The logic $\mathcal{L}^2$ completely characterizes event bisimulation on SNLMPs. I.e. $\mathcal{R}(\mathcal{L}^2) = \sim_e$.*

For the next result, we need to interpret logics $\mathcal{L}^0$ and $\mathcal{L}^1$ on SNLMP, but this is easy since $\langle a \rangle \psi$ of $\mathcal{L}^1$ corresponds to $\langle \{a\} \rangle \psi$ in $\mathcal{L}^2$ and $\langle a \rangle_q \varphi$ of $\mathcal{L}^0$ corresponds to $\langle \{a\} \rangle [\varphi]_{\geq q}$ in $\mathcal{L}^2$. The following theorem summarizes the results for bisimulations and logics in SNLMPs.

Theorem 29. $\sim_s \subseteq \sim_h \subsetneqq \sim_e = \mathcal{R}(\mathcal{L}^2) \subsetneqq \mathcal{R}(\mathcal{L}^1) \subsetneqq \mathcal{R}(\mathcal{L}^0)$.

The last inclusion is shown to be proper in Ex. 18. Besides we also showed that the inclusion $\mathcal{R}(\mathcal{L}^2) \subseteq \mathcal{R}(\mathcal{L}^1)$ is proper using SNLMP $\mathbf{S_2}$. The next example shows that inclusion $\sim_h \subseteq \sim_e$ is also proper.

Example 30. Consider the SNLMP $\mathbf{S_3}$ which is a variant of $\mathbf{S_2}$ where $V = (2.5, 3]$, f is measurable, and T is redefined as follows.

$$T(s) = \{a\} \times \{\delta_d \mid d \in [2, 3]\}$$
$$T(t) = \{a\} \times \{\delta_d \mid d \in [0, 1]\}$$
$$T(r) = T(f(r)) = \{(r, \delta_x) \mid r \in [0, 1]\backslash\{r\}\} \qquad \text{if } r \in [0, 1]$$
$$T(d) = \{(r, \delta_x) \mid r \in [0, 1]\} \qquad \text{if } d \in V$$
$$T(y) = \varnothing \qquad \text{otherwise.}$$

Note that the states r and $f(r)$ can perform any $[0, 1]$-labeled transition except for the r-labeled transition whenever $r \in [0, 1]$. Instead, every $d \in V$ can perform all $[0, 1]$-labeled transitions. Therefore, every pair of states in V are hit bisimilar, and every state $d \in V$ can be distinguished from states in $[0, 1] \cup [2, 3]\backslash V$ since $T(d)|_r \cap \Delta(S) = \{\delta_x\} \neq \varnothing = T(r)|_r \cap \Delta(S) = T(f(r))|_r \cap \Delta(S)$. Thus, $V \in \Sigma$ is $\sim_h$-closed and consequently $\delta_V = \{\delta_d \mid d \in V\} \in \Delta(\Sigma(\sim_h))$. From here we have that $T(s)|_a \cap \delta_V = \delta_V \neq \varnothing = T(t)|_a \cap \delta_V$, and therefore s and t are not hit bisimilar.

Now, take $\mathcal{F} = \{\{s, t\}, \{x\}, \{r, f(r)\}_{r\in[0,1]}\}$. It is not hard to prove that $\sigma(\mathcal{F})$ is an event bisimulation. Hence $s \sim_e t$. $\qquad\square$

Contrarily to what happens in NLMPs with example $\mathbf{S_2}$, the example above questions the hit bisimulation (rather than the event bisimulation) as it seems to distinguish sets of null measure. In fact any definable scheduler starting from state t has an *"almost surely equivalent"* scheduler starting from s (modulo state renaming).

Also, the question of whether state bisimilarity is strictly finer than hit bisimilarity on SNLMPs remains open. Notice that Ex. 20 is not a valid counterexample in the realm of SNLMPs because $\mathbf{S_1}$ is not an SNLMP.

6   Concluding Remarks

In this paper, we have presented the basic theory of LMPs and its extensions with internal non-determinism, namely NLMPs and SNLMPs. Much more research on this subject has been done. For instance, pseudometrics that behave like bisimulation in the limit have been defined and different kind of approximations for LMPs have been studied [4, 11, 13, 23, 24, 39, etc.].

When it comes to give semantics to languages or symbolic models that includes stochastic continuous behavior, NLMPs showed to be useful. Stochastic automata [15, 16] provide a symbolic framework to model soft real-timed systems. They can be seen as a non-deterministic extension of generalized semi-Markov processes that are amenable to composition. The semantics of a stochastic automaton naturally arises as a (structured) NLMP [48]. In a similar manner NLMPs have been used to give semantics to more complex models, particularly stochastic hybrid automata [27, 31]. As a consequence, NLMPs are also the concrete underlying semantics of a process algebra like SPADES [15, 16] and modeling languages like MODEST and HMODEST [3, 31]. These languages have been used to analyze real case studies (e.g. [30, 32]). [48] presents also mappings from pGCL [36] and abstract probabilistic automata [19] into NLMPs.

We have made use of the concept of schedulers to introduce SNLMPs. In fact, we have formally defined schedulers on SNLMPs and used them to define trace distribution semantics. See [48] for these results. We remark that a still overdue result in the setting of LMPs and NLMPs is a correspondence execution theorem which states that if two states are bisimilar (in any of the senses defined here), they share the same probabilistic execution structure and hence they are also trace distribution equivalent.

Desharnais et al. [22] followed a different approach to extend LMPs with some kind of internal non-determinism. Rather than explicitly introducing the branching set of probability measures as in NLMPs, they relax the requirements on the LMP by only asking that $\tau_a(s)$ is a super-additive function on Σ (instead of a sub-probability measure). They call this new model *infLMP*. An infLMP can be understood as a partially specified system where a possible implementation is an LMP in which its transition probability function is greater than or equal to the transition super-additive function of the infLMP. It would be interesting to draw conclusions whether NLMPs can capture infLMPs or not. A first (but inconclusive) approach to this relation is reported in [48].

Finally, the results in [42, 43] show that the generality of the models immediately leads to unwanted results. It seems reasonable to restrict only to standard Borel spaces. Confining to standard Borel spaces is not as restricting as it seems since most natural problems arise in this setting. For example, we have that the underlying semantics of stochastic (hybrid) automata is given in terms of an NLMP on standard Borel spaces, and in the case of stochastic automata, such NLMP is also image finite. Recall that stochastic automata and similar models are used to give semantics to stochastic process algebras and specification languages, see e.g. [3, 5, 6, 15, 16, 31]. Moreover, LMP-like models restricted to standard Borel spaces have been studied in [26].

References

1. Ash, R., Doléans-Dade, C.: Probability & Measure Theory. Academic Press (2000)
2. Baier, C., Katoen, J.: Principles of Model Checking. The MIT Press (2008)
3. Bohnenkamp, H., D'Argenio, P., Hermanns, H., Katoen, J.P.: MoDeST: A compositional modeling formalism for real-time and stochastic systems. IEEE Trans. Softw. Eng. 32(10), 812–830 (2006)
4. Bouchard-Côté, A., Ferns, N., Panangaden, P., Precup, D.: An approximation algorithm for labelled markov processes: towards realistic approximation. In: Proc. of QEST 2005, pp. 54–62. IEEE Computer Society (2005)
5. Bravetti, M.: Specification and Analysis of Stochastic Real-Time Systems. Ph.D. thesis, Università di Bologna, Padova, Venezia (2002)
6. Bravetti, M., D'Argenio, P.R.: Tutte le algebre insieme: Concepts, discussions and relations of stochastic process algebras with general distributions. In: Baier, C., Haverkort, B.R., Hermanns, H., Katoen, J.-P., Siegle, M. (eds.) Validation of Stochastic Systems. LNCS, vol. 2925, pp. 44–88. Springer, Heidelberg (2004)

7. Budde, C.: No determinismo completamente medible en procesos probabilísticos continuos. Master's thesis, FaMAF, Universidad Nacional de Córdoba (2012)
8. Cattani, S.: Trace-based Process Algebras for Real-Time Probabilistic Systems. Ph.D. thesis, University of Birmingham (2005)
9. Cattani, S., Segala, R., Kwiatkowska, M., Norman, G.: Stochastic transition systems for continuous state spaces and non-determinism. In: Sassone, V. (ed.) FOSSACS 2005. LNCS, vol. 3441, pp. 125–139. Springer, Heidelberg (2005)
10. Celayes, P.: Procesos de Markov Etiquetados sobre Espacios de Borel Estándar. Master's thesis, FaMAF, Universidad Nacional de Córdoba (2006)
11. Chaput, P., Danos, V., Panangaden, P., Plotkin, G.: Approximating labelled markov processes again! In: Kurz, A., Lenisa, M., Tarlecki, A. (eds.) CALCO 2009. LNCS, vol. 5728, pp. 145–156. Springer, Heidelberg (2009)
12. Clarke, E., Grumberg, O., Peled, D.: Model Checking. MIT Press (1999)
13. Danos, V., Desharnais, J.: Labelled markov processes: Stronger and faster approximations. In: Proc. of 18th LICS, pp. 341–350. IEEE Computer Society (2003)
14. Danos, V., Desharnais, J., Laviolette, F., Panangaden, P.: Bisimulation and cocongruence for probabilistic systems. Inf. & Comp. 204, 503–523 (2006)
15. D'Argenio, P.: Algebras and Automata for Timed and Stochastic Systems. Ph.D. thesis, Department of Computer Science, University of Twente (1999)
16. D'Argenio, P., Katoen, J.P.: A theory of stochastic systems, Part I: Stochastic automata, and Part II: Process algebra. Inf. & Comp. 203(1), 1–38, 39–74 (2005)
17. D'Argenio, P., Sánchez Terraf, P., Wolovick, N.: Bisimulations for non-deterministic labelled Markov processes. Mathematical. Structures in Comp. Sci. 22(1), 43–68 (2012)
18. D'Argenio, P., Wolovick, N., Sánchez Terraf, P., Celayes, P.: Nondeterministic labeled Markov processes: Bisimulations and logical characterization. In: Proc. of QEST 2009, pp. 11–20. IEEE Computer Society (2009)
19. Delahaye, B., Katoen, J.-P., Larsen, K.G., Legay, A., Pedersen, M.L., Sher, F., Wąsowski, A.: Abstract probabilistic automata. In: Jhala, R., Schmidt, D. (eds.) VMCAI 2011. LNCS, vol. 6538, pp. 324–339. Springer, Heidelberg (2011)
20. Desharnais, J.: Labeled Markov Process. Ph.D. thesis, McGill University (1999)
21. Desharnais, J., Edalat, A., Panangaden, P.: Bisimulation for labelled Markov processes. Inf. & Comp. 179(2), 163–193 (2002)
22. Desharnais, J., Laviolette, F., Turgeon, A.: A logical duality for underspecified probabilistic systems. Inf. Comput. 209(5), 850–871 (2011)
23. Desharnais, J., Gupta, V., Jagadeesan, R., Panangaden, P.: Approximating labelled markov processes. Inf. Comput. 184(1), 160–200 (2003)
24. Desharnais, J., Gupta, V., Jagadeesan, R., Panangaden, P.: Metrics for labelled markov processes. Theor. Comput. Sci. 318(3), 323–354 (2004)
25. Doberkat, E.E.: Kleisli morphisms and randomized congruences for the Giry monad. Journal of Pure and Applied Algebra 211(3), 638–664 (2007)
26. Doberkat, E.E.: Stochastic relations. Foundations for Markov transition systems. Studies in Informatics Series. Chapman & Hall/CRC (2007)
27. Fränzle, M., Hahn, E., Hermanns, H., Wolovick, N., Zhang, L.: Measurability and safety verification for stochastic hybrid systems. In: Caccamo, M., Frazzoli, E., Grosu, R. (eds.) Proc. of HSCC 2011, pp. 43–52. ACM (2011)
28. Giry, M.: A categorical approach to probability theory. In: Categorical Aspects of Topology and Analysis. LNM, vol. 915, pp. 68–85. Springer (1981)
29. van Glabeek, R.: The linear time–branching time spectrum I. The semantics of concrete, sequential processes. In: Bergstra, J.A., Ponse, A., Smolka, S.A. (eds.) Handbook of Process Algebra, pp. 3–99. North-Holland (2001)

30. Baró Graf, H., Hermanns, H., Kulshrestha, J., Peter, J., Vahldiek, A., Vasudevan, A.: A verified- wireless safety critical hard real-time design. In: Proc. of WOWMOM 2011, pp. 1–9. IEEE (2011)
31. Hahn, E., Hartmanns, A., Hermanns, H., Katoen, J.P.: A compositional modeling and analysis framework for stochastic hybrid systems. Formal Methods in System Design 43(2), 191–232 (2013)
32. Hartmanns, A., Hermanns, H.: Modelling and decentralised runtime control of self-stabilising power micro grids. In: Margaria, T., Steffen, B. (eds.) ISoLA 2012, Part I. LNCS, vol. 7609, pp. 420–439. Springer, Heidelberg (2012)
33. Hennessy, M., Milner, R.: Algebraic laws for nondeterminism and concurrency. J. ACM 32(1), 137–161 (1985)
34. Kechris, A.: Classical Descriptive Set Theory. Graduate Texts in Mathematics, vol. 156. Springer (1995)
35. Larsen, K., Skou, A.: Bisimulation through probabilistic testing. Inf. & Comp. 94(1), 1–28 (1991)
36. McIver, A., Morgan, C.: Abstraction, Refinement and Proof for Probabilistic Systems. Monographs in Computer Science. Springer (2005)
37. Milner, R.: Communication and Concurrency. Prentice Hall (1989)
38. Naimpally, S.: What is a hit-and-miss topology? Topological Comment 8(1) (2003)
39. Panangaden, P.: Labelled Markov Processes. Imperial College Press (2009)
40. Parma, A., Segala, R.: Logical characterizations of bisimulations for discrete probabilistic systems. In: Seidl, H. (ed.) FOSSACS 2007. LNCS, vol. 4423, pp. 287–301. Springer, Heidelberg (2007)
41. Puterman, M.: Markov Decision Processes: Discrete Stochastic Dynamic Programming. Wiley-Interscience (1994)
42. Sánchez Terraf, P.: Unprovability of the logical characterization of bisimulation. Inf. & Comp. 209(7), 1048–1056 (2011)
43. Sánchez Terraf, P.: Bisimilarity is not Borel. CoRR arXiv:1211.0967 (2012)
44. Segala, R.: Modeling and Verification of Randomized Distributed Real-Time Systems. Ph.D. thesis, Massachusetts Institute of Technology (1995)
45. Strulo, B.: Process Algebra for Discrete Event Simulation. Ph.D. thesis, Department of Computing, Imperial College, University of London (1993)
46. Vardi, M.: Automatic verification of probabilistic concurrent finite-state programs. In: 26th FOCS, pp. 327–338. IEEE (1985)
47. Viglizzo, I.: Coalgebras on Measurable Spaces. Ph.D. thesis, Department of Mathematics, Indiana University (2005)
48. Wolovick, N.: Continuous probability and nondeterminism in labeled transition systems. Ph.D. thesis, Universidad Nacional de Córdoba (2012)

On Abstraction of Probabilistic Systems

Christian Dehnert[1], Daniel Gebler[2], Michele Volpato[3], and David N. Jansen[3]

[1] Software Modeling and Verification Group,
RWTH Aachen University, Ahornstraße 55, 52056 Aachen, Germany
[2] Department of Computer Science, VU University Amsterdam,
De Boelelaan 1081a, 1081 HV Amsterdam, The Netherlands
[3] Institute for Computing and Information Sciences,
Radboud University, Heyendaalseweg 135, 6500 GL Nijmegen, The Netherlands

Abstract. Probabilistic model checking extends traditional model checking by incorporating quantitative information about the probability of system transitions. However, probabilistic models that describe interesting behavior are often too complex for straightforward analysis. Abstraction is one way to deal with this complexity: instead of analyzing the ("concrete") model, a simpler ("abstract") model that preserves the relevant properties is built and analyzed. This paper surveys various abstraction techniques proposed in the past decade. For each abstraction technique we identify in what sense properties are preserved or provide alternatively suitable boundaries.

1   Introduction

The advent of large-scale, distributed, dependable systems requires formal specification and verification methods which capture both *qualitative* and *quantitative* properties of systems. Performance and dependability evaluation of distributed systems therefore demands to use formal models and methods where both aspects are represented. Labeled transition systems (LTS) allow to capture qualitative (functional) aspects of software and hardware systems. To model the mentioned quantitative phenomena, one uses a probabilistic formalism, typically some extension of Markov chains. For instance, exchanging messages between distributed systems typically suffers from a failure probability. Hence, interesting properties of real systems often express that some functional behavior can be guaranteed to happen with at least some given probability or, dually, some bad behavior appears with at most some given probability. Probabilistic models, such as Markov chains and Markov decision processes, allow to model and reason over both qualitative (functional) and quantitative (non-functional) aspects.

The properties of probabilistic systems are typically specified in temporal logics such as the probabilistic computation tree logic (PCTL) [25, 6]. Model checking is a method to verify those properties. However, it suffers from the state space explosion problem, which means that the number of reachable states of the model under investigation is too large. In models specified compositionally, it is often exponential in the number of components of the model. Additionally,

A. Remke and M. Stoelinga (Eds.): ROCKS Autumn School 2012, LNCS 8453, pp. 87–116, 2014.

probabilistic model checking relies on expensive numerical methods, making the problem even more pressing. Consequently, it is generally of crucial importance to simplify the model prior to verification. However, given the complexity of typical models, this procedure needs to be both automated and efficient.

In this paper, we present selected *abstraction* techniques for probabilistic systems. Intuitively, abstraction removes details from concrete models that are not relevant to the property of interest. In many cases, only abstraction makes the analysis of the model feasible or at least speeds up verification considerably. Verification of realistic models requires the application of aggressive abstraction techniques (c.f. [10]). We present the following abstraction techniques:

Multi-valued abstraction allows to partition the state space and to abstract transition probabilities by intervals. Both positive and negative verification results in the abstract model carry over to the concrete model. However, in the abstract model some properties may evaluate to "unknown" if the abstract model does not allow a conclusive evaluation of the property.

Counterexample-guided abstraction refinement (CEGAR) uses counterexamples for the abstract model obtained from a model checker to refine the abstraction. By (dis)proving realizability of the abstract counterexample in the concrete model, this allows for an automatic abstraction-refinement technique that proves or refutes properties. We survey the probabilistic CEGAR algorithm introduced by [27] and [45, Section 7].

Game-based abstraction provides the means to abstraction while maintaining the separation between nondeterminism present in the concrete model and nondeterminism introduced by the abstraction. We present probabilistic game-based and menu-based abstraction, which employ two-player games where opponent and defender take different roles in resolving the nondeterminism. The resulting game allows to give distinct upper and lower bounds on reachability properties. This interval can also be understood as a measure of the quality of the abstraction.

Organization of the paper. Section 2 provides a survey on the available literature and the tools for the discussed subject. Section 3 introduces the formal framework, i.e. probabilistic models, probabilistic temporal logics and probabilistic games. Multi-valued abstraction is covered in Section 4. Probabilistic CEGAR is surveyed in Section 5 and, finally, game-based abstraction techniques are presented in Section 6. We summarize and conclude in Section 7.

2   Related Work

2.1   Literature

Abstraction is of immense importance for the analysis of large probabilistic systems. Consequently, the field has been studied extensively. One of the most popular techniques is *bisimulation minimization* [4]. Here, the states of the abstract system represent equivalence classes of an equivalence relation on the states, a bisimulation, such that the abstract system is guaranteed to preserve certain properties. [18] investigates several kinds of strong and weak bisimulations regarding the minimality of the quotient system with respect to the number of states, the number of transitions and transition fan-out. In [32], the authors show that (strong) bisimulation can, in practice, lead to significant savings in memory and runtime of explicit state probabilistic model checking. Approaches that compute the bisimulation quotients on a symbolic representation of the abstract state space are proposed in both [47] and [14], where the former uses multi-terminal binary decision diagrams and the latter focuses on a representation of the state space in terms of predicates. [37, 17] compute a bisimulation based on symmetry in the models that is easier to compute than strong bisimulation, but may produce larger quotients. [15] proposes an abstraction technique for probabilistic automata based on may and must modalities inspired by modal transition systems [41]. [12, 13] pioneered the use of an abstraction-refinement approach for probabilistic systems that tries to prove a reachability property on a very coarse abstraction of the system. If the verification fails, the system is successively refined until a conclusive answer can be given. While probabilistic *CE-GAR* [27, 8] uses counterexamples obtained from a (probabilistic) model checker to refine the abstract system, the game-based techniques [35, 45] typically rely on disagreeing strategies for the individual players to make the abstraction more precise when required. *Magnifying-lens abstraction* [1] uses a similar scheme, but rather considers the concrete states contained in an abstract state in each step and thus "magnifies" the state.

In infinite state probabilistic models, typically almost the whole probability mass is concentrated in a finite subset of the states. *Sliding-window abstraction* [26] is a technique to abstract from an infinite state space by "hiding" irrelevant states (in the sense that they possess a negligible amount of probability mass) in a way similar to the view through a window. Over time, as different states become relevant, the window slides to different areas of the state space.

Often, probabilistic models arise from the parallel composition of several components. *Assume-guarantee* verification [36, 21, 39] aims at proving a property of the composed model without actually building a representation of the full model by verifying the components in isolation. As the interaction between the components is typically essential to prove a given property, these techniques try to create small assumptions that can be proven on one component and suffice to establish the property on the other components. Note that some of the aforementioned techniques, for example bisimulation minimization, are also compositional in the sense that they can be applied to the individual components that are further subject to parallel composition.

Abstraction and refinement are closely related to *simulation* relations. A simulation relation is a relation between two models that shows a form of weak preservation: all properties expressible as positive formulas are preserved. In a probabilistic context, one usually chooses a *liveness* view on simulation: a probabilistic liveness property is a lower bound on the probability of some (good) behavior. One wants the concrete model to be at least as good as the abstract one, so every liveness property ensured by the abstract model should also hold in the concrete one. A good simulation relation shows a form of weak preservation, i. e., all liveness properties in some suitable logic are preserved. Simulation relations for probabilistic systems have been studied for systems without [5, 29] and with nondeterminism [43, 50]. Work in this area also explores systems with continuous state spaces and how such state spaces can be approximated by a finite Markov model [16].

2.2 Tools

Tools implementing one or several of the aforementioned abstraction techniques have been developed. Such tools not only served as prototypical implementations for evaluations in the literature, but are still available and in use. SIGREF [47] is a tool implementing bisimulation minimization for systems represented as (variants of) binary decision diagrams that is, for example, applied in performance analysis of AADL models [7]. Another tool, PASS [23], employs a mixture of probabilistic CEGAR and the game-based approaches to provide lower and upper bounds for both minimal and maximal reachability probabilities. Finally, PRISM-games [9], extends the well-known probabilistic model checker PRISM [38] by an engine for probabilistic games.

3   Preliminaries

This section briefly introduces the basic notions and definitions. All material is standard and the interested reader is pointed to the original literature [4] and the referred material therein.

3.1   Markov Models

Markov models are similar to transition systems in that they comprise states and transitions between these states. In discrete-time Markov chains, each state is associated with a discrete probability distribution over successor states according to which the next state is chosen. Let $Dist(S)$ be the set of *discrete probability distributions* over a set S, i. e., the set of functions $\mu\colon S \to [0,1]$ such that $\sum_{s \in S} \mu(s) = 1$.

Definition 1 (Discrete-time Markov chain (DTMC)). *A discrete-time Markov chain is a tuple* $\mathcal{D} = (S, \mathbf{P}, s_{init}, AP, L)$ *where*

- *S is a countable, non-empty set of states,*
- *$\mathbf{P}\colon S \times S \to [0,1]$ is the transition probability function that assigns to each pair (s, s') of states the probability $\mathbf{P}(s, s')$ of moving from state s to s' in one step such that $\mathbf{P}(s, \cdot) \in Dist(S)$,*
- *$s_{init} \in S$ is the initial state,*
- *AP is a set of atomic propositions, and*

- $L: S \rightarrow 2^{AP}$ is the labeling function that assigns a (possibly empty) set of atomic propositions $L(s) \subseteq AP$ to a state $s \in S$.

Let $\mathbf{P}(s, S) = \sum_{s' \in S} \mathbf{P}(s, s')$ be the probability to move from state s to some state $s' \in S$. A path in a DTMC is an infinite sequence of states of the form $\omega = s_1 s_2 \ldots$ such that $\mathbf{P}(s_i, s_{i+1}) > 0$ for all $i \geq 1$. Let $Path^{\mathcal{D}}$ denote the set of paths in the DTMC $\mathcal{D}$ and $Path_{\text{fin}}^{\mathcal{D}}$ denote the set of finite prefixes of all paths. For $\omega \in Path_{\text{fin}}^{\mathcal{D}}$, by $last(\omega)$ we refer to the last state of the finite path. A probability measure $Pr^{\mathcal{D}}$ on the set $Path^{\mathcal{D}}$ can be defined as unique extension of the measure on the respective cones [25].

As the behavior of a DTMC is purely probabilistic, it is not well suited to model concurrent systems. *Markov decision processes* add nondeterministic behavior to DTMCs by allowing (external) nondeterministic choice over probability distributions in each state.

Definition 2 (Markov decision process (MDP)). *A* Markov decision process *is a tuple* $\mathcal{M} = (S, Act, \mathbf{P}, s_{init}, AP, L)$ *where*

- *S, s_{init}, AP and L are as for DTMCs (see Definition 1),*
- *Act is a finite set of actions,*
- *$\mathbf{P}: S \times Act \times S \rightarrow [0, 1]$ is the transition probability function that specifies the probability to move from s to s' with action $\alpha \in Act$ such that $\mathbf{P}(s, \alpha, \cdot) \in Dist(S)$ or $\mathbf{P}(s, \alpha, \cdot)$ is the constant zero function.*

Let $Act(s)$ denote the set of enabled actions in state s, i.e., the actions α that satisfy $\sum_{s' \in S} \mathbf{P}(s, \alpha, s') = 1$. For simplicity, we require that the MDP has no deadlock states, i.e., $Act(s) \neq \emptyset$ for all states $s \in S$. We denote the set of distributions available at state s by $Steps(s) = \{\mathbf{P}(s, \alpha, \cdot) \in Dist(S) \mid \alpha \in Act(s)\}$. In each state s, first some enabled action $\alpha \in Act(s)$ is chosen nondeterministically. Then, the probabilistic choices given by $\mathbf{P}(s, \alpha, \cdot)$ yield the successor state s'. Thus, the set $Path^{\mathcal{M}}$ is given by all sequences of the form $\omega = s_1 \alpha_1 s_2 \alpha_2 \ldots$, where $s_i \in S$ and $\alpha_i \in Act$, such that $\mathbf{P}(s_i, \alpha_i, s_{i+1}) > 0$ for all $i \geq 1$. Similarly to DTMCs, we define $Path_{\text{fin}}^{\mathcal{M}}$ as the set of finite prefixes, ending with a state, of all paths and use $last(\cdot)$ accordingly.

Schedulers provide means to resolve the nondeterministic choices of MDPs. The most general class of schedulers uses the complete trajectory up to the current state and resolves the nondeterminism to a probabilistic choice (history-dependent randomized schedulers). An important subclass are schedulers which depend only on the current state (not on the history) and which resolve the nondeterministic choice to a deterministic choice (memoryless deterministic schedulers). Formally they are given by a function $\sigma: S \rightarrow Act$ such that $\sigma(s) \in Act(s)$. They are powerful enough to reason over probabilistic reachability properties. The resolution of the nondeterministic choices in an MDP by a scheduler σ leads to an (infinite) DTMC $\mathcal{D}_{\sigma}^{\mathcal{M}} = (S^+, \mathbf{P}', s_{init}, AP, L')$ where $\mathbf{P}'(\omega, \omega s') = \mathbf{P}(last(\omega), \sigma(\omega), \omega s')$ and $L'(\omega) = L(last(\omega))$. Intuitively, the behavior of this DTMC corresponds to the behavior of the MDP under the scheduler σ. Not surprisingly, the probability measure over the infinite paths of $\mathcal{M}$ under σ, denoted $Pr_{\sigma}^{\mathcal{M}}$, is thus given by the measure over the DTMC $\mathcal{D}_{\sigma}^{\mathcal{M}}$.

In order to express properties over these models, probabilistic extensions of common logics are used. While *probabilistic computation tree logic* (PCTL) [3] is the most prominent one, a probabilistic interpretation of linear temporal logic [44] that can enforce probability bounds on the set of paths satisfying a regular LTL formula is also popular. We will, however, focus our attention to PCTL in the further course of the paper. Formulae in this logic are given by the following grammar.

Definition 3 (Probabilistic computation tree logic (PCTL)). *The syntax of PCTL state formulae over a set of atomic propositions AP is given by the following rules:*

$$\Phi ::= true \mid a \mid \Phi \wedge \Phi \mid \neg \Phi \mid P_{\bowtie p}(\varphi)$$

where $a \in AP$, φ is a PCTL path formula, $\bowtie \in \{<, \leq, >, \geq\}$ and $p \in [0, 1]$.
    PCTL path formulae are defined by the grammar:

$$\varphi ::= X\,\Phi \mid \Phi\,U\,\Phi \mid \Phi\,U^{\leq k}\,\Phi$$

where Φ is a state formula and $k \in \mathbb{N}$.

PCTL keeps the basic structure of CTL [19] and replaces the path quantifiers by the single new operator $P_{\bowtie p}(\varphi)$. Intuitively, this formula holds, if the probability mass of all paths in the model that satisfy φ conforms to $\bowtie p$. To improve readability, we will use the abbreviations $false$ for $\neg true$ and $\Phi_1 \vee \Phi_2$ for $\neg(\neg\Phi_1 \wedge \neg\Phi_2)$. For DTMCs the interpretation of a PCTL formula is straightforward. However, for MDPs the probability mass of all paths that satisfy a given path formula depends on the resolution of nondeterminism. A PCTL formula holds in a state of an MDP if it holds for all possible schedulers. Fortunately, there is no algorithmic need to optimize over all (infinitely many) schedulers, because it can be proven that only finitely many schedulers have to be considered [6]. More specifically, memoryless schedulers attain minimal and maximal probabilities for next-step and unbounded-until path formulae whereas k step-bounded schedulers are sufficient for bounded-until formulae with time-bound k. The minimal and maximal probabilities of an MDP $\mathcal{M}$ satisfying a given PCTL path formula φ are denoted $Pr_{min}^{\mathcal{M}}(\varphi)$ and $Pr_{max}^{\mathcal{M}}(\varphi)$, respectively:

$$Pr_{min}^{\mathcal{M}}(\varphi) = \inf_{\sigma} Pr_{\sigma}^{\mathcal{M}}(\varphi) \qquad Pr_{max}^{\mathcal{M}}(\varphi) = \sup_{\sigma} Pr_{\sigma}^{\mathcal{M}}(\varphi).$$

Sometimes, we need to restrict ourselves to a fragment $PCTL_{reach}$ of PCTL that only expresses *probabilistic reachability* properties. A probabilistic reachability property P is a PCTL formula of the form $\Phi = P_{\bowtie p}(true \; U \; \Phi_F)$, which we abbreviate as $P_{\bowtie p}(\Diamond \Phi_F)$, where $p \in [0, 1]$, $\bowtie \in \{<, \leq, >, \geq\}$ and Φ_F is a propositional logic formula over atomic propositions. Note that the truth value of Φ_F can be determined for each state in isolation, which is why we will treat F like an atomic proposition and assume that states are labeled accordingly.

Yet, sometimes it is necessary to further restrict this subset of PCTL to the set $PCTL_{safe}$ of *probabilistic safety* properties, which are PCTL formulae that apply negation only to literals and only use comparison operators from $\{<, \leq\}$. While these restrictions seem very severe, many problems can be reduced to reachability problems, and safety properties allow for expressing very interesting properties of a system in practice, e. g., "the probability to reach a set of states F is less than 0.5".

3.2 Probabilistic Two-Player Games

While MDPs extend DTMCs with nondeterministic choices, some abstractions rely on the separation of the nondeterminism introduced in the abstraction and the nondeterminism present in the original model. Probabilistic, or stochastic, two-player games are a natural formalism for this.

Definition 4 (Probabilistic game). *A probabilistic two-player game is a tuple* $\mathcal{G} = ((V, E), v_{init}, (V_1, V_2, V_p), \delta)$ *where*

- *(V, E) is a directed graph with edge set $E \subseteq V_1 \times V_2 \cup V_2 \times V_p \cup V_p \times V_1$,*
- *$v_{init} \in V_1$ is the initial vertex,*
- *(V_1, V_2, V_p) is a partition of V where V_1 is the set of vertices of player 1, V_2 is the set of vertices of player 2, and elements of V_p are probabilistic vertices,*
- *$\delta\colon V_p \to Dist(V_1)$ is a function that maps each probabilistic vertex to a probability distribution specifying its successor vertices such that $\delta(v_p)(v_1) > 0$ implies $(v_p, v_1) \in E$.*

A play in this game is an infinite sequence $\omega = v_{1,1}v_{2,1}v_{p,1}v_{1,2}v_{2,2}v_{p,2}\ldots$, where $v_{1,i} \in V_1, v_{2,i} \in V_2$ and $v_{p,i} \in V_p$, such that $(v_{1,i}, v_{2,i}), (v_{2,i}, v_{p,i}) \in E$ and $\delta(v_{p,i})(v_{1,i+1}) > 0$ for all $i \geq 1$. Let $Play^{\mathcal{G}}$ denote the set of all plays in $\mathcal{G}$, $Play^{\mathcal{G}}_{fin}$ the finite prefixes thereof and $last(\omega_{fin})$ refers to the last vertex of the finite prefix ω_{fin} of a play. Furthermore, $\omega(i)$ denotes the ith vertex of ω.

Intuitively, the behavior of a probabilistic game is as follows. Initially, starting in $v_{1,1} \in V_1$, player 1 nondeterministically chooses a successor vertex $v_{2,1} \in V_2$ belonging to player 2. Player 2 reacts by choosing a successor state $v_{p,1} \in V_p$. Then, the next vertex is chosen according to the probability distribution in v_p and it is again player 1's turn. Thus, in order to resolve the nondeterminism, a scheduler is needed for each of the players. In the context of games, these are called strategies for player 1 and 2, respectively. Just like for MDPs, these strategies are functions that map a finite prefix of a play to a possible choice. Formally, strategies for the players are given by functions $\sigma_1\colon (V_1 V_2 V_p)^* V_1 \to V_2$ and $\sigma_2\colon (V_1 V_2 V_p)^* V_1 V_2 \to V_p$, respectively, such that $\sigma_1(\omega v_1) = v_2$ and $\sigma_2(\omega v_1 v_2) = v_p$ implies $(v_1, v_2) \in E$ and $(v_2, v_p) \in E$, respectively. If two strategies, σ_1 for player 1 and σ_2 for player 2, are fixed, then, given a vertex v, the sets of finite and infinite plays starting in v which follow those strategies are denoted as $Play^{\sigma_1, \sigma_2}_{fin}(v)$ and $Play^{\sigma_1, \sigma_2}(v)$ respectively. In such a play, all nondeterministic choices are resolved by the strategies and the remaining behavior is purely probabilistic. Hence, a probability measure, denoted $Prob^{\sigma_1, \sigma_2}_v$, can be defined over the resulting model as in [11].

Given a set $F \subseteq V_1$ of target vertices, let

$$p_v^{\sigma_1,\sigma_2}(F) = Prob_v^{\sigma_1,\sigma_2}(\{\omega \in Play^{\sigma_1,\sigma_2}(v) \mid \exists i \in \mathbb{N}, \omega(i) \in F\})$$

be the probability for reaching a vertex in F if the game is played according to the strategies σ_1 and σ_2. For the case where the two players play adversarially, we define the optimal reachability probabilities as

$$p_v^{+-}(F) = \sup_{\sigma_1} \inf_{\sigma_2} p_v^{\sigma_1,\sigma_2}(F)$$
$$p_v^{-+}(F) = \inf_{\sigma_1} \sup_{\sigma_2} p_v^{\sigma_1,\sigma_2}(F).$$

Accordingly, for the opposite case in which the two players cooperate, we have the optimal reachability probabilities

$$p_v^{--}(F) = \inf_{\sigma_1,\sigma_2} p_v^{\sigma_1,\sigma_2}(F)$$
$$p_v^{++}(F) = \sup_{\sigma_1,\sigma_2} p_v^{\sigma_1,\sigma_2}(F).$$

These probabilities can be computed, e. g., using value iteration [11].

3.3 Probabilistic Programs

The main motivation for abstracting models prior to verification is the hope that this will reduce the time and memory needed for verification. As building the full model is often the a time and memory consuming step in the verification procedure, applying the abstraction after that step has little potential to improve the overall performance. Hence, many successful abstraction techniques avoid building the concrete model by employing a symbolic representation. A common model for succinctly representing Markov models are probabilistic programs, which are also able to finitely represent possibly infinite Markov models. Let $BExpr_{Var}$ denote the set of boolean expressions over a set of variables Var and $b \in BExpr_{Var}$ be an element of such a set. We denote by $[\![b]\!]$ the set of all valuations of variables in Var under which b evaluates to true.

Definition 5 (Probabilistic program). *A probabilistic program is a tuple* $\mathcal{P} = (Var, \Sigma, s_{init}, C)$ *where*

- *$Var = \{v_1, \ldots, v_n\}$ is a finite set of variables,*
- *$\Sigma = \Sigma(v_1) \times \ldots \times \Sigma(v_n)$ is the state space of the program, where $\Sigma(v)$ denotes the (possibly infinite) domain of the variable $v \in Var$,*
- *$s_{init} \in \Sigma$ is the initial state,*
- *C is a finite set of guarded commands of the form $c = g \rightarrow p_1 : u_1 \oplus \ldots \oplus p_m : u_m$ where*
  - *$g \in BExpr_{Var}$ is the guard of the command,*
  - *probabilities $p_i \in [0,1]$, such that $\sum_{1 \leq i \leq m} p_i = 1$,*
  - *update functions $u_i \colon \Sigma \rightarrow \Sigma$ such that $u_i \neq u_j$ for $i \neq j$.*

$[a]$ $x + y \leq 1 \longrightarrow 0.5 : x' = x + 1 \oplus 0.5 : x' = x + 1 \wedge y' = y + 1;$
$[b]$ $1 \leq x + y \leq 2 \longrightarrow 0.8 : x' = 2 \wedge y' = x - 1 \oplus 0.2 : x' = 2;$
$[c]$ $x = 2 \longrightarrow 1 : x' = x \wedge y' = y;$

Fig. 1. A probabilistic program $\mathcal{P}$

*Additionally, without loss of generality, we assume that for every state $s \in \Sigma$
there is a command $c \in C$ such that $s \in [\![g]\!]$ where g is the guard of c.*

Intuitively, the state space of a probabilistic program is the set of all valuations
of its variables. The commands then define the probabilistic transitions between
these states. A guarded command is enabled in all states satisfying its guard
g, i.e., the states $s \in [\![g]\!]$, which we write as $s \models g$. If a command is enabled
in some state, that state possesses an outgoing probability distribution that is
given by the probabilities and the update functions. Given a command $c = g \rightarrow$
$p_1 : u_1 \oplus \ldots \oplus p_m : u_m$ and a variable valuation $s \in \Sigma$ such that $s \models g$, s has a
transition with probability p_i to the state $s_i = u_i(s)$ for all $i \in \{1, \ldots, m\}$.

The semantics of a probabilistic program is a DTMC or an MDP, depending
on whether there exists a state that satisfies multiple guards. If there is no such
state, each state has exactly one command that is enabled and the resulting
model is a DTMC. If a state satisfies multiple guards, this corresponds to a
nondeterministic choice between multiple commands in that particular state and,
hence, the model is an MDP.

Example 1. Consider the probabilistic program $\mathcal{P}$ depicted in Figure 1 with three
commands a, b and c over two integer variables x and y with range $[0, 2]$. The
semantics of this program is the MDP $\mathcal{M}$ shown in Figure 2 where we assume
that only state $\langle 2, 1 \rangle$ is of interest and thus labeled with the special atomic
proposition F indicated by the double circle around the state. ∎

Note that a set of predicates $\Pi = \{b_1, \ldots, b_k\} \subseteq BExpr_{Var}$ induces a (finite)
partitioning Q of the state space, where for $q \in Q$, $s_i \in q$ and $s_j \in q$ if and
only if $s_i \models p \Leftrightarrow s_j \models p$ for all $p \in \Pi$. Stated differently, the partition is given
by the sets of states that satisfy exactly the same predicates of Π. Satisfiability
solvers that support richer theories, such as linear integer arithmetic, can be
used to reason over probabilistic programs and build abstractions w.r.t. a set of
predicates directly from such a representation [45]. That is, abstractions may be
built without building the full model first. In the further course of this paper,
we will only consider partitions that respect the labeling of states in the model.
That is, a partition Q is viable if for each $a \in AP$ and $q \in Q$ we have that
either all or no $s \in q$ are labeled with a. For a given partition Q of a set S and
a probability distribution $\mu \in Dist(S)$, we denote by $\bar{\mu}$ the lifted distribution
$\bar{\mu} \in Dist(Q)$ defined by $\bar{\mu}(q) = \sum_{s \in q} \mu(s)$ for all $q \in Q$.

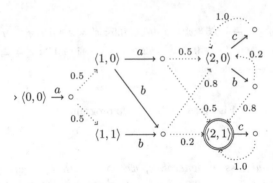

Fig. 2. The MDP $\mathcal{M}$ induced by the program $\mathcal{P}$

Example 2. Reconsider the probabilistic program from Figure 1 and let the set of predicates Π be given as $\Pi = \{x < 2, x \geq 2 \wedge y \geq 1\}$. Π induces the partition

$$Q = \{\underbrace{\{\langle 0,0\rangle, \langle 1,0\rangle, \langle 1,1\rangle\}}_{A}, \underbrace{\{\langle 2,0\rangle\}}_{B}, \underbrace{\{\langle 2,1\rangle\}}_{C}\}$$

of the state space of $\mathcal{M}$, because, e.g., for all $s \in A$ we have that $s \models x < 2$ and $s \not\models x \geq 2 \wedge y \geq 1$.   ∎

3.4   MDP Quotienting

Given an MDP $\mathcal{M} = (S, Act, \mathbf{P}, s_{init}, AP, L)$ and a partition Q of its state space S that respects its labeling, a first idea is to construct a simple abstract system by merging the states of $\mathcal{M}$ according to Q. A state of the abstract system, thus, corresponds to a block of Q. In order to still keep the behavior of the original system, the transition probability function must over-approximate the transition probability function of the concrete model. This can be achieved by giving an abstract state q the joint behavior of all contained states $s \in q$. Put differently, if there is a state $s \in q$ and an action $\alpha \in Act$ such that α is enabled in s and associated with the distribution $\mu \in Dist(S)$, then there must be an action α' that is enabled in q and associated with $\bar{\mu} \in Dist(Q)$. Note that this possibly involves a renaming of the action name. This is necessary, because the action α may be enabled in several states in q, but can only be associated with a single probability distribution in the quotient system. Formally, the resulting MDP is given by $\mathcal{M}/Q = (Q, Act', \mathbf{P}/Q, AP, L/Q)$ where $\mathbf{P}/Q$ is such that $(\mathbf{P}/Q)(q, \alpha', \cdot) = \bar{\mu}$ if and only if there exists an $s \in q$ such that $\mathbf{P}(s, \alpha, \cdot) = \mu$ and $(L/Q)(q) = L(s)$ for an $s \in q$. Note that the labeling is well-defined, because of our requirement for partitions to only group states with the same labeling.

Example 3. Consider the MDP $\mathcal{M}$ depicted in Figure 2 and the partition Q from Example 2. The quotient MDP $\mathcal{M}/Q$ is shown in Figure 3. Note that from block A of the partition we now have the union of all (lifted) distributions

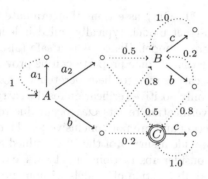

Fig. 3. The MDP $\mathcal{M}/Q$

available in the states contained in A and that we preserved the labeling of the states contained in each block. Furthermore, we needed to rename the two distributions labeled with a to a_1 and a_2, respectively.

∎

As can be seen from the example, this abstraction mixes the nondeterminism present in the original model and the nondeterminism introduced by the abstraction. Consequently, the minimal probability for satisfying any given PCTL path formula in the abstract MDP $\mathcal{M}/Q$ is a lower bound for the corresponding probability in $\mathcal{M}$. A similar result holds for the maximal probability. Formally, we have the following theorem.

Theorem 1. *Let $\mathcal{M}$ be an MDP, Q be a partition of its state space and φ be a PCTL path formula, then*

$$Pr_{min}^{\mathcal{M}/Q}(\varphi) \leq Pr_{min}^{\mathcal{M}}(\varphi) \leq Pr_{max}^{\mathcal{M}}(\varphi) \leq Pr_{max}^{\mathcal{M}/Q}(\varphi)$$

Stated differently, this means that the abstraction only guarantees that the extremal probabilities for satisfying φ are in between the extremal probabilities obtained from the abstract MDP. This can lead to very coarse results as illustrated by the next example.

Example 4. For $\mathcal{M}/Q$ of Figure 3 and $\varphi = \Diamond F$, we have $Pr_{min}^{\mathcal{M}/Q}(\varphi) = 0$ and $Pr_{max}^{\mathcal{M}/Q}(\varphi) = 1$, which gives no information about the real values of $Pr_{min}^{\mathcal{M}}(\varphi)$ and $Pr_{max}^{\mathcal{M}}(\varphi)$ which can lie anywhere in between. In fact, the correct values are 0.2 and 1, respectively. ∎

4 Multi-valued Abstraction

Multi-valued abstraction aims to partition the state space combined with an abstraction of transition probabilities to sets of transition probabilities such that both positive and negative verification results in the abstract model carry over to the concrete model [20]. Only assertions that evaluate to other values than true or false in the abstract model, typically called indefinite values, are non-conclusive for the concrete model. Hence, it is a safe (also called conservative) abstraction in the sense that if some property can be proven or disproven in the abstract model, then it carries over to the concrete model. In contrast, for MDP quotients (Section 3.4) only positive verification results carry over to the concrete model while negative verification results may occur due to over-approximation in the quotient abstraction and are not conclusive for the concrete model.

In this section we provide a survey of the three-valued abstraction technique presented in [33]. We consider abstractions of DTMCs and properties expressed by PCTL. Three-valued abstraction of models without probabilistic choice [28, 40] yields abstract models that over- and under-approximate transitions in the concrete model by *may* and *must* transitions. This concept generalizes naturally for DTMCs to transitions equipped with intervals in the abstract model where upper and lower bounds of the intervals represent accordingly the over- and under-approximation of the abstract probabilistic transitions in the concrete model. We will show that abstract states simulate the concrete states by an adapted notion of probabilistic simulation [31]. Furthermore, we demonstrate that an appropriate three-valued semantics of PCTL provides that affirmative and negative verification results on abstract DTMCs carry over to the concrete model.

As running example we will consider the DTMC presented in Figure 4a. This DTMC corresponds to the MDP of Figure 2 after abstracting from transition labels. The colour of states represents the state labelling L.

4.1 Three-Valued Abstraction

We start by introducing abstract DTMCs. A state in the abstract DTMC represents a set of concrete states. Transitions between abstract states are equipped with an interval of probabilities instead of a concrete probability. The lower and upper bound of the intervals represent the lowest and highest probability in the concrete model. Let $\mathbb{B}_3$ denote the three-valued domain with carrier $\{\bot, ?, \top\}$ and order $\bot < ? < \top$.

Definition 6 (Abstract DTMC (ADTMC)). *An* abstract DTMC *is a tuple* $\mathcal{M} = (S, \mathbf{P}^l, \mathbf{P}^u, L, \mu_0)$ *where*

- S *is a countable set of states,*
- $\mathbf{P}^l, \mathbf{P}^u : S \times S \to [0, 1]$ *are probabilistic transition functions with*
  - $\mathbf{P}^l(s, s') \leq \mathbf{P}^u(s, s')$ *for all* $s, s' \in S$ *and*
  - $\mathbf{P}^l(s, S) \leq 1 \leq \mathbf{P}^u(s, S),$
- $L : S \times AP \to \mathbb{B}_3$ *evaluates an atomic proposition for a given state and*
- $\mu_0 \in Dist(S)$ *is the initial distribution.*

We write $s \xrightarrow{[a,b]} s'$ for $\mathbf{P}^l(s, s') = a$ and $\mathbf{P}^u(s, s') = b$. This transition may happen with any (nondeterministically chosen) probability in the interval $[a, b]$.

Furthermore, the validity of atomic propositions, denoted by L, may now evaluate to the indefinite value $? \in \mathbb{B}_3$. Thus, ADTMCs can be described by MDPs, where the distributions reachable from s are given by $\{\mu \in Dist(S) \mid \mu(s') \in [\mathbf{P}^l(s, s'), \mathbf{P}^u(s, s')]\}$. It is clear that every DTMC is also an ADTMC if $\mathbf{P}^l(s, s') = \mathbf{P}^u(s, s')$ for all $s, s' \in S$ and $L(s, p) \in \{\top, \bot\}$ for all $s \in S, p \in AP$.

We proceed by defining the abstraction of an ADTMC based on some partitioning of its state space. Because every DTMC is also an ADTMC this directly gives us a notion to abstract a DTMC to an ADTMC.

Definition 7 (Abstraction of ADTMC). *Let* $\mathcal{M} = (S, \mathbf{P}^l, \mathbf{P}^u, L, \mu_0)$ *be an ADTMC and* Q *be a finite partitioning of* S. *The abstraction of* $\mathcal{M}$ *with respect to* Q *is an ADTMC* $(Q, \widetilde{\mathbf{P}}^l, \widetilde{\mathbf{P}}^u, \widetilde{L}, \mu_0)$ *such that for any* $q, q' \in Q$ *we have*

- $\widetilde{\mathbf{P}}^l(q, q') = \inf_{s \in q} \mathbf{P}^l(s, q')$,
- $\widetilde{\mathbf{P}}^u(q, q') = \min(\sup_{s \in q} \mathbf{P}^u(s, q'), 1)$
- $\widetilde{L}(q, a) = \begin{cases} \top & \text{if } L(s, a) = \top \text{ for all } s \in q \\ \bot & \text{if } L(s, a) = \bot \text{ for all } s \in q \\ ? & \text{otherwise} \end{cases}$

We denote by $\mathcal{M}/Q$ the ADTMC that arises from abstracting $\mathcal{M}$ by Q. The definition of the upper bound $\widetilde{\mathbf{P}}^u(q, q')$ of the probabilistic transition between q and q' needs to be bounded by 1 because $\mathbf{P}^u(s, q') = \sum_{s' \in q'} \mathbf{P}^u(s, s')$ may exceed 1. Every abstraction leads again to an ADTMC [33, Lemma 1].

Example 5. Figure 4b represents the ADTMC after grouping states $\langle 1, 0 \rangle$ and $\langle 1, 1 \rangle$ of the DTMC in Figure 4a into a single abstract state. The probabilistic transition between state $(\langle 1, 0 \rangle, \langle 1, 1 \rangle)$ and $\langle 2, 0 \rangle$ is equipped with the interval $[0.5, 0.8]$ which represents exactly the minimal and maximal probabilities of reaching state $\langle 2, 0 \rangle$ from some of the states $\{\langle 1, 0 \rangle, \langle 1, 1 \rangle\}$.   ∎

The notion of abstraction on ADTMCs is closely related to forward simulation [31]. In detail, for any ADTMC $\mathcal{M}$ and partition Q we have that $\mathcal{M}$ is simulated by $\mathcal{M}/Q$ [33, Theorem 1]. The three-valued abstraction technique can be adapted to CTMCs without technical difficulties when applying prior uniformization (i. e. all states have equal residence time).

4.2   Reachability Analysis and Model Checking

In the following section we investigate how logical properties, in detail reachability analysis, can be verified on abstract models. The nondeterminism introduced by intervals is resolved using schedulers which lead also to a natural notion of induced DTMC from an ADTMC by a specific scheduler. Interestingly, extreme schedulers, which are schedulers that resolve the probabilities in the intervals to one of the boundaries, suffice to compute maximal/minimal reachability properties ([33, Theorem 2]).

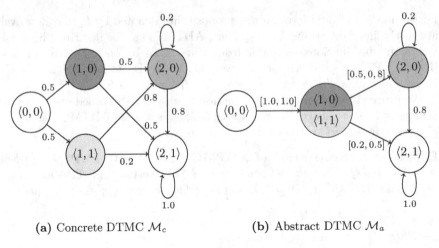

(a) Concrete DTMC $\mathcal{M}_c$ (b) Abstract DTMC $\mathcal{M}_a$

Fig. 4. Example for three-valued abstraction

The classical interpretation of PCTL is over a two-valued truth domain $\{\bot, \top\}$. ADTMCs group states together such that some PCTL properties are no longer strictly true or false. For instance, consider the reachability property $\Phi = P_{\geq 0.7}(\Diamond \bullet)$ which evaluates to true if the probability to reach a state with label $\bullet$ is at least 0.7. Let us consider the state $(\langle 1, 0 \rangle, \langle 1, 1 \rangle)$ of the ADTMC $\mathcal{M}_a$. The interval of probabilities $[0.5, 0.8]$ to reach state $\langle 2, 0 \rangle$ allows for probabilities that are greater than 0.7 but also for probabilities that are less than 0.7. Hence, the property Φ evaluates to the indefinite value ? in state $(\langle 1, 0 \rangle, \langle 1, 1 \rangle)$. On the other hand, $P_{\geq 0.9}(\Diamond \bullet)$ evaluates to $\bot$ in $(\langle 1, 0 \rangle, \langle 1, 1 \rangle)$ because for none of the realizable probabilities the property can become true. Similarly, $P_{\geq 0.3}(\Diamond \bullet)$ evaluates in $(\langle 1, 0 \rangle, \langle 1, 1 \rangle)$ to $\top$ because for all realizable probabilities the property becomes true.

To summarize, PCTL properties evaluate in the abstract states of ADTMCs to $\mathbb{B}_3$. The semantics differs from the two-valued semantics mainly by the fact that the evaluation of $P_{\bowtie p}(\Diamond \varphi)$ is split up in the case $P_{<p}(\Diamond \varphi), P_{\leq p}(\Diamond \varphi)$ and case $P_{>p}(\Diamond \varphi), P_{\geq p}(\Diamond \varphi)$. For the first case, $P_{<p}(\Diamond \varphi)$ (resp. $P_{\leq p}(\Diamond \varphi)$) evaluates to $\top$ if in all realizable probabilistic choices φ is reachable by strictly less than p (resp. at most p). $P_{<p}(\Diamond \varphi)$ (resp. $P_{\leq p}(\Diamond \varphi)$) evaluates to $\bot$ if in all realizable probabilistic choices φ is reachable by at least p (resp. strictly more than p). The reasoning for the second case is analogous. In all other cases the property evaluates to ?. It was shown that abstraction preserves validity of PCTL formulae [33, Thm. 3,5 & Cor. 1]. This paves the way for three-valued abstraction-based model checking.

5   Counterexample-Guided Abstraction Refinement

Proposed in 2000 [10], *counterexample-guided abstraction refinement* (CEGAR) quickly became a very successful technique for qualitative verification of safety properties that proceeds in an iterative manner: starting with an initially coarse overapproximation of the concrete model, it tries to add precision to the parts of the model where required. It does so by analyzing information obtained from the model checking process on the abstract model. The key idea is the following: if the abstract model violates the safety property Φ, it must be possible for a model checker to extract a reason for this, a so-called *counterexample*. This is then analyzed with respect to its realizability in the original model. If it is in fact realizable, we can conclude that the original model also violates Φ. On the other hand, if the counterexample is not realizable, the verification result on the abstract model does not carry over to the concrete model. In this case, the abstraction introduced the *spurious* behavior and the abstract models needs to be refined. As it is known that the counterexample was indeed spurious, it also carries information about the reason why the abstraction introduced this behavior, which can be exploited to refine the current partition. The overall approach is sketched in Figure 5. It is easy to see that for finite models the

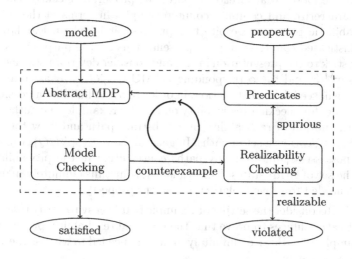

Fig. 5. A schema of the CEGAR loop

CEGAR loop will only be traversed a finite number of times until a decision can be made, assuming that the predicate synthesis always splits at least one abstract state. For infinite models, however, the procedure may not terminate.

We start by introducing the notion of a counterexample for safety properties. Then, we present the core procedure of the CEGAR loop [27]; in particular we discuss the counterexample analysis using satisfiability solvers, and the synthesis of predicates. Note that while [27] also treats systems with arbitrary (even infinitely) many initial states, we restrict our attention to systems with one initial state for the sake of simplicity.

5.1   Counterexamples for Safety Properties in MDPs

In the traditional qualitative model checking setting, safety properties express that a certain set of "bad" states F must not be reachable. A probabilistic safety property $\mathcal{P}_{\trianglelefteq p}(\Diamond F)$ with $\trianglelefteq \in \{<, \leq\}$ establishes an upper bound p on the probability to reach the bad states F.

Example 6. Consider for example the MDP in Figure 2 and the probabilistic safety property $\Phi = \mathcal{P}_{<1}(\Diamond F)$. Obviously, there are schedulers that violate Φ, namely all schedulers σ that pick action b in $\langle 2, 0 \rangle$ and some available action in the other states. ∎

Given a (deterministic memoryless) scheduler σ that violates the safety property Φ, σ can be called a counterexample for Φ. Likewise, the DTMC $\mathcal{M}_\sigma$ resulting from the application of σ on $\mathcal{M}$ can be called a counterexample. A lot of work has been conducted on finding more succinct representations of counterexamples to make them effectively useful in the context of debugging the system. Several approaches [48, 2, 30] revolve around the identification of a small subsystem of the concrete model that already violates the property. Recently, [49] showed how to characterize and compute counterexamples in terms of the commands of a probabilistic program. Despite this progress, it is yet unclear how to use these sophisticated counterexample representations for the purpose of CEGAR. Hence, we stick to the presentation in [27] and consider deterministic memoryless schedulers $\sigma^{\mathcal{M}/Q}$ and the corresponding DTMC $(\mathcal{M}/Q)_{\sigma^{\mathcal{M}/Q}}$ as counterexamples. As we will see in the next section, in order to avoid building a possibly huge concrete model for checking realizability of a counterexample, the authors of [27] resort to the notion of realizability of an abstract path and view the abstract DTMC $(\mathcal{M}/Q)_{\sigma^{\mathcal{M}/Q}}$ as a set of paths. In general, however, $(\mathcal{M}/Q)_{\sigma^{\mathcal{M}/Q}}$ is cyclic and thus possesses infinitely many paths contributing to the probability mass reaching the set of "bad" states. Unfortunately, sometimes infinitely many paths are in fact needed to prove violation of a safety property.

Example 7. Reconsider the setting of Example 6. It is easy to verify that no finite number of paths suffices to prove that the maximal reachability probability is 1. In this example all paths, i. e., infinitely many, are needed to witness the violation of the safety property. ∎

Luckily, it can be shown that this phenomenon only occurs when the comparison operator is strict [24]. In order words, if a property that uses only smaller-or-equal comparisons is violated, then there is always a finite set of paths whose probability mass exceeds the bound. In this case, following the ideas of Han and Katoen [24], a minimal set of paths that exceeds the probability bound p can be efficiently obtained by a reduction to a graph problem. Hence, from now on we assume that we can obtain, one by one, a finite set of paths in decreasing probability order whose accumulated probability mass exceeds the given bound.

5.2   Realizability of a Counterexample

Figure 5 illustrates that counterexample analysis is at the heart of the CEGAR approach. It must be possible to determine whether a counterexample in the abstract system carries over to the concrete model without building it. In order to do this, we first need to define what it means for a counterexample to be realizable. For the remainder of this section, we will assume that the concrete MDP $\mathcal{M} = (S, Act, \mathbf{P}, s_{init}, AP, L)$ is given as a probabilistic program such that no two commands have the same label. Furthermore, for the current partition Q of the state space the abstract MDP $\mathcal{M}^{\#} := \mathcal{M}/Q = (Q, Act', \mathbf{P}/Q, AP, L/Q)$ has been built and proven to violate a reachability property $\Phi = \mathcal{P}_{\leq p}(\Diamond F)$ by a model checker. In addition, a counterexample $\sigma^{\#}$ is provided as a witness for the violation by the model checker.

For checking realizability of the abstract scheduler $\sigma^{\#}$, the idea is to check whether a similarly behaving scheduler σ on $\mathcal{M}$ will exhibit the same violating behavior. Formally, the concretization of a counterexample is defined as follows.

Definition 8 (Concretization and realizability of a counterexample).

(i) *The concretization* $\gamma(\sigma^{\#})$ *of a counterexample* $\sigma^{\#}$ *is defined as the scheduler* σ *for* $\mathcal{M}$ *such that for all* $s \in S$

$$\sigma(s) = \begin{cases} \sigma^{\#}(q) & s \in q \wedge s \models g \\ \bot & \text{otherwise} \end{cases}$$

where $q \in Q$ *and* g *is the guard of the command associated with the command* $\sigma^{\#}(q)$, *which is given by the probabilistic program for* $\mathcal{M}$.

(ii) *A counterexample* $\sigma^{\#}$ *is called* realizable *if the probability of reaching a state in* F *in* $\mathcal{M}_{\gamma(\sigma^{\#})}$ *exceeds the given bound, i.e.,* $Prob^{\mathcal{M}}_{\gamma(\sigma^{\#})}(\Diamond F) > p$ *and* spurious *otherwise.*

Intuitively, the concretisation $\gamma(\sigma^{\#})$ of a scheduler $\sigma^{\#}$ is a scheduler for the concrete MDP $\mathcal{M}$ that chooses an action a in a concrete state s iff $\sigma^{\#}$ chooses a in the abstract state $q \in Q$ containing s and a is available in s. If a is not available in s, because s fails to satisfy the guard of the command, the concretization will just stop in s, which is indicated by $\bot$. Now, a counterexample is realizable if the concretization induces enough probability mass on the concrete model to violate the bound p of the probabilistic safety property under the concrete scheduler.

Example 8. Reconsider the MDP from Figure 2 and let the safety property be given as $\Phi = \mathcal{P}_{\leq 0.6}(\Diamond F)$. Obviously, $\mathcal{M}^{\#}$ (Figure 3) does not satisfy Φ, because the scheduler $\sigma^{\#}$ that picks a_2 in A, b in B and c in C achieves a probability of 1. The concretization $\gamma(\sigma^{\#})$, thus, has also to choose a in all states of A and so on. However, $\langle 1, 1 \rangle$ fails to satisfy the guard of b, so $\gamma(\sigma^{\#})(\langle 1, 1 \rangle) = \bot$.   ∎

However, given the definition of realizability of a counterexample, it remains to show how to actually perform the realizability check without having the concrete model at hand. As previously mentioned, the authors of [27] resort to viewing the abstract DTMC induced by the abstract counterexample as a set of paths reaching a state in F. This enables to check the realizability of paths in isolation rather than a "full" counterexample at once.

Realizability of a Path. Intuitively, a path in the abstract counterexample DTMC $\mathcal{M}_{\sigma^{\#}}^{\#}$ is *realizable*, if there exists a path in $\mathcal{M}$ that (i) starts in the initial state and ends in F, (ii) does not visit F before the last state (iii) chooses the same actions and updates as the abstract path and (iv) is in a concrete state $s_i \in q_i$ whenever the abstract path is in $q_i \in Q$. Note that this implies that all states along the path satisfy the appropriate guards. The following definition captures this formally:

Definition 9 (Concretization of a path).

Let $\omega^{\#} = q_0 \alpha_1 q_1 \alpha_2 \ldots q_n \in Path_{fin}^{\mathcal{M}_{\sigma^{\#}}^{\#}}$ be a finite path prefix.

(i) *The* concretization $\gamma(\omega^{\#})$ *of* $\omega^{\#}$ *is given by*

$$\{\omega \in Path_{fin}^{\mathcal{M}_{\gamma(\sigma^{\#})}} \mid \omega = s_0 \alpha_1 s_1 \alpha_2 \ldots s_n \text{ with } s_i \in q_i \text{ for } 0 \leq i \leq n\}$$

(ii) $\omega^{\#}$ *is called* realizable *if* $\gamma(\omega^{\#}) \neq \emptyset$.

Example 9. Let the setting be the same as the one in Example 8. Furthermore, consider the path prefix $\omega^{\#} = A a_2 C$ in $\mathcal{M}_{\sigma^{\#}}^{\#}$. By inspection, we can see that $\gamma(\omega^{\#}) = \emptyset$, because it is impossible to reach a state in F from the initial state of $\mathcal{M}$ within one step. In fact, no finite path prefix in $\mathcal{M}_{\sigma^{\#}}^{\#}$ possesses a non-empty concretization. ∎

The existence of an element in the concretization of the abstract path is formulated as a query to an SMT solver. The actual formula $\varphi_\gamma(\omega^{\#})$ for the path $\omega^{\#}$ can be constructed using repeated applications of the weakest precondition operator and additional information obtainable from the path. This formula may then be dispatched to a standard SMT solver supporting linear integer arithmetic, as, for example, Z3 or MathSat. For details, we refer to [27].

Algorithmically checking realizability of a counterexample. Now that we have presented a method for determining whether a path is realizable, we can check the realizability of the counterexample as follows. Given $\mathcal{M}_{\sigma^{\#}}^{\#}$, we start extracting finite paths $\omega^{\#} = q_0 \alpha_1 q_1 \alpha_2 \ldots q_n$ with $q_i \notin F$ for $0 \leq i < n$ and $q_n \in F$ in decreasing probability order. Each of these paths is individually checked for realizability. All realizable paths are added to a set Ω^+ whereas unrealizable paths are added to Ω^- and we denote by p_{Ω^+} and p_{Ω^-}, respectively, the sum of the probabilities of the paths in these sets. If we get to a point where $p_{\Omega^+} > p$, we know that enough probability mass of the abstract counterexample is also present in the concrete model to exceed the bound p. This directly implies that the counterexample is in fact realizable and we can conclude that $\mathcal{M}$ violates the given safety property as well. Conversely, we need a criterion to stop looking for new paths in $\mathcal{M}_{\sigma^{\#}}^{\#}$ if there is no hope of ever exceeding p. As the model checker that provided the abstract counterexample computed the probability $p_{\sigma^{\#}}(\lozenge F)$ of reaching F in the abstract model, we can at any point determine whether there is enough probability mass left in the abstract counterexample that is potentially realizable and suffices to exceed p. If $p_{\Omega^+} + (p_{\sigma^\#}(\lozenge F) - p_{\Omega^-}) \leq p$, we can conclude that even if all remaining paths in $\mathcal{M}^\#_{\sigma^\#}$ were realizable, they could still not exceed the bound p and thus, the abstract counterexample was determined to be spurious. Formally, we get the following result:

Lemma 1 (Termination criterion). *Let $\epsilon = p_{\sigma^\#}(\lozenge F) - p_{\Omega^-}$ be the probability of paths reaching F in the abstract model that were not yet proven to be (un)realizable.*

(i) $p_{\Omega^+} > p$ implies that $\sigma^\#$ is realizable.
(ii) $p_{\Omega^+} + \epsilon \leq p$ implies that $\sigma^\#$ is spurious.

Stated differently, the result is only inconclusive if $p_{\Omega^+} \in (p - \epsilon, p]$. In this case, the next most likely abstract path is considered until the result is in fact conclusive.

Example 10. Reconsider Example 8. The model checker initially returns the counterexample $\sigma^\#$ along with the reachability probability $p_{\sigma^\#}(\lozenge F) = 1$. Let the first path prefix that is found in $\mathcal{M}^\#_{\sigma^\#}$ by the procedure be $\omega^\# = Aa_2C$. As shown in Example 9, this path prefix is found to be unrealizable. It is then added to Ω^- and its probability is added to p_{Ω^-}, which then becomes 0.5. Now, however, ϵ becomes 0.5, which means that there is at most a probability mass of 0.5 left in $\mathcal{M}^\#_{\sigma^\#}$ that might be realizable. As this does not suffice to exceed the bound $p = 0.6$ of Φ, the counterexample is known to be spurious. If this was not directly the case, the procedure would check the next path prefix for realizability and carry on. ∎

5.3 Predicate Synthesis

Suppose the decision procedure previously described determines that a given counterexample in the abstract model is spurious. This means that the abstraction falsely introduced behavior that was not present in the original model. In other words, we need to refine the abstract model to rule out this spurious behavior. As the abstract model is built using predicate abstraction, this corresponds to introducing additional predicates. Since the formulae $\varphi_\gamma(\omega^\#)$ do not involve quantitative aspects, but are similar to the non-probabilistic case, standard techniques, such as predicate interpolation [42, 45], may be used to obtain predicates to rule out the source of spuriousness.

Example 11. After the counterexample from Example 10 was found to be spurious, the predicate $x = 0 \wedge y = 0$ could be added to rule out the possibility to reach F within one step in the quotient model $\mathcal{M}^\#$. ∎

6 Game-Based Abstraction

The success of *counterexample-guided abstraction refinement* inspired another abstraction-refinement framework for MDPs. First, observe that probabilistic

CEGAR (as presented in Section 5) can only (dis)prove probabilistic safety properties. Intuitively, this is because the concrete MDP is again abstracted to an MDP. Doing so, however, merges the nondeterminism of the concrete model with the nondeterminism introduced by the abstraction. Effectively this means that the minimal and maximal reachability probabilities in the abstract MDP are lower and upper bounds, respectively, for the corresponding reachability probabilities in the concrete model (see Theorem 1). Rather than merging the two sources of nondeterminism, the two abstraction techniques presented in this section keep them separated by using probabilistic games [11] as their underlying abstract model. This way, they are able to provide lower and upper bounds on both minimal and maximal reachability probabilities. Hence, they are applicable to the broader class of probabilistic reachability properties.

6.1 Idea

Reconsider Example 4. In the abstract MDP, the minimal and maximal probability to a reach a state in F are 0 and 1, respectively. This, however, means that the reachability probilities in the concrete model may lie anywhere in between those values providing no information at all. This phenomenon stems from the fact that, in the abstract state A, both the nondeterministic choices of state $\langle 0,1 \rangle$ and the transitions emanating from $\langle 0,0 \rangle$ are enabled.

The key idea of game-based abstraction [34] is the follwing. Instead of using an MDP as the "target" of the abstraction, the concrete MDP is mapped to an abstract probabilistic game. This way, the two sources of nondeterminism can be assigned to the different players and, thus, kept separate. In other words, one player is responsible for resolving the nondeterminism of the abstraction while the other governs the nondeterminism of the original model. Depending on whether the two players both try to maximize or minimize the probability to reach the target states or they take an adversarial role, the resulting value of the game is a lower or upper bound on the minimal or maximal reachability probability, respectively. If these bounds are precise enough for proving or refuting a given property, a conclusive answer for satisfaction of the property on the concrete model can be given. In the other case, at least one block of the abstraction can be refined based on the strategies of the players. The resulting game then yields more precise results and, similarly to CEGAR, the procedure may be iterated until the obtained bounds are precise enough. The approach is sketched in Figure 6. Note that in practice this approach can be implemented in a fully symbolic way (just like CEGAR) by using SMT solvers that avoids building the concrete model $\mathcal{M}$ altogether. However, for the sake of simplicity, our presentation will abstract from this.

6.2 Simple Game-Based Abstraction

We will now present how to obtain an appropriate (abstract) probabilistic game from an MDP. Just like for MDP quotienting, a set of predicates is used to partition the state space S of the concrete MDP into blocks of states

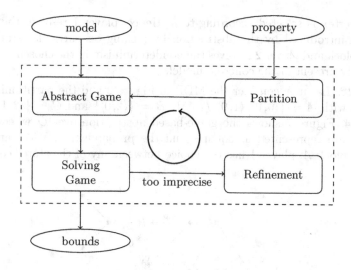

Fig. 6. A schema of refinement loop of game-based abstraction

$Q = \{S_1, \dots, S_n\}$. Recall that $\bar{\mu} \in Dist(Q)$ denotes the lifting of $\mu \in Dist(S)$ to $Dist(Q)$.

Definition 10 (Simple probabilistic game-based abstraction). *Given an MDP* $\mathcal{M} = (S, Act, \mathbf{P}, s_{init}, AP, L)$ *and a partition* Q *of* S, *the* simple probabilistic game-based abstraction *of* $\mathcal{M}$ *over* Q *is the probabilistic game*

$$\mathcal{G}_{\mathcal{M}}^{Q} = ((V = V_1 \cup V_2 \cup V_p, E), v_{init}, (V_1, V_2, V_p), \delta)$$

where

- $V_1 = Q$ *are player 1's vertices,*
- $V_2 = \{v_2 \subseteq Dist(Q) \mid v_2 = \{\bar{\mu} \mid \mu \in Steps(s)\}$ *for some* $s \in S\}$ *are player 2's vertices,*
- $V_p = \{v_p \in Dist(Q) \mid v_p \in \{\bar{\mu} \mid \mu \in Steps(s)\}$ *for some* $s \in S\}$ *are the probabilistic vertices,*
- $v_{init} = q \in Q$ *such that* $s_{init} \in q$ *is the initial vertex,*
- $\delta : V_p \to Dist(V)$ *is the identity function,*

and the set of edged E *is given by*

$$E = \{(v_1, v_2) \mid v_1 \in V_1 \text{ and } v_2 = \{\bar{\mu} \mid \mu \in Steps(s) \text{ for some } s \in v_1\}$$
$$\cup \{(v_2, v_p) \mid v_2 \in V_2 \text{ and } v_p \in v_2\}$$
$$\cup \{(v_p, v_1) \mid v_p = \bar{\mu} \text{ with } \bar{\mu}(v_1) > 0\}.$$

Intuitively, the game proceeds as follows. In each $v_1 \in V_1$, player 1 picks a concrete state $s \in v_1$ and moves to the corresponding player 2 vertex. Then, player 2 chooses a (lifted) probability distribution $\bar{\mu}$ available in s. Finally, the next player 1 vertex is selected according to $\bar{\mu}$. Hence, player 1 resolves the nondeterminism introduced by the abstraction by picking on particular state in an abstract block and player 2 resolves the nondeterminism in the chosen state that was already present in the concrete model.

Example 12. Let us reconsider the MDP in Figure 2 and the partitioning $Q = \{A, B, C\}$ with $A = \{\langle 0,0 \rangle, \langle 1,0 \rangle, \langle 1,1 \rangle\}$, $B = \{\langle 2,0 \rangle\}$ and $C = \{\langle 2,1 \rangle\}$ from Example 4. Figure 7 shows the game-based abstraction over Q where player 2 vertices are represented as squares and the probabilistic vertices are small circles. In block A, player 1 has the choice between any of the states contained

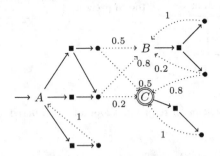

Fig. 7. Game-based abstraction $\mathcal{G}_{\mathcal{M}}^Q$

in A. Say he chooses to pick state $\langle 1,0 \rangle$ and, hence, moves to the topmost successor vertex of A. Then, player 2 can choose between the two distributions available in $\langle 1,0 \rangle$ and move to the probabilistic vertex corresponding to the lifted distribution. After the successor state was determined according to the probability distribution, it is again player 1's turn.   ∎

By solving the game $\mathcal{G}_{\mathcal{M}}^Q$ using numerical methods, such as value iteration [11], bounds for both minimal and maximal reachability probability in the concrete model are defined as follows. Suppose both players try to minimize the probability to reach a target state in F. Then, the reachability probability $p_{v_{init}}^{--}(F)$ is the minimal reachability probability for F in the MDP quotient $\mathcal{M}/Q$ and is thus, by Theorem 1, a lower bound for the minimal reachability probability in $\mathcal{M}$. Suppose, on the other hand, that the player controlling the abstraction (player 1) tries to maximize while player 2 still minimizes over the nondeterministic choices available in the state selected by player 1. The probability $p_{v_{init}}^{+-}(F)$ is an upper bound for the minimal reachability probability in $\mathcal{M}$. Formally, we have the following result.

Theorem 2 (Correctness of game-based abstraction [34]). *Let $\mathcal{G}_{\mathcal{M}}^Q$ be the game-based abstraction for an MDP $\mathcal{M}$ with state space S and Q a partition of S. Then for all $v \in V_1$ and $s \in v$*

$$p_v^{--}(F) \leq p_s^-(F) \leq p_v^{+-}(F), \tag{1}$$

$$p_v^{-+}(F) \leq p_s^+(F) \leq p_v^{++}(F). \tag{2}$$

Example 13. For the abstraction in Example 12 we obtain $p^{--}_{v_{init}}(\{C\}) = 0$. Unlike the MDP abstraction $\mathcal{M}/Q$, we can, however, obtain a better upper bound on the minimal reachability probability than 1. Observe that, if only player 1 tries to maximize the probability value, he does not choose the state $\langle 0, 0 \rangle$ (the bottommost choice emanating from A) but any of the other states. Then, player 2 can not completely avoid reaching C any more, but has to go to C with a probability of at least 0.2. Indeed, solving the game yields $p^{+-}_{v_{init}}(\{C\}) = 0.2$. Hence, the minimal reachability probabilty in $\mathcal{M}$ is determined to lie in the interval $[0, 0.2]$ providing more precise information than the MDP quotient over the same partition Q. ∎

As indicated in Figure 6, after obtaining bounds by solving a game, the partition Q may need to be refined in order to obtain more precise results. We will show how the refinement may be done in a way that guarantees termination of the procedure for finite models $\mathcal{M}$.

Refinement. Recall that, given the goals of players 1 and 2, solving a game not only comprises computing the extremal reachability probability for the game, but also produces memoryless deterministic strategies for the two players that together achieve the computed probability. Suppose we obtained the bounds $[l, u]$ for the minimal reachability probability by solving the game $\mathcal{G}^Q_\mathcal{M}$ (twice). Further assume that the bounds were imprecise, i.e., $l < u$. Then, two pairs of memoryless deterministic strategies (σ^l_1, σ^l_2) and (σ^u_1, σ^u_2) are generated such that:

$$p^{\sigma^l_1 \sigma^l_2}_{v_{init}}(F) = l \text{ and } p^{\sigma^u_1 \sigma^u_2}_{v_{init}}(F) = u.$$

Since $l \neq u$, there is at least one $v_1 \in V_1$ where the two player 1 strategies disagree, i.e., $\sigma^l_1(v_1) \neq \sigma^u_1(v_1)$. Intuitively, this means that player 1 chose different concrete states contained in v_1 depending on whether he wanted to minimize or maximize the reachability probability. Consequently, v_1 can be split to narrow down the choices of player 1 in the resulting vertices. A possible way to achieve this, is to split v_1 into blocks v^l_1, v^u_1 and v^r_1 where

$$v^l_1 = \{s \in v_1 \mid \sigma^l_1(v_1) = \{\overline{\mu} \mid \mu \in Steps(s)\}\}$$
$$v^u_1 = \{s \in v_1 \mid \sigma^u_1(v_1) = \{\overline{\mu} \mid \mu \in Steps(s)\}\}$$
$$v^r_1 = v_1 \setminus (v^l_1 \cup v^u_1).$$

Of course, there may be several vertices that can be split according to this criterion and it is not clear which or how many blocks should be refined in order to get more precise bounds that are able to prove or disprove the property at hand. This refinement method is called *strategy-based*. There exist other refinement techniques, for example *value-based* refinement, which are not covered here. For details, we refer to [34].

Example 14. As pointed out in Example 13, player 1 chooses state $\langle 0, 0 \rangle \in A$ or either of the states $\langle 1, 0 \rangle, \langle 1, 1 \rangle \in A$ if he wants to minimize or maximize, respectively, the reachability probability in $\mathcal{G}_{\mathcal{M}}^{Q}$. Consequently, A is split into blocks $A_1 = \{\langle 0,0 \rangle\}$ and $A_2 = \{\langle 1,0 \rangle, \langle 1,1 \rangle\}$. The resulting game over the partition $Q' = (Q \setminus A) \cup \{A_1, A_2\}$ is depicted in Figure 8. Solving the refined game determines the minimal reachability probability to be in the interval $[0.2, 0.2]$.

∎

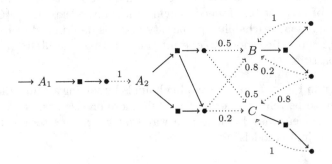

Fig. 8. Game-based abstraction after the refinement of block A

6.3 Menu-Based Abstraction

While game-based abstraction yields good results for many examples, the constructed game can become very large. The reason is that the game representation may need one player 2 vertex for each combination of (lifted) probability distributions available in some block $q \in Q$. The worst-case appears if all the states contained in a particular block happen to have different (combinations of) lifted probability distributions. Roughly speaking, in the context of MDPs given by a probabilistic program, the game may become large if there exist many states in which different combinations of guarded commands are enabled. In this case, constructing and solving the game might be very expensive.

Example 15. The game $\mathcal{G}_{\mathcal{M}}^{Q}$ in Example 12 has three player 2 vertices reachable in one step from player 1 vertex A, even though block A contained only three states of the concrete model. ∎

Menu-based abstraction [46, 45] aims to overcome this by considering commands of the probabilistic program in isolation. That is, it builds a possibly smaller game than game-based abstraction that might produce coarser probability approximations in the hope that it can be constructed and solved more easily. Instead of letting player 1 pick a concrete state out of a given block and move to the vertex representing the enabled commands at this particular state, it lets player 1 choose a command. This means that the choice of a concrete state is still open (among all states that have the chosen command enabled). Consequently, in the successor vertex, player 2 has the choice between all possible realizations of the chosen command in all states of the block.

Definition 11 (Menu game). *Given an MDP* $\mathcal{M} = (S, Act, \mathbf{P}, s_{init}, AP, L)$ *and a partition Q of S, the* menu-based abstraction *of $\mathcal{M}$ over Q is the probabilistic game*

$$\widehat{\mathcal{G}}_{\mathcal{M}}^Q = ((V = V_1 \,\dot\cup\, V_2 \,\dot\cup\, V_p, E), v_{init}, (V_1, V_2, V_p), \delta)$$

where

- $V_1 = Q \cup \{\bot\}$ *are player 1's vertices,*
- $V_2 = \{(v_1, a) \mid v_1 \in V_1, a \in Act(v_1)\}$ *are player 2's vertices,*
- $V_p = \{\mathbf{P}(s, a, \cdot) \mid s \in S, a \in Act(s)\} \cup \{v_p^\bot\}$ *are the probabilistic vertices,*
- $v_{init} = B \in P$ *such that $s_{init} \in B$ is the initial vertex, and*
- $\delta : V_p \to Dist(V)$ *is the identity function,*

and the set of edges E is given by

$$
\begin{aligned}
E = {} & \{(v_1, v_2) \mid v_1 \in V_1, v_2 = (v_1, a) \in V_2, a \in Act(v_1)\} \\
& \cup \{(v_2, v_p) \mid v_2 = (v_1, a) \in V_2, \exists s \in v_1 : v_p = \mathbf{P}(s, a, \cdot)\} \\
& \cup \{(v_2, v_p^\bot), (v_p^\bot, \bot) \mid v_2 = (v_1, a) \in V_2, \exists s \in V_1 : a \notin Act(s)\} \\
& \cup \{(v_p, v') \mid v_p \in V_p, v' \in V_1 : v_p(v') > 0\}
\end{aligned}
$$

where $v_p^\bot \in Dist(S \cup \{\bot\})$ is defined by $v_p^\bot(v) = 1$ iff $v = \bot$.

To distinguish the probabilities obtained in the game-based abstraction $\mathcal{G}_{\mathcal{M}}^Q$ from the ones obtained in the menu-based abstraction $\widehat{\mathcal{G}}_{\mathcal{M}}^Q$, we will denote the latter by $\widehat{p}_v^{\circ_1 \circ_2}(F)$ with $\circ_1, \circ_2 \in \{-, +\}$ and $\widehat{p}_v^{\sigma_1 \sigma_2}(F)$.

Starting in the initial vertex, player 1 chooses one of the commands that are enabled in at least one state in the current block. Then, player 2 implicitly chooses a state from the current block by choosing a probability distribution that is (i) created by the chosen command and (ii) is available at some state in the block. Finally, the successor vertices are given by that distribution and the play is once again in a vertex owned by player 1.

Example 16. Reconsider the MDP $\mathcal{M}$ in Figure 2 and the partition Q from Example 2. The resulting menu game is shown in Figure 9. Note that the labeling of player 1's choices with commands is added to illustrate the correspondence, but is not actually part of the game itself.   ∎

As for game-based abstraction, we can state the correctness of the abstraction in the sense that the reachability probabilities obtained from the game are lower and upper bounds for the reachability probabilities in the original MDP.

Theorem 3 (Correctness of menu-based abstraction [45]). *Let $\widehat{\mathcal{G}}_{\mathcal{M}}^Q$ be the menu game for an MDP $\mathcal{M}$ with state space S and Q a partition of S. Then for all $v \in V_1$ and $s \in v$*

$$\widehat{p}_v^{--}(F \cup \{\bot\}) \le p_s^-(F) \le \widehat{p}_v^{-+}(F \cup \{\bot\}), \tag{3}$$

$$\widehat{p}_v^{+-}(F) \le p_s^+(F) \le \widehat{p}_v^{++}(F). \tag{4}$$

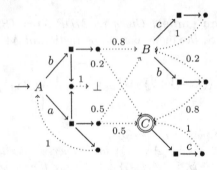

Fig. 9. The menu game $\widehat{\mathcal{G}}_{\mathcal{M}}^{Q}$

For minimal reachability objectives, $\perp$ becomes a target vertex in addition to the given set F. Intuitively, this is because it corresponds to the case where player 2 selects a concrete state in the abstract vertex in which the command previously selected by player 1 is not enabled. This would, however, result in a lower bound of 0 for the minimum reachability probability in all vertices that contain some state in which at least one action is not enabled. Instead, it should be implicitly forbidden for player 2 to choose $\perp$ in this case, which is done by assigning the worst (with respect to the goal of player 2) possible value to it. Conversely, if player 2 tries to maximize its value, the construction would be incorrect if $\perp$ was not considered a target vertex as well, which is illustrated by the next example.

Example 17. Reconsider the probabilistic program from Figure 1 with the difference that the guard of command b is strengthened to $1 < x+y \leq 2$. Note that the menu game $\widehat{\mathcal{G}}_{\mathcal{M}'}^{Q}$ for the resulting modified MDP $\mathcal{M}'$ is equal to $\widehat{\mathcal{G}}_{\mathcal{M}}^{Q}$ from Example 16 even though b is now disabled in $\langle 1, 0 \rangle$ in $\mathcal{M}'$ and the minimum reachability probability of reaching F from $\langle 1, 0 \rangle$ is 0.5 then. Now, suppose $\perp$ was not considered a target state in the menu game. If player 1 chooses the command b in A, the best player 2 can do to maximize the probability to eventually reach F is to choose the state that has b enabled and achieve a probability of 0.2. This is, however, not an upper bound for the minimal reachability probability for all states in block A, as $\langle 1, 0 \rangle \in A$ has a minimal reachability probability of 0.5 in $\mathcal{M}'$. Intuitively, it is not legal for player 1 to pick a command that is not enabled in all states. ∎

The menu game can be solved in the same fashion that the previous games were solved, e.g., using value iteration. As stated by the correctness theorem, this will result in lower and upper bounds for both the minimal and the maximal reachability probability with respect to the given set of target states.

Compared to game-based abstraction, it should be noted that it is no longer the case that player 1 resolves the nondeterminism introduced by the abstraction and player 2 the nondeterminism of the original model, but that the two have swapped roles. In the game-based abstraction setting, player 1 determined whether the resulting probability was a lower or upper bound while player 2 could control whether the result was an approximation of the minimal or maximal reachability probability in the original MDP. This is exactly reversed in the context of menu games, which is reflected in the previous correctness theorem by swapping the goals of the two players (compared to game-based abstraction).

Also, there are examples for which game-based abstraction produces tighter bounds than menu-based abstraction if the same partition Q of the state space is used for building the games. Technically, this happens because game-based abstraction constructs a game representation of the best transformer on the partition induced by the predicates whereas menu-based abstraction represents an abstract transformer that does not necessarily coincide with the best transformer [45].

Example 18. Reconsider the menu game $\widehat{\mathcal{G}}_{\mathcal{M}}^{Q}$ from Example 16. For the menu-based abstraction, the lower bounds for minimum and maximum reachability are both 0 and, likewise, the upper bounds are both 1, effectively yielding no information. As shown in Example 13, this is coarser than the bounds obtained via game-based abstraction (using the same partition Q). ∎

This immediately raises the issue of termination if menu-based abstraction is to be used in a refinement loop. Fortunately, the following result can be established.

Theorem 4 (Refinability). *Any finite partition Q can be refined to a finite partition Q' on which the reachability probabilities in the menu game approximate the reachability properties in the original MDP at least as precisely as the game-based abstraction over Q. Formally, for every $s \in S$, let $v \in Q$, $v' \in Q'$ such that $s \in v$ and $s \in v'$, then:*

$$p_{v,Q}^{--}(F) \leq \widehat{p}_{v',Q'}^{--}(F \cup \{\bot\}) \leq p_s^-(F) \leq \widehat{p}_{v',Q'}^{-+}(F \cup \{\bot\}) \leq p_{v,Q}^{+-}(F) \quad (5)$$

$$p_{v,Q}^{-+}(F) \leq \widehat{p}_{v',Q'}^{+-}(F) \leq p_s^+(F) \leq \widehat{p}_{v',Q'}^{++}(F) \leq p_{v,Q}^{++}(F). \quad (6)$$

Note that the players swapped roles, so the order of the superscripts of the reachability probabilities is important.

Extensions. For models that involve parametric transition probabilities depending on state variables, the usual game construction possibly produces games of infinite size, because infinitely many probability distributions might be available in a block. Recently, [22] proposed to solve this problem by constructing *constraint Markov games,* an extension of probabilistic games that is able to deal with *variable probabilities,* instead. Intuitively, the idea is to avoid introducing a game vertex for every available distribution by shifting the selection of the distribution into a different level of non-determinism in the game.

7   Conclusion

We have described three successful techniques for the abstraction of probabilistic systems. Which one is most useful in a concrete situation? Multi-valued abstraction seems to be the simplest method: one stays within the model of MDPs, so analysis of the abstract model can use mainstream model checkers. However, the disadvantage is that some questions cannot be answered. In those cases the abstraction-refinement frameworks demonstrate their strengths. CEGAR overcomes a part of the weakness of multi-valued abstraction, by providing a direction in which to refine a model if model checking on the abstract model has led to a spurious counterexample. Game-based techniques do not rely on MDPs as their underlying abstract model but rather use probabilistic games. This way, they can provide lower bounds on both minimal and maximal reachability probabilities.

References

[1] de Alfaro, L., Roy, P.: Magnifying-lens abstraction for Markov decision processes. In: Damm, W., Hermanns, H. (eds.) CAV 2007. LNCS, vol. 4590, pp. 325–338. Springer, Heidelberg (2007)

[2] Aljazzar, H., Leue, S.: Directed explicit state-space search in the generation of counterexamples for stochastic model checking. IEEE Trans. Software Eng. 36(1), 37–60 (2010)

[3] Aziz, A., Singhal, V., Balarin, F., Brayton, R.K., Sangiovanni-Vincentelli, A.L.: It usually works: The temporal logic of stochastic systems. In: Wolper, P. (ed.) CAV 1995. LNCS, vol. 939, pp. 155–165. Springer, Heidelberg (1995)

[4] Baier, C., Katoen, J.-P.: Principles of model checking. MIT Press, Cambridge (2008)

[5] Baier, C., Katoen, J.-P., Hermanns, H., Wolf, V.: Comparative branching-time semantics for Markov chains. Information and Computation 200(2), 149–214 (2005)

[6] Bianco, A., de Alfaro, L.: Model checking of probabilistic and nondeterministic systems. In: Thiagarajan, P.S. (ed.) FSTTCS 1995. LNCS, vol. 1026, pp. 499–513. Springer, Heidelberg (1995)

[7] Bozzano, M., Cimatti, A., Katoen, J.-P., Nguyen, V.Y., Noll, T., Roveri, M.: Safety, dependability and performance analysis of extended aadl models. Comput. J. 54(5), 754–775 (2011)

[8] Chadha, R., Viswanathan, M.: A counterexample-guided abstraction-refinement framework for markov decision processes. ACM Trans. Comput. Log. 12(1), 1 (2010)

[9] Chen, T., Forejt, V., Kwiatkowska, M., Parker, D., Simaitis, A.: Prism-games: A model checker for stochastic multi-player games. In: Piterman, N., Smolka, S.A. (eds.) TACAS 2013. LNCS, vol. 7795, pp. 185–191. Springer, Heidelberg (2013)

[10] Clarke, E., Grumberg, O., Jha, S., Lu, Y., Veith, H.: Counterexample-guided abstraction refinement. In: Emerson, E.A., Sistla, A.P. (eds.) CAV 2000. LNCS, vol. 1855, pp. 154–169. Springer, Heidelberg (2000)

[11] Condon, A.: The complexity of stochastic games. Information and Computation 96(2), 203–224 (1992)

[12] D'Argenio, P.R., Jeannet, B., Jensen, H.E., Larsen, K.G.: Reachability analysis of probabilistic systems by successive refinements. In: de Luca, L., Gilmore, S. (eds.) PAPM-PROBMIV 2001. LNCS, vol. 2165, pp. 39–56. Springer, Heidelberg (2001)

[13] D'Argenio, P.R., Jeannet, B., Jensen, H.E., Larsen, K.G.: Reduction and refinement strategies for probabilistic analysis. In: Hermanns, H., Segala, R. (eds.) PAPM-PROBMIV 2002. LNCS, vol. 2399, pp. 57–76. Springer, Heidelberg (2002)

[14] Dehnert, C., Katoen, J.-P., Parker, D.: SMT-based bisimulation minimisation of markov models. In: Giacobazzi, R., Berdine, J., Mastroeni, I. (eds.) VMCAI 2013. LNCS, vol. 7737, pp. 28–47. Springer, Heidelberg (2013)

[15] Delahaye, B., Katoen, J.-P., Larsen, K.G., Legay, A., Pedersen, M.L., Sher, F., Wąsowski, A.: Abstract probabilistic automata. In: Jhala, R., Schmidt, D. (eds.) VMCAI 2011. LNCS, vol. 6538, pp. 324–339. Springer, Heidelberg (2011)

[16] Desharnais, J., Gupta, V., Jagadeesan, R., Panangaden, P.: Approximating labelled Markov processes. Information and Computation 184(1), 160–200 (2003)

[17] Donaldson, A.F., Miller, A.: Symmetry reduction for probabilistic model checking using generic representatives. In: Graf, S., Zhang, W. (eds.) ATVA 2006. LNCS, vol. 4218, pp. 9–23. Springer, Heidelberg (2006)

[18] Eisentraut, C., Hermanns, H., Schuster, J., Turrini, A., Zhang, L.: The quest for minimal quotients for probabilistic automata. In: Piterman, N., Smolka, S.A. (eds.) TACAS 2013. LNCS, vol. 7795, pp. 16–31. Springer, Heidelberg (2013)

[19] Emerson, E.A., Clarke, E.M.: Using branching time temporal logic to synthesize synchronization skeletons. Science of Computer Programming 2(3), 241–266 (1982)

[20] Fecher, H., Leucker, M., Wolf, V.: Don't know in probabilistic systems. In: Valmari, A. (ed.) SPIN 2006. LNCS, vol. 3925, pp. 71–88. Springer, Heidelberg (2006)

[21] Feng, L., Kwiatkowska, M.Z., Parker, D.: Compositional verification of probabilistic systems using learning. In: QEST, pp. 133–142. IEEE Computer Society (2010)

[22] Ferrer Fioriti, L.M., Hahn, E.M., Hermanns, H., Wachter, B.: Variable probabilistic abstraction refinement. In: Chakraborty, S., Mukund, M. (eds.) ATVA 2012. LNCS, vol. 7561, pp. 300–316. Springer, Heidelberg (2012)

[23] Hahn, E.M., Hermanns, H., Wachter, B., Zhang, L.: Pass: Abstraction refinement for infinite probabilistic models. In: Esparza, J., Majumdar, R. (eds.) TACAS 2010. LNCS, vol. 6015, pp. 353–357. Springer, Heidelberg (2010)

[24] Han, T., Katoen, J.-P.: Counterexamples in probabilistic model checking. In: Grumberg, O., Huth, M. (eds.) TACAS 2007. LNCS, vol. 4424, pp. 72–86. Springer, Heidelberg (2007)

[25] Hansson, H., Jonsson, B.: A logic for reasoning about time and reliability. Formal Aspects of Computing 6(5), 512–535 (1994)

[26] Henzinger, T.A., Mateescu, M., Wolf, V.: Sliding window abstraction for infinite Markov chains. In: Bouajjani, A., Maler, O. (eds.) CAV 2009. LNCS, vol. 5643, pp. 337–352. Springer, Heidelberg (2009)

[27] Hermanns, H., Wachter, B., Zhang, L.: Probabilistic CEGAR. In: Gupta, A., Malik, S. (eds.) CAV 2008. LNCS, vol. 5123, pp. 162–175. Springer, Heidelberg (2008)

[28] Huth, M., Jagadeesan, R., Schmidt, D.: Modal transition systems: A foundation for three-valued program analysis. In: Sands, D. (ed.) ESOP 2001. LNCS, vol. 2028, pp. 155–169. Springer, Heidelberg (2001)

[29] Jansen, D.N., Song, L., Zhang, L.: Revisiting weak simulation for substochastic Markov chains. In: Joshi, K., Siegle, M., Stoelinga, M., D'Argenio, P.R. (eds.) QEST 2013. LNCS, vol. 8054, pp. 209–224. Springer, Heidelberg (2013)

[30] Jansen, N., Ábrahám, E., Katelaan, J., Wimmer, R., Katoen, J.-P., Becker, B.: Hierarchical counterexamples for discrete-time markov chains. In: Bultan, T., Hsiung, P.-A. (eds.) ATVA 2011. LNCS, vol. 6996, pp. 443–452. Springer, Heidelberg (2011)

[31] Jonsson, B., Larsen, K.G.: Specification and refinement of probabilistic processes. In: Proc. LICS 1991, pp. 266–277. IEEE Comp. Soc. Pr. (1991)

[32] Katoen, J.-P., Kemna, T., Zapreev, I., Jansen, D.N.: Bisimulation minimisation mostly speeds up probabilistic model checking. In: Grumberg, O., Huth, M. (eds.) TACAS 2007. LNCS, vol. 4424, pp. 87–101. Springer, Heidelberg (2007)

[33] Katoen, J.-P., Klink, D., Leucker, M., Wolf, V.: Three-valued abstraction for probabilistic systems. JLAP 81(4), 356–389 (2012)

[34] Kattenbelt, M., Kwiatkowska, M., Norman, G., Parker, D.: A game-based abstraction-refinement framework for markov decision processes. Formal Methods in System Design 36(3), 246–280 (2010)

[35] Kattenbelt, M., Kwiatkowska, M.Z., Norman, G., Parker, D.: A game-based abstraction-refinement framework for markov decision processes. Formal Methods in System Design 36(3), 246–280 (2010)

[36] Komuravelli, A., Păsăreanu, C.S., Clarke, E.M.: Assume-guarantee abstraction refinement for probabilistic systems. In: Madhusudan, P., Seshia, S.A. (eds.) CAV 2012. LNCS, vol. 7358, pp. 310–326. Springer, Heidelberg (2012)

[37] Kwiatkowska, M., Norman, G., Parker, D.: Symmetry reduction for probabilistic model checking. In: Ball, T., Jones, R.B. (eds.) CAV 2006. LNCS, vol. 4144, pp. 234–248. Springer, Heidelberg (2006)

[38] Kwiatkowska, M., Norman, G., Parker, D.: Prism 4.0: Verification of probabilistic real-time systems. In: Gopalakrishnan, G., Qadeer, S. (eds.) CAV 2011. LNCS, vol. 6806, pp. 585–591. Springer, Heidelberg (2011)

[39] Kwiatkowska, M., Norman, G., Parker, D., Qu, H.: Assume-guarantee verification for probabilistic systems. In: Esparza, J., Majumdar, R. (eds.) TACAS 2010. LNCS, vol. 6015, pp. 23–37. Springer, Heidelberg (2010)

[40] Larsen, K.G., Thomsen, B.: A modal process logic. In: Proc. LICS 1988, pp. 203–210. Los Alamitos, Calif (1988)

[41] Larsen, K.G., Thomsen, B.: A modal process logic. In: LICS, pp. 203–210 (1988)

[42] McMillan, K.L.: Lazy abstraction with interpolants. In: Ball, T., Jones, R.B. (eds.) CAV 2006. LNCS, vol. 4144, pp. 123–136. Springer, Heidelberg (2006)

[43] Segala, R., Lynch, N.: Probabilistic simulations for probabilistic processes. Nordic Journal of Computing 2, 250–273 (1995)

[44] Vardi, M.Y.: Automatic verification of probabilistic concurrent finite-state programs. In: FOCS, pp. 327–338. IEEE Comp. Soc. Pr., Washington, DC (1985)

[45] Wachter, B.: Refined probabilistic abstraction. Ph.D. thesis, Universität des Saarlandes, Saarbrücken (2011)

[46] Wachter, B., Zhang, L.: Best probabilistic transformers. In: Barthe, G., Hermenegildo, M. (eds.) VMCAI 2010. LNCS, vol. 5944, pp. 362–379. Springer, Heidelberg (2010)

[47] Wimmer, R., Herbstritt, M., Hermanns, H., Strampp, K., Becker, B.: Sigref- a symbolic bisimulation tool box. In: Graf, S., Zhang, W. (eds.) ATVA 2006. LNCS, vol. 4218, pp. 477–492. Springer, Heidelberg (2006)

[48] Wimmer, R., Jansen, N., Ábrahám, E., Becker, B., Katoen, J.-P.: Minimal critical subsystems for discrete-time markov models. In: Flanagan, C., König, B. (eds.) TACAS 2012. LNCS, vol. 7214, pp. 299–314. Springer, Heidelberg (2012)

[49] Wimmer, R., Jansen, N., Vorpahl, A., Ábrahám, E., Katoen, J.-P., Becker, B.: High-level counterexamples for probabilistic automata. In: Joshi, K., Siegle, M., Stoelinga, M., D'Argenio, P.R. (eds.) QEST 2013. LNCS, vol. 8054, pp. 39–54. Springer, Heidelberg (2013)

[50] Zhang, L.: Decision algorithms for probabilistic simulations. Ph.D. thesis, Universität des Saarlandes, Saarbrücken (2009)

Computing Behavioral Relations
for Probabilistic Concurrent Systems

Daniel Gebler[1], Vahid Hashemi[2,3], and Andrea Turrini[4]

[1] Department of Computer Science, VU University Amsterdam,
De Boelelaan 1081a, NL-1081 HV Amsterdam, The Netherlands
[2] Max Planck Institute for Informatics, 66123 Saarbrücken, Germany
[3] Department of Computer Science
Saarland University, 66123 Saarbrücken, Germany
[4] State Key Laboratory of Computer Science, Institute of Software,
Chinese Academy of Sciences, 100190 Beijing, China

Abstract. Behavioral equivalences and preorders are fundamental notions to formalize indistinguishability of transition systems and provide means to abstraction and refinement. We survey a collection of models used to represent concurrent probabilistic real systems, the behavioral equivalences and preorders they are equipped with and the corresponding decision algorithms. These algorithms follow the standard refinement approach and they improve their complexity by taking advantage of the efficient algorithms developed in the optimization community to solve optimization and flow problems.

1 Introduction

1.1 Probabilistic Systems

Probability, time, and *nondeterminism.* These are three main characteristics of several real-world applications. *Probability* occurs every time the behavior of the applications is not unique, either by construction or by physical properties. For example, distributed algorithms like the Zeroconf protocol or cryptographic protocols like SSL are based on random choices to break symmetry or to insert uncertainty in order to achieve their goals. Each time a message is transmitted on the network, in fact, transmission protocols have to manage the corruption of the messages, as well as their loss, as the effect of the interference with other concurrent transmissions or physical properties of the transmission medium. For instance, simultaneous transmissions on the same channel of a wireless network lead to the collision of the sent messages and their corruption.

Beside probabilities, these systems often have another source of uncertainty, namely *nondeterminism,* that appears whenever an event may occur with unpredictable behavior; for instance, the event of a host starting the transmission in a wireless network.

Time governs the evolution of the system: with the time passing, the system performs and reacts to actions and correspondingly changes its state, according

A. Remke and M. Stoelinga (Eds.): ROCKS Autumn School 2012, LNCS 8453, pp. 117–155, 2014.
© Springer-Verlag Berlin Heidelberg 2014

to its goals. Time can be considered as a discrete component (e.g., a program running on a computer performs one operation at each tick of the digital clock) or as a continuum behavior (e.g., the arrival and service of customers at the information desk).

To study the properties of such real-world applications, several models have been proposed by researchers: the basic model in the discrete time domain is the discrete time Markov chains (*DTMCs*) model [26,52], where the time is discrete (i.e., the system performs one operation per clock tick) and only probability determines the reached states. The continuous-time counterpart is known as the continuous-time Markov chains (*CTMCs*) [3,5] model, where exponentially distributed sojourn times distributions control the evolution of the system.

DTMCs and *CTMCs* are purely probabilistic, and they have been extended with nondeterminism to permit different operations or behaviors from a specific state. This extension results to Markov decision processes (*MDPs*) [10,31,32,44] and continuous-time Markov decision processes (*CTMDPs*) [7,11,31,44,54], respectively. These models, despite being widely used to represent and study real systems, are not fully compositional, that is, there is no guarantee that complex systems can be obtained by composing smaller components while preserving the intended behavior. This property is rather important as it is usually much easier to model and study (a set of) small systems and then combine them together rather than a single large system. Moreover, in the real world, usually applications and protocols involve several parties each one composed by modules working together in parallel. Two models have been proposed to achieve such compositional property: the probabilistic automata (*PAs*) model [47, 48] for discrete time systems and the interactive Markov chains (*IMCs*) [28] model for continuous-time systems. Recently one model has been proposed to unify and merge all such models in a single framework: the Markov automata (*MAs*) model [17, 22, 23]. This formalism is suitable for studying systems featuring continuous-time based behaviors as well as probabilistic and nondeterministic choices. Moreover, the Markov automata model provides the semantics to every generalized stochastic Petri net (GSPN) [19], a popular modelling formalism for performance and dependability analysis.

1.2   Comparing System Behaviors

Given a real world system we want to analyze, for instance by verifying whether it satisfies a set of properties, we can model it in several ways. This analysis is commonly known as *model checking*. In particular, we can decide to model it as a *DTMC* or as a *PA* whenever we are interested in its properties as a discrete time system; alternatively, if we want to study its behavior in continuous time, we can use *CTMCs* or *IMCs*. The choice of the model framework depends on the properties we are interested in and the details we want to consider.

Once the model framework has been chosen, the real system can be represented by several different models: for example, we can use different names for the states, we can encode probabilistic choices as sequences of events or as single events, we can detail or abstract from particular details, and so on. It is clear that these choices affect the resulting model whose size may vary even if all these models represent the same real system.

A possible way to abstract away from this modelling details is to use the so called *simulation* and *bisimulation* relations that allow us to declare that two models are similar or equivalent whenever they are related, respectively. Intuitively, a system S_1 simulates a system S_2 if $S1$ is able to mimic whatever S_2 can do; the bisimulation requires that also S_2 simulates S_1. Usually, a simulation (or bisimulation) is defined as a binary relation over the states of the model and for each pair (s_1, s_2), if s_1 can perform a step, then s_2 has to match such step via its own steps in order to reach states that are related to the states reached from s_1. Depending on the steps s_2 is allowed to perform, simulation relations can be classified as *strong* (s_2 has to match with exactly one step) or as *weak* (s_2 is free to perform an arbitrary additional number of internal steps). Computing such simulation relations is rather easy by using classical refinement algorithms, provided that we have a procedure for deciding the existence of the matching step from s_2 given a step from s_1. As we will see in Section 6, such procedure is the only part that has to be changed in order to decide different simulations and it is also the bottleneck of the computation and the main source of the complexity of the decision procedure.

We are interested in systems related by a simulation relation since also the properties they satisfy are related, so we can check whether the real world system satisfies a given property by verifying it in one of the similar models: the theory ensures us that the evaluation of the property does not depend on the specific model we consider to represent the real world system. When we consider the bisimulation relation, among all possible bisimilar models there is a unique minimal model (up to isomorphism) that represents the original system [21]: the *quotient* model. The quotient is the model with the minimum number of states and transitions still behaving as the system we want to analyze; this minimality mitigates the state explosion problem of the model checking [8, 14, 34] as well as it helps in reducing the computational effort needed to verify whether the desired properties are fulfilled. Moreover, the computation of the quotient automaton is independent on the properties we want to check, thus even if it may be rather time consuming, the overall gain it provides to the following model checking phase may justify it.

1.3   Optimization Problems

Optimization or mathematical programming uses mathematical techniques to find the best solution among a set of given alternatives. More precisely, an optimization problem asks for maximizing or minimizing a real valued function for which the variables take values from a permissible set. It includes many diverse areas such as decision theory [42], flow network optimization [1] and so on. Flow network optimization is a subclass of linear programming that has application in a number of domains such as computer science, logistics, transportation systems. Although flow network based models are not as wide as models that can be formulated mathematically using linear or integer programming, they can be solved very quickly which enables them to be a powerful tool for decision making [1].

1.4 Probabilistic Systems vs. Optimization

To a casual observer, flow and optimization problems seem rather unrelated to probabilistic concurrent systems. In fact, as we have seen, the former aim to optimize problems like resource allocation or goods transportation and distribution while the latter model systems that run in parallel where the behavior depends on probabilistic events as well like random failures, errors, and choices needed to break symmetry. To a careful observer, flow and optimization problems and probabilistic concurrent systems are not so unrelated, since the probability mass concentrated in the initial state can be seen as a liquid that flows and distributes in the network representing the possible evolution of the system. To highlight this connection, in this survey we consider a selection of papers [29,30,55,57] that, together with other works in concurrency literature such as [2,4,13,15,20,21,43,45], make use of flow and optimization problems to decide or solve efficiently the challenges of probabilistic concurrent systems.

Organization of the Paper. After the mathematical preliminaries in Section 2, we present in Section 3 the discrete and continuous-time models, followed in Section 4 by the simulation and bisimulation relations defined on them. We recall in Section 5 the theory about networks and flow problems that are widely used in Section 6 to efficiently compute simulations and bisimulations. We conclude the paper in Section 7.

2 Mathematical Preliminaries

2.1 Functions and Relations

Given a set X and $\perp \notin X$, we denote by $X_\perp$ the set $X \cup \{\perp\}$.

Let X, Y be two finite sets, $f\colon X \to \mathbb{R}$ and $g\colon X \times Y \to \mathbb{R}$ be two functions. For $X' \subseteq X$, we denote by $f(X')$ the value $f(X') = \sum_{x \in X'} f(x)$; for $x \in X$ and $Y' \subseteq Y$, $g(x, Y') = \sum_{y \in Y'} g(x, y)$ and similarly, for $y \in Y$ and $X' \subseteq X$, $g(X', y) = \sum_{x \in X'} g(x, y)$. Finally, we define for each $x \in X$ and $y \in Y$ the functions $g(x, \cdot)\colon Y \to \mathbb{R}$ and $g(\cdot, y)\colon X \to \mathbb{R}$ as $g(x, \cdot)(y') = g(x, y')$ for each $y' \in Y$ and $g(\cdot, y)(x') = g(x', y)$ for each $x' \in X$, respectively. Given two functions $f, g\colon X \to \mathbb{R}$ and $p \in \mathbb{R}$, we denote by $p \cdot f\colon X \to \mathbb{R}$ the function $(p \cdot f)(x) = p \cdot f(x)$ for each $x \in X$ and $f + g\colon X \to \mathbb{R}$ the function $(f + g)(x) = f(x) + g(x)$ for each $x \in X$.

For a function $f\colon X \to \mathbb{R}^{\geq 0}$, we denote by $\mathrm{Supp}(f)$ the *support* set $\mathrm{Supp}(f) = \{ x \in X \mid f(x) > 0 \}$.

Given a relation $\mathcal{R} \subseteq X \times Y$ and the sets $X' \subseteq X$ and $Y' \subseteq Y$, we define $\mathcal{R}(X') = \{ y \in Y \mid \exists x \in X'.x \mathrel{\mathcal{R}} y \}$ and $\mathcal{R}^{-1}(Y') = \{ x \in X \mid \exists y \in Y'.x \mathrel{\mathcal{R}} y \}$.

Given a relation $\mathcal{R} \subseteq X \times X$, we call $\mathcal{R} \cap \mathcal{R}^{-1}$ the *kernel* of $\mathcal{R}$ and we denote by $\mathcal{R}_\perp \subseteq X_\perp \times X_\perp$ the relation $\mathcal{R} \cup \{ (\perp, x) \mid x \in X_\perp \}$.

2.2 Probability Distributions

For a set X, denote by $\mathrm{Disc}(X)$ the set of discrete probability distributions over X, and by $\mathrm{SubDisc}(X)$ the set of discrete sub-probability distributions over X. Since a discrete sub-probability distribution $\rho \in \mathrm{SubDisc}(X)$ can be seen as a function $\rho\colon X \to [0,1]$, we adopt the same terminology and operations. Given $\rho \in \mathrm{SubDisc}(X)$, we denote by $\rho(\bot)$ the value $1 - \rho(X)$ where $\bot \notin X$, and by $|\rho|$ the size $|\mathrm{Supp}(\rho)|$. We extend ρ to a probability distribution $\rho_\bot \in \mathrm{Disc}(X_\bot)$ by defining $\rho_\bot(\bot) = 1 - \rho(X)$ and $\rho_\bot(x) = \rho(x)$ for each $x \in X$. We denote by δ_x, where $x \in X_\bot$, the *Dirac* distribution such that $\delta_x(y) = 1$ for $y = x$, 0 otherwise. For a sub-probability distribution ρ, we also write $\rho = \{\,(x, p_x) \mid x \in X\,\}$ where p_x is the probability of x. We say that ρ is *stochastic* if $\rho(X) = 1$ and *absorbing* if $\rho(\bot) = \delta_\bot$. We sometimes refer to $\rho(X)$ as the *mass* of ρ.

The lifting $\mathcal{L}(\mathcal{R}) \subseteq \mathrm{Disc}(X) \times \mathrm{Disc}(X)$ [34] of a relation $\mathcal{R} \subseteq X \times X$ to distributions is defined as: for $\rho_1, \rho_2 \in \mathrm{Disc}(X)$, $\rho_1\, \mathcal{L}(\mathcal{R})\, \rho_2$ holds if there exists a *weighting function* $w\colon X \times X \to [0,1]$ such that

1. for each $(x_1, x_2) \in X \times X$, $w(x_1, x_2) > 0$ implies $x_1\, \mathcal{R}\, x_2$,
2. for each $x_1 \in X$, $w(x_1, X) = \rho_1(x_1)$, and
3. for each $x_2 \in X$, $w(X, x_2) = \rho_2(x_2)$.

This definition of lifting has been proposed for discrete systems [34,50] and it is indeed equivalent [55] to the definition based on $\mathcal{R}$-closure introduced by [18] for non-discrete systems: the lifting $\mathcal{L}(\mathcal{R}) \subseteq \mathrm{Disc}(X) \times \mathrm{Disc}(X)$ of a relation $\mathcal{R} \subseteq X \times X$ is defined as: for $\rho_1, \rho_2 \in \mathrm{Disc}(X)$, $\rho_1\, \mathcal{L}(\mathcal{R})\, \rho_2$ holds if for each $X' \subseteq X$, $\rho_1(X') \leq \rho_2(\mathcal{R}(X'))$.

Extending the lifting to sub-distributions is rather easy [57]: for $\rho_1, \rho_2 \in \mathrm{SubDisc}(X)$, $\rho_1\, \mathcal{L}(\mathcal{R})\, \rho_2$ holds if there exists a *weighting function* $w\colon X_\bot \times X_\bot \to [0,1]$ such that

1. for each $(x_1, x_2) \in X_\bot \times X_\bot$, $w(x_1, x_2) > 0$ implies $x_1\, \mathcal{R}_\bot\, x_2$,
2. for each $x \in X_\bot$, $w(x, X_\bot) = \rho_1(x)$, and
3. for each $x \in X_\bot$, $w(X_\bot, x) = \rho_2(x)$.

3 The Models

We now introduce the formal models for probabilistic concurrent systems we consider in this survey paper. We first recall the discrete time models and then the continuous-time models. In this work we consider only finite models, i.e., systems such that states, actions, and transition relations are finite.

3.1 Discrete Time Models

The first model we consider is the labelled substochastic discrete time Markov chain model where each state enables only a transition that may reach several states, each one with a given probability. The status of the system is represented by a set AP of atomic propositions that are true in the given state.

Definition 1 (Substochastic discrete time Markov chain [8, 33]). *A labelled substochastic Discrete Time Markov Chain (sDTMC) $\mathcal{S}$ is a tuple $\mathcal{S} = (S, \bar{s}, \mathbf{P}, L)$ where S is a finite set of states, $\bar{s}$ is the start state, $\mathbf{P} \colon S \times S \to [0, 1]$ is a probability matrix such that $\mathbf{P}(s, \cdot) \in \mathrm{SubDisc}(S)$ for all $s \in S$, and $L \colon S \to 2^{AP}$ is a labeling function.*

Given a state s and the associated distribution $\mu_s = \mathbf{P}(s, \cdot) \in \mathrm{SubDisc}(S)$, we call (s, μ_s) a *transition* and we say that (s, μ_s) is enabled by s and that μ_s is the target of (s, μ_s).

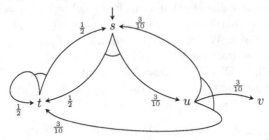

Fig. 1. An example of substochastic discrete time Markov chain

Figure 1 shows an example of a *sDTMC*, where s is the initial state, denoted by the short incoming arrow. For each state, we represent the enabled transition by a set of arrows grouped by an arc and pointing to the target states, each one decorated with the corresponding probability. For example, the transition enabled by s reaches t and u with probability $\frac{1}{2}$ and $\frac{3}{10}$, respectively. As usual in this kind of representation of the model, to keep the picture clear we have omitted the arrows reaching states with probability 0. For instance, from s there should be also an arrow reaching v with probability 0. As labels of the states, we take $AP = S = \{s, t, u, v\}$ and we let $L(z) = z$ for each state $z \in S$. Note that the transitions from both s and u have as target a sub-probability distribution that is not a probability distribution. In fact, for the transition (s, μ_s) the mass of μ_s is $\frac{8}{10}$ and the transition (u, μ_u) the mass of μ_u is $\frac{9}{10}$.

We call a state s *stochastic (absorbing)* if the distribution $\mathbf{P}(s, \cdot)$ is stochastic (absorbing) respectively. For the *sDTMC* in Figure 1, t is stochastic, v is absorbing while both s and u are neither stochastic nor absorbing. If we restrict the states of a *sDTMC* to be either stochastic or absorbing, we obtain a discrete time Markov chain:

Definition 2 (Discrete time Markov chain [26, 52]). *A labelled Discrete Time Markov Chain (DTMC) $\mathcal{D}$ is a labelled sDTMC $\mathcal{D} = (S, \bar{s}, \mathbf{P}, L)$ such that for each state $s \in S$, $\mathbf{P}(s, S) \in \{0, 1\}$.*

Figure 2 shows an example of a *DTMC*. It is actually the *sDTMC* in Figure 1 where probability distributions have been normalized to have mass 1.

These two models are suitable for systems exhibiting only probabilistic behaviors, that is, they are not able to represent systems where different transitions

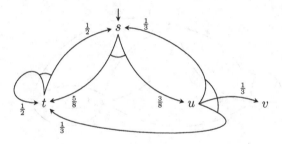

Fig. 2. An example of discrete time Markov chain

are available from the states. For instance, the system that is in a particular state may react differently to different stimuli and this can be modeled by performing different transitions leading to different distributions over the states of the system. We call this capacity nondeterminism that is encoded, together with probability, by the following two discrete time models: Markov decision processes and probabilistic automata. In order to have a uniform approach, for probabilistic automata we adopt the notation of [57] instead of the one used in [47, 48].

Definition 3 (Probabilistic automaton [47, 48]). *A* Probabilistic Automaton *(PA) $\mathcal{P}$ is a tuple $\mathcal{P} = (S, \bar{s}, \Sigma, \rightarrow, L)$ where S is a finite set of* states, *$\bar{s}$ is the* start state, *Σ is a finite set of* actions, *$\rightarrow \subseteq S \times \Sigma \times \mathrm{Disc}(S)$ is a finite* probabilistic transition relation, *and $L \colon S \rightarrow 2^{AP}$ is a labeling function.*

The set Σ is divided in two sets H and E of *internal* (*hidden*) and *external* actions, respectively. We remark that the definition of probabilistic automata we are presenting here is different from the original one given by Segala in [48] named *simple probabilistic automata,* but currently known as just probabilistic automata. In fact, in such work (simple) probabilistic automata are defined as follows (cf. [48, Section 3.1]): A Probabilistic Automaton (PA) $\mathcal{P}$ is a tuple $(S, \bar{s}, \Sigma, \rightarrow)$ where S is a countable set of *states*, $\bar{s} \in S$ is the *start state*, Σ is a countable set of *actions*, and $\rightarrow \subseteq S \times \Sigma \times \mathrm{Disc}(S)$ is a *probabilistic transition relation*. The main difference with Definition 3 is that in [48] there is no labeling function. This difference can be easily bridged by defining L as $L(s) = \emptyset$ for each $s \in S$. Figure 3 shows an example of a *PA* where H $= \{\tau\}$ and E $= \{a, b\}$.

In a probabilistic automaton $\mathcal{P}$ we can distinguish between two kinds of nondeterminism: external and internal nondeterminism. We say that a state s exhibits external nondeterminism if there exist two different actions a and b such that $(s, a, \mu_a) \in \rightarrow$ and $(s, b, \mu_b) \in \rightarrow$ for some $\mu_a, \mu_b \in \mathrm{Disc}(S)$. For instance, this is the case for the state v of the *PA* in Figure 3 since we have the two transitions (v, a, δ_s) and (v, b, δ_u). On the other hand, we say that a state s exhibits internal nondeterminism if there exist an action a and two different distributions $\mu_1, \mu_2 \in \mathrm{Disc}(S)$ such that $(s, a, \mu_1) \in \rightarrow$ and $(s, a, \mu_2) \in \rightarrow$. This happens for the state u that enables two different transitions both with action b. Note that

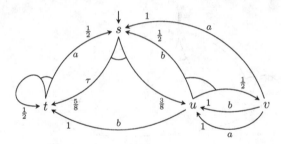

Fig. 3. An example of probabilistic automaton

a state may exhibit both internal and external nondeterminism (as happens for v) or none of them (see states s and t).

3.2   Continuous-Time Models

We now consider the continuous-time counterparts of the previous models, where state transitions are governed by the passing of the time. Essentially, they are defined as the discrete time models except for the probability distributions that are replaced by transition rates, i.e., the speed of transition firing.

The first model we recall is about continuous-time Markov chains that are just discrete time Markov chains where the probability matrix is replaced by the rate matrix.

Definition 4 (Continuous-time Markov chain [5, 44, 55]).
A labelled Continuous-Time Markov Chain (CTMC) $\mathcal{C}$ is a tuple $\mathcal{C} = (S, \bar{s}, \mathbf{R}, L)$ such that S, $\bar{s}$, and L are defined as for DTMC and $\mathbf{R}\colon S \times S \to \mathbb{R}^{\geq 0}$ is the rate matrix.

Note that the usual definition of CTMCs, such as the one in [3], requires that $\mathbf{R}\colon S \times S \to \mathbb{R}$ where for each $s \in S$, $\mathbf{R}(s, s') \geq 0$ for each $s' \neq s$ and $\mathbf{R}(s, s) = -\sum_{s' \neq s} \mathbf{R}(s, s')$. As remarked in [5], allowing self loops neither alters the transient nor the steady-state behavior of the CTMC, but it allows the usual interpretation of the linear-time CSL operators like next-step and until.

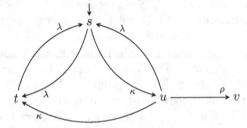

Fig. 4. An example of continuous-time Markov chain

Figure 4 shows an example of a *CTMC*. Greek letters λ, κ, and ρ are the rates governing the speed of the firing of the transitions. So, for example, the λ on the transition from s to t means that $\mathbf{R}(s,t) = \lambda$. We omitted the transitions with rate 0 to keep the picture clear.

The probability of performing a transition and reaching a given state can be computed as follows: starting from the state s, the probability of performing a transition within time t is $1 - e^{-\mathbf{R}(s,S) \cdot t}$ and the probability of reaching the state s' with this transition is $(1 - e^{-\mathbf{R}(s,S) \cdot t}) \cdot \frac{\mathbf{R}(s,s')}{\mathbf{R}(s,S)}$.

This allows us to consider the *DTMC* embedded into a *CTMC* that captures the system behavior after abstracting away the time:

Definition 5 (Embedded DTMC [5, 55, 57]). *Let $\mathcal{C}$ be a CTMC. The embedded DTMC $\mathcal{D}$ of $\mathcal{C}$ is defined by $emb(\mathcal{C}) = (S, \bar{s}, \mathbf{P}, L)$ where for each $s, s' \in S$, $\mathbf{P}(s, s')$ is defined as $\mathbf{P}(s, s') = \frac{\mathbf{R}(s,s')}{\mathbf{R}(s,S)}$ if $\mathbf{R}(s, S) > 0$, and $\mathbf{P}(s, s') = 0$ otherwise.*

Similarly to *CTMC*s and *DTMC*s, the continuous-time counterparts of *PA*s, called continuous-time probabilistic automata (*CTPA*), are obtained by replacing the transition relation with a rate matrix.

We call a function $r \colon S \to \mathbb{R}^{\geq 0}$ a *rate function* and we denote the set of all rate functions by $Rate(S)$. Given the rate function r, we call $r(S)$ the *exit rate*. Given $\mathbf{R}$ and a state s of a *CTMC* $\mathcal{C}$, we call $\mathbf{R}(s, \cdot) \colon S \to \mathbb{R}^{\geq 0}$ the *rate function* associated with s and we usually denote it by r_s.

Definition 6 (Continuous-time probabilistic automaton [11, 37, 44]). *A Continuous Time Probabilistic Automaton (CTPA) $\mathcal{CP}$ is a tuple $\mathcal{CP} = (S, \bar{s}, \Sigma, \mathbf{R}, L)$, where S is a finite set of states, $\bar{s}$ is the start state, Σ is a finite set of actions, $\mathbf{R} \subseteq S \times \Sigma \times Rate(S)$ is a finite rate matrix, and $L \colon S \to 2^{AP}$ is a labeling function.*

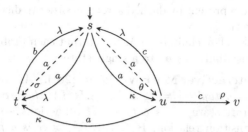

Fig. 5. An example of continuous-time probabilistic automaton

Figure 5 shows an example of *CTPA*; arrows emanating from a state with the same label and shape belong to the same transition. For instance, state s enables two transitions (s, a, r) and (s, a, r') with $r, r' \in Rate(S)$ such that $r(t) = \sigma$, $r(u) = \theta$, and $r(s) = r(v) = 0$ and $r'(t) = \lambda$, $r'(u) = \kappa$, and $r'(s) = r'(v) = 0$, respectively.

3.3 Mixed Discrete and Continuous-Time Models

We now present two models that merge continuous-time and discrete time behavior, the interactive Markov chains and the Markov automata. They exhibit continuous-time behavior like *CTMCs*, where transitions are fired by the passage of the time, as well as discrete time behavior like labelled transitions systems where transitions are fired by actions. These two models are especially suited for compositional reasoning over continuous-timed systems due to the separation of action and Markovian transitions and the maximal progress assumption, that is, if a state enables both timed transitions and internally labelled transitions, then the latter take precedence and the former are ignored.

Definition 7 (Markov automaton [17,22,23]). *A Markov Automaton (MA) $\mathcal{MA}$ is a tuple $\mathcal{MA} = (S, \bar{s}, \Sigma, \rightarrow, \mathbf{R}, L)$ where S is a finite set of states, $\bar{s}$ is the* start state, *Σ is a finite set of* actions, *$\rightarrow \subseteq S \times \Sigma \times \mathrm{Disc}(S)$ is a finite probabilistic transition relation, $\mathbf{R} \subseteq S \times \mathbb{R}^{\geq 0} \times S$ is a finite set of* timed transitions, *and $L \colon S \rightarrow 2^{AP}$ is a labeling function.*

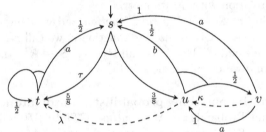

Fig. 6. An example of Markov automaton

Figure 6 shows an example of a *MA*. As for the *CTMC* in Figure 4, we use Greek letters λ and κ for the rates governing the speed of the firing of the transitions that we represent by dashed arrows in order to distinguish them from probabilistic transitions.

An interactive Markov chain is an *MA* such that each probabilistic transition leads to a Dirac distribution, i.e., to a single state:

Definition 8 (Interactive Markov chain [28]). *An Interactive Markov Chain (IMC) $\mathcal{I}$ is a tuple $\mathcal{I} = (S, \bar{s}, \Sigma, \rightarrow, \mathbf{R}, L)$ where S is a finite set of states, $\bar{s}$ is the* start state, *Σ is a finite set of* actions, *$\rightarrow \subseteq S \times \Sigma \times S$ is an interactive transition relation, $\mathbf{R} \subseteq S \times \mathbb{R}^{\geq 0} \times S$ is a finite set of* timed transitions, *and $L \colon S \rightarrow 2^{AP}$ is a labeling function.*

Figure 7 shows an example of an *IMC*. In particular, *IMC* can be seen as the merger of labelled transitions systems and *CTMCs* while *MA* can be seen as the merger of *PAs* and *CTMCs*. In fact, each model is an instance of the *MA* model with specific restrictions on $\rightarrow$ and $\mathbf{R}$ (cf. [22, Section 3]). As for probabilistic automata, the original definitions do not involve the labeling function L that we have added for uniformity. Again, the original model can be recovered by defining $L(s) = \emptyset$ for each $s \in S$.

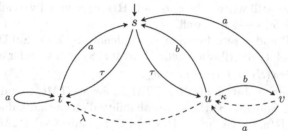

Fig. 7. An example of interactive Markov chain

3.4 Terminology and Notation

In the remaining of the paper we adopt the following terminology and notation, given in the context of probabilistic automata [27, 29, 47, 48, 57].

We refer to each instance of the discrete and continuous-time models as *automaton* and we denote it by $\mathcal{A}$, that is, we use the term (discrete time) automaton and $\mathcal{A}$ for the *sDTMC* $\mathcal{S}$, the *DTMC* $\mathcal{D}$, and the *PA* $\mathcal{P}$ as well as the term (continuous-time) automaton and $\mathcal{A}$ for the *CTMC* $\mathcal{C}$ and the *CTPA* $\mathcal{CP}$.

Given a *PA* $\mathcal{P}$, we let s,t,u,v, and their variants with indices range over S; a, b range over actions; and τ range over internal actions. A transition $tr = (s, a, \mu) \in \rightarrow$, also denoted by $s \xrightarrow{a} \mu$, is said to *leave* from state s, to be *labelled* by a, and to *lead* to the *target* distribution μ, also denoted by μ_{tr}. We denote by $src(tr)$ the *source* state s and by $act(tr)$ the *action* a. We also say that s enables action a, that action a is enabled from s, and that (s, a, μ) is enabled from s. Finally, we let $s \longrightarrow = \{\, tr \in \rightarrow \mid src(tr) = s \,\}$ be the set of transitions enabled by s and $\xrightarrow{a} = \{\, tr \in \rightarrow \mid act(tr) = a \,\}$ be the set of transitions with label a.

An *execution fragment* of a *PA* $\mathcal{P}$ is a finite or infinite sequence of alternating states and actions $\alpha = s_0 a_1 s_1 a_2 s_2 \ldots$ starting from a state s_0, also denoted by $first(\alpha)$, and, if the sequence is finite, ending with a state denoted by $last(\alpha)$, such that for each $i > 0$ there exists a transition $(s_{i-1}, a_i, \mu_i) \in \rightarrow$ such that $\mu_i(s_i) > 0$. The *length* of α, denoted by $len(\alpha)$, is the number of occurrences of actions in α. If α is infinite, then $len(\alpha) = \infty$. Denote by $frags(\mathcal{P})$ the set of execution fragments of $\mathcal{P}$ and by $frags^*(\mathcal{P})$ the set of finite execution fragments of $\mathcal{P}$. An execution fragment α is a *prefix* of an execution fragment α', denoted by $\alpha \leqslant \alpha'$, if the sequence α is a prefix of the sequence α'. The *trace* $trace(\alpha)$ of α is the sub-sequence of external actions of α; we denote by ε the empty trace and we define $trace(a) = a$ for $a \in \mathbf{E}$ and $trace(a) = \varepsilon$ for $a \in \mathbf{H}$.

We extend the above terminology to the other models introduced so far, when applicable; in particular, we use $\rightarrow$ to denote the transition relations $\mathbf{P}$ and $\mathbf{R}$ of *DTMCs* and *sDTMCs*, and of *CTMCs* and *CTPAs*, respectively. For instance, given a *DTMC* $\mathcal{D}$, a state s, and a probability distribution μ, we still call (s, τ, μ) a transition, denoted by $s \xrightarrow{\tau} \mu$, also written $(s, \tau, \mu) \in \mathbf{P}$, provided that $\mu = \mathbf{P}(s, \cdot)$. Note that here τ denotes just a step since it is not an actual action labeling the transition. Similarly, given a *CTPA* $\mathcal{CP}$, a state s, an action a, and a rate function r, we still write $(s, a, r) \in \rightarrow$ if $\mathbf{R}(s, a) = r$ and we call (s, a, r) a transition, denoted by $s \xrightarrow{a} r$ as well.

For a *CTPA* $\mathcal{CP}$ and a rate function r, we denote by $\mu_r \in \mathrm{SubDisc}(S)$ the induced sub-probability distribution defined by: if $r(S) > 0$, then for each $s \in S$, $\mu_r(s) = \frac{r(s)}{r(S)}$, and if $r(S) = 0$, then $\mu_r = \delta_\perp$.

We adopt a similar notation also for *CTMCs*: for a *CTMC* $\mathcal{C}$ and a state s, we denote by $\mu_{r_s} \in \mathrm{SubDisc}(S)$ the sub-probability distribution induced by the rate function $r_s = \mathbf{R}(s, \cdot)$, i.e., $\mu_{r_s} = \mathbf{P}(s, \cdot)$ for the embedded *DTMC* $emb(\mathcal{C})$.

Given an automaton $\mathcal{A}$ and a state s, we denote by $post(s)$ the set of successors of the state s, that is, $post(s) = \mathrm{Supp}(\mathbf{P}(s, \cdot))$ if $\mathcal{A}$ is a *DTMC* or a *sDTMC*, and $post(s) = \{\, s' \in S \mid \mathbf{R}(s, s') > 0 \,\}$ if $\mathcal{A}$ is a *CTMC*. For a *sDTMC* S and a state s, we denote by $post_\perp(s)$ the set $post_\perp(s) = \mathrm{Supp}(\mu_\perp)$ where $\mu = \mathbf{P}(s, \cdot)$. Similarly, we denote by $pre(s)$ the set of predecessors of the state s, that is, $pre(s) = \{\, s' \in S \mid \mathbf{P}(s', s) > 0 \,\}$ if $\mathcal{A}$ is a *DTMC* or a *sDTMC*, and $pre(s) = \{\, s' \in S \mid \mathbf{R}(s', s) > 0 \,\}$ if $\mathcal{A}$ is a *CTMC*. Finally, we denote by $reach(s)$ the states that are reachable with positive probability from s, that is, $reach(s) = \{\, t \in S \mid \exists \alpha \in frags^*(\mathcal{A}).last(\alpha) = t \,\}$.

4   Simulations and Bisimulations

We recall now the main behavioral preorders and equivalences that are used for the models presented in Section 3. These relations allow us to relate system that are syntactically different, for instance because they use different names for the states, but exhibit equivalent behaviors. Moreover, they allow to reduce the size of the automata without changing their properties. This is especially useful to mitigate the state space explosion problem that usually happens in model checking [8, 14, 34]. An empirical investigation to show the effectiveness of such behavioral relations minimization is performed in [36]. This study indicates that for traditional model checking, huge state space reductions (up to logarithmic) may be acquired. It is worthwhile to mention that the definition of such relations is based on a single automaton; however, as we will see, they are usually used to relate two automata $\mathcal{A}_1$ and $\mathcal{A}_2$. This technical problem is easily solved by taking the disjoint union of the two automata, that is, the automaton whose set of states is the disjoint union of the sets of states of $\mathcal{A}_1$ and $\mathcal{A}_2$, and whose other components are the union of the corresponding components of $\mathcal{A}_1$ and $\mathcal{A}_2$.

4.1   Strong Simulation and Bisimulation

The first relations we introduce are the strong simulation and bisimulation, that are the natural extension to probabilistic systems of the homonymous relations for labelled transition systems [39].

Definition 9 (Strong simulation for discrete time probabilistic automata [9, 50, 56, 57]). *Let $\mathcal{A}$ be a discrete time probabilistic automaton. A relation $\mathcal{R}$ on S is a strong simulation if, for each pair of states $s, t \in S$ such that $s \, \mathcal{R} \, t$,*

- $L(s) = L(t)$ *and*
- *if* $s \xrightarrow{a} \mu_s$ *for some probability distribution* μ_s, *then there exists* μ_t *such that* $t \xrightarrow{a} \mu_t$ *and* $\mu_s \mathcal{L}(\mathcal{R}) \mu_t$.

We say that the discrete time automaton $\mathcal{A}_2$ *strongly simulates* $\mathcal{A}_1$ *if there exists a strong simulation* $\mathcal{R}$ *on the disjoint union* $S_1 \uplus S_2$ *such that* $\bar{s}_1 \mathcal{R} \bar{s}_2$ *and we say that the state* t *strongly simulates the state* s *if there exists a strong simulation* $\mathcal{R}$ *such that* $s \mathcal{R} t$. *We denote the coarsest strong simulation, called strong similarity, by* $\precsim$.

In the remaining of the paper and similarly for the following simulations, we refer to the second condition (*if* $s \xrightarrow{a} \mu_s$ *for some probability distribution* μ_s, *then there exists* μ_t *such that* $t \xrightarrow{a} \mu_t$ *and* $\mu_s \mathcal{L}(\mathcal{R}) \mu_t$) as the *step condition* since it ensures that from two similar states s and t, each transition (or *step*) from s is matched by a transition/step from t and the reached states are still related according to the lifting of the reached distributions.

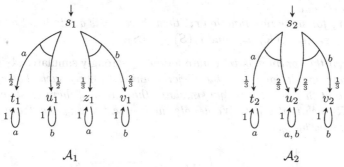

Fig. 8. Two *PA*s with $L(x) = \emptyset$ for each state x such that $\mathcal{A}_1 \precsim \mathcal{A}_2$. The single transition from u_2 with action a, b is just a compact form for the two transitions (u_2, a, δ_{u_2}) and (u_2, b, δ_{u_2}).

Figure 8 shows two probabilistic automata $\mathcal{A}_1$ and $\mathcal{A}_2$ such that $\mathcal{A}_1 \precsim \mathcal{A}_2$. In fact, consider the relation $\mathcal{R} = \{(s_1, s_2), (t_1, t_2), (u_1, u_2), (v_1, v_2), (z_1, u_2)\}$; it is rather easy to verify that $\mathcal{R}$ satisfies Definition 9: it is trivial to verify the step condition for the pairs (t_1, t_2), (u_1, u_2), (v_1, v_2), and (z_1, u_2). The only interesting case is the pair (s_1, s_2); the transition $s_1 \xrightarrow{a} \mu_{1a}$ with $\mu_{1a} = \{(t_1, \frac{1}{2}), (u_1, \frac{1}{2})\}$ is matched by s_2 via the transition $s_2 \xrightarrow{a} \mu_{2a}$ with $\mu_{2a} = \{(t_2, \frac{1}{3}), (u_2, \frac{2}{3})\}$ such that $\mu_{1a} \mathcal{L}(\mathcal{R}) \mu_{2a}$. The weighting function [34,56,57] w_a justifying $\mu_{1a} \mathcal{L}(\mathcal{R}) \mu_{2a}$ is defined as follows:

$$w_a(x_1, x_2) = \begin{cases} \frac{1}{3} & \text{if } x_1 = t_1 \text{ and } x_2 = t_2, \\ \frac{1}{6} & \text{if } x_1 = t_1 \text{ and } x_2 = u_2, \\ \frac{1}{2} & \text{if } x_1 = u_1 \text{ and } x_2 = u_2, \text{ and} \\ 0 & \text{otherwise.} \end{cases}$$

Similarly, the transition $s_1 \xrightarrow{b} \mu_{1b}$ with $\mu_{1b} = \{(z_1, \frac{1}{3}), (v_1, \frac{2}{3})\}$ is matched by s_2 via the transition $s_2 \xrightarrow{b} \mu_{2b}$ with $\mu_{2b} = \{(u_2, \frac{2}{3}), (v_2, \frac{1}{3})\}$ such that $\mu_{1b} \mathcal{L}(\mathcal{R}) \mu_{2b}$.

The weighting function w_b justifying $\mu_{1b} \, \mathcal{L}(\mathcal{R}) \, \mu_{2b}$ is:

$$w_b(x_1, x_2) = \begin{cases} \frac{1}{3} & \text{if } x_1 = z_1 \text{ and } x_2 = u_2, \\ \frac{1}{3} & \text{if } x_1 = v_1 \text{ and } x_2 = u_2, \\ \frac{1}{3} & \text{if } x_1 = v_1 \text{ and } x_2 = v_2, \text{ and} \\ 0 & \text{otherwise.} \end{cases}$$

The definition of strong simulation for continuous-time automata is almost the same, except for the fact that we require that t can move stochastically *faster* than s, i.e., t has a rate higher than s:

Definition 10 (Strong simulation for continuous-time probabilistic automata [9,56,57]). *Let $\mathcal{A}$ be a continuous-time probabilistic automaton. A relation $\mathcal{R}$ on S is a strong simulation if, for each pair of states $s, t \in S$ such that $s \, \mathcal{R} \, t$,*

- *$L(s) = L(t)$ and*
- *if $s \xrightarrow{a} r_s$ for some rate function r_s, then there exists a rate function r_t such that $t \xrightarrow{a} r_t$, $\mu_{r_s} \, \mathcal{L}(\mathcal{R}) \, \mu_{r_t}$, and $r_s(S) \leq r_t(S)$.*

We say that the continuous-time automaton $\mathcal{A}_2$ strongly simulates $\mathcal{A}_1$ if there exists a strong simulation $\mathcal{R}$ on the disjoint union $S_1 \uplus S_2$ such that $\bar{s}_1 \, \mathcal{R} \, \bar{s}_2$ and we say that the state t strongly simulates the state s if there exists a strong simulation $\mathcal{R}$ such that $s \, \mathcal{R} \, t$. We denote the coarsest strong simulation, called strong similarity, by $\precsim$.

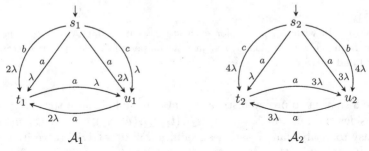

Fig. 9. Two *CTPAs* with $L(s_1) = L(s_2) = \{s\}$ and $L(x) = \emptyset$ for each remaining state x such that $\mathcal{A}_1 \precsim \mathcal{A}_2$

Figure 9 shows two continuous time probabilistic automata $\mathcal{A}_1$ and $\mathcal{A}_2$ such that $\mathcal{A}_1 \precsim \mathcal{A}_2$. The relation $\mathcal{R} = \{(s_1, s_2), (t_1, t_2), (u_1, u_2), (t_1, u_2), (u_1, t_2)\}$ indeed justifies $\mathcal{A}_1 \precsim \mathcal{A}_2$: consider for instance the pair (t_1, t_2); the rate function r_{t_1} induces the probability distribution $\mu_{r_{t_1}} = \delta_{u_1}$ and the overall rate $r_{t_1}(S) = \lambda$. For t_2, we have the rate function r_{t_2} that induces the probability distribution $\mu_{r_{t_2}} = \delta_{u_2}$ and the overall rate $r_{t_1}(S) = 3\lambda$, thus $r_{t_1}(S) \leq r_{t_2}(S)$. Since $(u_1, u_2) \in \mathcal{R}$, then $\delta_{u_1} \, \mathcal{L}(\mathcal{R}) \, \delta_{u_2}$ is trivially satisfied, hence the step condition is satisfied. A similar argument shows that the step condition is satisfied for the pairs (u_1, u_2), (t_1, u_2), and (u_1, t_2).

Now, consider the pair (s_1, s_2): we distinguish the case of the transitions with label b and c and the transition with label a, all from s_1. The transition from s_1 with label b induces the probability distribution $\mu_{r_{s_1}^b} = \delta_{t_1}$ and the overall rate $r_{s_1}^b(S) = 2\lambda$. For s_2, we have the rate function $r_{s_2}^b$ that induces the probability distribution $\mu_{r_{s_2}^b} = \delta_{u_2}$ and the overall rate $r_{s_2}^b(S) = 4\lambda$, thus $r_{s_1}^b(S) \leq r_{s_2}^b(S)$. Since $(t_1, u_2) \in \mathcal{R}$, then $\delta_{t_1} \mathcal{L}(\mathcal{R}) \delta_{u_2}$ trivially holds, hence the step condition is satisfied. The case for the label c is similar.

The last step condition we have to check involves the transition with label a from s_1. The rate function $r_{s_1}^a$ induces the probability distribution $\mu_{r_{s_1}^a} = \{(t_1, \frac{\lambda}{3\lambda}), (u_1, \frac{2\lambda}{3\lambda})\}$ and the overall rate $r_{s_1}^a(S) = 3\lambda$. For s_2, we have the rate $r_{s_2}^a$ that induces the probability distribution $\mu_{r_{s_2}^a} = \{(t_2, \frac{\lambda}{4\lambda}), (u_2, \frac{3\lambda}{4\lambda})\}$ and the overall rate $r_{s_2}^a(S) = 4\lambda$. Obviously, $r_{s_1}^a(S) \leq r_{s_2}^a(S)$; $\mu_{r_{s_1}^a} \mathcal{L}(\mathcal{R}) \mu_{r_{s_2}^a}$ is justified by the weighting function w defined as

$$w(x_1, x_2) = \begin{cases} \frac{1}{4} & \text{if } x_1 = t_1 \text{ and } x_2 = t_2, \\ \frac{1}{12} & \text{if } x_1 = t_1 \text{ and } x_2 = u_2, \\ \frac{2}{3} & \text{if } x_1 = u_1 \text{ and } x_2 = u_2, \text{ and} \\ 0 & \text{otherwise} \end{cases}$$

The definition of strong bisimulation and strong bisimilarity, denoted by $\sim$, is obtained by requiring $\mathcal{R}$ to be a symmetric relation.

Definition 11 (Strong bisimulation [38]). *Let $\mathcal{A}$ be a discrete time or a continuous-time probabilistic automaton. A relation $\mathcal{R}$ on S is a strong bisimulation if $\mathcal{R}$ is symmetric and a strong simulation.*

We denote the coarsest strong bisimulation, called strong bisimilarity, by $\sim$.

Other definitions of strong bisimulation require $\mathcal{R}$ to be an equivalence relation but it is easy to show that such definitions are equivalent to Definition 11.

Finally, only strong bisimulation on *IMC*s has been defined [28], and it is the expected merge of the bisimulation for *CTMC*s and labelled transition systems:

Definition 12 (Strong bisimulation for *IMC*s [28]). *Let $\mathcal{I}$ be a IMC. An equivalence relation $\mathcal{R}$ on S is a strong bisimulation if, for each pair of states $s, t \in S$ such that $s \mathcal{R} t$,*

- *$L(s) = L(t)$,*
- *if $s \xrightarrow{a} s'$ for some $s' \in S$ and $a \in \Sigma$, then there exists t' such that $t \xrightarrow{a} t'$ and $s' \mathcal{R} t'$, and*
- *if s does not enable a transition with label τ, then for each $\mathcal{C} \in S/\mathcal{R}$, $\gamma(s, \mathcal{C}) = \gamma(t, \mathcal{C})$ where $\gamma(v, \mathcal{C}) = \sum_{\{\lambda \in \mathbb{R}^{\geq 0} | v \xrightarrow{\lambda} v', v' \in \mathcal{C}\}} \lambda$.*

We say that the IMC $\mathcal{A}_2$ strongly bisimulates $\mathcal{A}_1$ if there exists a strong bisimulation $\mathcal{R}$ on the disjoint union $S_1 \uplus S_2$ such that $\bar{s}_1 \mathcal{R} \bar{s}_2$ and we say that the state t strongly bisimulates the state s if there exists a strong bisimulation $\mathcal{R}$ such that $s \mathcal{R} t$. We denote the coarsest strong bisimulation, called strong bisimilarity, by $\sim$.

A simulation (and a bisimulation) can be seen as a game where in each round the challenger, or attacker, s proposes a transition, or step, that has to be matched by the defender t. The two states s and t are strong (bi-)similar if the defender is always able to match the challenging transitions proposed by the attacker, that is, the game can be played forever.

4.2 Strong Probabilistic Simulation and Bisimulation

The fact that (continuous-time) probabilistic automata may exhibit internal non-determinism, i.e., a state can enable different transitions with the same label, allows us to define the probabilistic counterpart of strong simulation and bisimulation where each transition proposed by the challenger is matched by some convex combination of the defender's enabled transitions.

Given a PA $\mathcal{P}$, a state $s \in S$, an action $a \in \Sigma$, and a distribution $\mu \in \text{Disc}(S)$, we say that there exists a *combined transition* $s \xrightarrow{a}_C \mu$ if there exists a finite set I of indexes, a family $\{p_i\}_{i \in I} \subseteq [0, 1]$ such that $\sum_{i \in I} p_i = 1$, and a family $\{s \xrightarrow{a} \mu_i\}_{i \in I} \subseteq \rightarrow$ such that $\mu = \sum_{i \in I} p_i \cdot \mu_i$.

Definition 13 (Strong probabilistic simulation for PAs [49, 50]). *Let $\mathcal{A}$ be a PA. A relation $\mathcal{R}$ on S is a strong probabilistic simulation if, for each pair of states $s, t \in S$ such that $s \mathcal{R} t$,*

- *$L(s) = L(t)$ and*
- *if $s \xrightarrow{a} \mu_s$ for some probability distribution μ_s, then there exists μ_t such that $t \xrightarrow{a}_C \mu_t$ and $\mu_s \, \mathcal{L}(\mathcal{R}) \, \mu_t$.*

We say that the PA $\mathcal{P}_2$ strongly probabilistically simulates $\mathcal{P}_1$ if there exists a strong probabilistic simulation $\mathcal{R}$ on the disjoint union $S_1 \uplus S_2$ such that $\bar{s}_1 \mathcal{R} \bar{s}_2$ and we say that the state t strongly probabilistically simulates the state s if there exists a strong probabilistic simulation $\mathcal{R}$ such that $s \mathcal{R} t$. We denote the coarsest strong probabilistic simulation, called strong probabilistic similarity, by $\lesssim_p$.

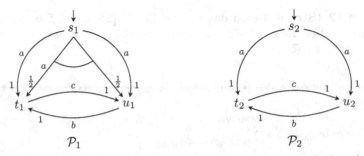

Fig. 10. Two PAs with $L_i(x_i) = \{x\}$ for each state x_i, $i = 1, 2$ such that $\mathcal{A}_1 \lesssim_p \mathcal{A}_2$

Figure 10 shows two PAs $\mathcal{A}_1$ and $\mathcal{A}_2$ such that $\mathcal{A}_1 \lesssim_p \mathcal{A}_2$. The relation justifying $\mathcal{A}_1 \lesssim_p \mathcal{A}_2$ is $\mathcal{R} = \{ (x_1, x_2) \mid x \in \{s, t, u\} \}$. All cases are trivial, except for the pair (s_1, s_2) and the transition $s_1 \xrightarrow{a} \mu 1$ with $\mu_1 = \{(t_1, \frac{1}{2}), (u_1, \frac{1}{2})\}$. This transition is matched by s_2 via the combined transition $s_2 \xrightarrow{a} \mu 2$ with $\mu_2 = \{(t_2, \frac{1}{2}), (u_2, \frac{1}{2})\}$. Such combined transition is obtained by taking transitions $s_2 \xrightarrow{a} \delta_{t_2}$ and $s_2 \xrightarrow{a} \delta_{u_2}$ both with probability $\frac{1}{2}$.

The definition of combined transition for *CTPAs* requires to consider for the convex combination only transitions with the same exit rate, in order to obtain a combined transition that is still exponentially distributed (see [57, Example 2.17] for more details).

Given a *CTPA* $\mathcal{CP}$, a state $s \in S$, an action $a \in \Sigma$, and a rate function $r: S \to \mathbb{R}^{\geq 0}$, we say that there exists a *combined transition* $s \xrightarrow{a}_C r$ if there exists a finite set I of indexes, a family $\{p_i\}_{i \in I} \subseteq [0, 1]$ such that $\sum_{i \in I} p_i = 1$, and a family $\{s \xrightarrow{a} r_i\}_{i \in I} \subseteq \mathbf{R}$ such that $r_i(S) = r_j(S)$ for each $i, j \in I$ and $r = \sum_{i \in I} p_i \cdot r_i$.

As before, the definition of strong probabilistic simulation for *CTPAs* is the obvious continuous-time counterpart of the definition for *PAs*:

Definition 14 (Strong probabilistic simulation for *CTPAs* [9, 28, 56, 57]). *Let $\mathcal{A}$ be a CTPA. A relation $\mathcal{R}$ on S is a strong probabilistic simulation if, for each pair of states $s, t \in S$ such that $s \, \mathcal{R} \, t$,*

- *$L(s) = L(t)$ and*
- *if $s \xrightarrow{a} r_s$ for some rate r_s, then there exists r_t such that $t \xrightarrow{a}_C r_t$, $\mu_{r_s} \, \mathcal{L}(\mathcal{R})$ μ_{r_t}, and $r_s(S) \leq r_t(S)$.*

We say that the CTPA $\mathcal{CP}_2$ strongly probabilistically simulates $\mathcal{CP}_1$ if there exists a strong probabilistic simulation $\mathcal{R}$ on the disjoint union $S_1 \uplus S_2$ such that $\bar{s}_1 \, \mathcal{R} \, \bar{s}_2$ and we say that the state t strongly probabilistically simulates the state s if there exists a strong probabilistic simulation $\mathcal{R}$ such that $s \, \mathcal{R} \, t$. We denote the coarsest strong probabilistic simulation, called strong probabilistic similarity, by $\precsim_p$.

As for the strong case, the definition of strong probabilistic bisimulation and strong probabilistic bisimilarity, denoted by $\sim_p$, is obtained by requiring $\mathcal{R}$ to be a symmetric relation. Note that the two *PAs* in Figure 10 are actually strong probabilistic bisimilar, not just strongly probabilistic similar.

4.3 Weak Simulation and Bisimulation

Strong (probabilistic) simulations and bisimulations require that each transition proposed by the challenger is matched by the defender via a single (combined) transition. If we are not interested in internal computations, but just on the visible behavior, these relations are too restrictive. In order to abstract away internal steps, such relations have been relaxed to weak (probabilistic) simulations and bisimulations where the defender is able to match the challenging transition by performing several internal steps before and after having exhibited the same visible behavior, for instance, the same external action. The simplest example of weak transition is the one for labelled transition systems [39]: it is just the concatenation of arbitrarily many internal steps, the external transition (if we have to match an external challenging transition), and again arbitrarily many internal steps.

The definition of weak transition for probabilistic systems is not so easy as we have to take into account probabilistic choices. We first consider weak simulation and bisimulation for Markov chains and sDTMCs, and then for probabilistic automata. We are not aware of any definition of weak simulation and bisimulation for CTPAs where sequences of transitions are involved.

Markov Chains. Before presenting the weak simulation and bisimulation for Markov chains, we need to introduce some additional definition [55,57].

For a given pair of states (s_1, s_2) of the automaton $\mathcal{A}$ and functions $\gamma_i \colon S \to [0,1]$, we denote by U_i and V_i the sets $\{ u \in post(s_i) \mid \gamma_i(u) > 0 \}$ and $\{ v \in post(s_i) \mid \gamma_i(v) < 1 \}$, respectively. Essentially, U_i represents the states that can be reached with non-zero probability according to γ_i from s_i by performing one transition while V_i represents the states that cannot be reached with probability 1 according to γ_i from s_i by performing one transition. It is, however, worthwhile to mention that U_i and V_i are in general non-disjoint. The definition of weak simulation for DTMCs is not so immediate, because the "weak step" does not represent the fact that multiple transitions can be performed as in non-probabilistic settings like CCS and π-calculus [39,40] or in the other probabilistic models, as we will see later in the section, but that a single transition represents a *visible* or *stutter* step to a reached state z depending on whether z is in U or in V, respectively. More precisely we require for the visible steps (i.e., steps reaching states in U_i) that there exists a weighting function w for the conditional distributions $\frac{\mathbf{P}(s_1, \cdot)}{K_1}$ and $\frac{\mathbf{P}(s_2, \cdot)}{K_2}$ where K_i is essentially the probability to perform a visible step. The stutter steps (i.e., steps reaching states in V_i) must respect the weak bisimulations, that is, states in V_1 are weakly simulated by s_2 and s_1 is weakly simulated by all states in V_2, as depicted in Figure 11. Since a state t may belong to both U and V, the functions γ_i take care of distributing s_i over U_i and V_i. See [55, Section 4.3.1] for more details.

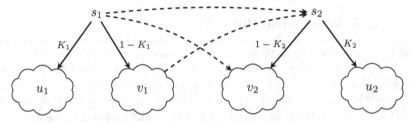

Fig. 11. Splitting of successor states in weak simulations for DTMCs

Definition 15 (Weak simulation for *DTMCs* [6, 9, 56, 57]). *Let $\mathcal{D}$ be a DTMC. A relation $\mathcal{R}$ on S is a* weak simulation *if, for each pair of states $s_1, s_2 \in S$ such that $s_1 \mathcal{R} s_2$,*

- $L(s_1) = L(s_2)$ *and*
- *there exist functions $\gamma_i \colon S \to [0,1]$ for $i \in \{1,2\}$ such that*
  1. *(a) $v_1 \mathcal{R} s_2$ for each $v_1 \in V_1$ and (b) $s_1 \mathcal{R} v_2$ for each $v_2 \in V_2$;*
  2. *there exists a weighting function $w \colon S \times S \to [0,1]$ such that*
     (a) $w(u_1, u_2) > 0$ implies $u_1 \in U_1$, $u_2 \in U_2$, and $u_1 \mathcal{R} u_2$,
     (b) if $K_1 > 0$ and $K_2 > 0$, then for all states $t \in S$,

$$K_1 \cdot w(t, U_2) = \mathbf{P}(s_1, t) \cdot \gamma_1(t) \ \text{ and } \ K_2 \cdot w(U_1, t) = \mathbf{P}(s_2, t) \cdot \gamma_2(t)$$

   where $K_i = \sum_{u_i \in U_i} \mathbf{P}(s_i, u_i) \cdot \gamma_i(u_i)$ for $i \in \{1,2\}$; and
  3. *for $u_1 \in U_1$ there exist an execution fragment $s_2 t_1 \ldots t_n u_2$ with positive probability such that $n \in \mathbb{N}$, $s_1 \mathcal{R} t_j$ for $0 < j \leq n$, and $u_1 \mathcal{R} u_2$.*

We say that the DTMC $\mathcal{D}_2$ weakly simulates $\mathcal{D}_1$ if there exists a weak simulation $\mathcal{R}$ on the disjoint union $S_1 \uplus S_2$ such that $\bar{s}_1 \mathcal{R} \bar{s}_2$ and we say that the state t weakly simulates the state s if there exists a weak simulation $\mathcal{R}$ such that $s \mathcal{R} t$. We denote the coarsest weak simulation, called weak similarity*, by $\precsim$.*

Figure 12 shows a *DTMC* for which $s_i \precsim t_j$ for $i, j \in \{1, 2, 3\}$. For each of these pairs we can select $U_1 = \emptyset$ and $V_2 = \emptyset$. Since $K_1 = 0$, we need to check only the Condition 1. However, since all the successor states of s_i are either empty or itself, this conditions holds trivially. It holds similarly that $v_1 \precsim v_2$.

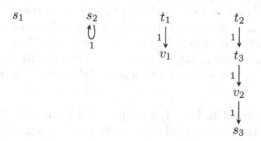

Fig. 12. A *DTMC* with $L(v_1) = L(v_2) = \{v\}$ and $L(x) = \emptyset$ for each other state x

The definition of weak simulation for *CTMCs* is similar, where condition (3) is replaced by $K_1 \cdot \mathbf{R}(s_1, S) \leq K_2 \cdot \mathbf{R}(s_2, S)$.

Similarly, the definition of weak simulation for *sDTMCs* is just a slight variation of the one for *DTMCs*, where we consider sub-distributions instead of distributions: for a given pair of states (s_1, s_2) of the *sDTMC* $\mathcal{S}$ and functions $\gamma_i \colon S_\perp \to [0,1]$, we change the definition of U_i and V_i as follows: U_i and V_i are the sets $\{\, u \in post_\perp(s_i) \mid \gamma_i(u) > 0 \,\}$ and $\{\, v \in post_\perp(s_i) \mid \gamma_i(v) < 1 \,\}$, respectively.

Definition 16 (Weak simulation for sDTMCs [6, 9, 56, 57]). *Let S be a sDTMC. A relation $\mathcal{R}$ on S is a* weak simulation *if, for each pair of states $s_1, s_2 \in S$ such that $s_1 \, \mathcal{R} \, s_2$,*

- $L(s_1) = L(s_2)$ *and*
- *there exist functions $\gamma_i : S_\perp \to [0, 1]$ for $i \in \{1, 2\}$ such that*
  1. *(a) $v_1 \, \mathcal{R} \, s_2$ for each $v_1 \in V_1 \backslash \{\perp\}$ and (b) $s_1 \, \mathcal{R} \, v_2$ for each $v_2 \in V_2 \backslash \{\perp\}$;*
  2. *there exists a function $w : S_\perp \times S_\perp \to [0, 1]$ such that*
     (a) $w(u_1, u_2) > 0$ implies $u_1 \in U_1$, $u_2 \in U_2$, and $u_1 \, \mathcal{R}_\perp \, u_2$,
     (b) if $K_1 > 0$ and $K_2 > 0$, then for all states $t \in S$,

$$K_1 \cdot w(t, U_2) = \mathbf{P}(s_1, t) \cdot \gamma_1(t) \text{ and } K_2 \cdot w(U_1, t) = \mathbf{P}(s_2, t) \cdot \gamma_2(t)$$

  where $K_i = \sum_{u_i \in U_i} \mathbf{P}(s_i, u_i) \cdot \gamma_i(u_i)$ for $i \in \{1, 2\}$; and
  3. *for $u_1 \in U_1 \backslash \{\perp\}$ there exist an execution fragment $s_2 t_1 \ldots t_n u_2$ with positive probability such that $n \in \mathbb{N}$, $s_1 \, \mathcal{R} \, t_j$ for $0 < j \leq n$, and $u_1 \, \mathcal{R} \, u_2$.*

We say that the sDTMC S_2 weakly simulates S_1 if there exists a weak simulation $\mathcal{R}$ on the disjoint union $S_1 \uplus S_2$ such that $\bar{s}_1 \, \mathcal{R} \, \bar{s}_2$ and we say that the state t weakly simulates the state s if there exists a weak simulation $\mathcal{R}$ such that $s \, \mathcal{R} \, t$. We denote the coarsest weak simulation, called weak similarity, by $\precsim$.

As for the strong bisimulation, the definition of weak bisimulation and weak bisimilarity, denoted by $\approx$, is obtained by requiring $\mathcal{R}$ to be a symmetric relation.

Remark 1. The definition of weak simulation for sDTMC we present here is neither sound nor complete for the liveness fragment of PCTL without the next operator [33]. To fix this problem, [33] proposes a new definition of weak simulation for sDTMC that is sound and conjectured to be complete. However, the associated technical report shows that completeness does not hold as well.

We have decided to maintain the definition from [55, 57] instead of switching to the definition proposed in [33] because the latter currently lacks of a published decision algorithm while such algorithm is available for the former.

Interactive Markov Chains. The definition of weak bisimulation for IMC is rather simple, since it is the obvious extension to the weak case of the strong bisimulation. Given an IMC $\mathcal{I}$, two state s and t, and an action a, we denote by $s \xrightarrow{a} t$ the sequence of transitions $s \xRightarrow{\tau} s' \xrightarrow{a} t' \xRightarrow{\tau} t$ for some state s' and t' where $s \xRightarrow{\tau} s'$ is the reflexive and transitive closure of $\xrightarrow{\tau}$, as defined for labelled transition systems [40]. For an IMC $\mathcal{I}$, we recall that $\gamma(v, \mathcal{C}) = \sum_{\{\lambda \in \mathbb{R}^{\geq 0} | v \xrightarrow{\lambda} v', v' \in \mathcal{C}\}} \lambda$.

Definition 17 (Weak bisimulation for IMCs [28]). *Let $\mathcal{I}$ be a IMC. An equivalence relation $\mathcal{R}$ on S is a* weak bisimulation *if, for each pair of states $s, t \in S$ such that $s \, \mathcal{R} \, t$,*

- $L(s) = L(t)$,

- if $s \stackrel{a}{\Longrightarrow} s'$ for some $s' \in S$ and $a \in \Sigma$, then there exists t' such that $t \stackrel{a}{\Longrightarrow} t'$ and $s' \mathcal{R} t'$, and
- if $s \stackrel{\tau}{\Longrightarrow} s'$ and s' does not enable a transition with label τ, then there exists t' such that t' does not enable a transition with label τ, $t \stackrel{\tau}{\Longrightarrow} t'$, and for each $\mathcal{C} \in S/\mathcal{R}$, $\gamma(s', \mathcal{C}^\tau) = \gamma(t', \mathcal{C}^\tau)$ where $\mathcal{C}^\tau = \{ u \mid \exists v \in \mathcal{C}.u \stackrel{\tau}{\Longrightarrow} v \}$.

We say that the IMC $\mathcal{A}_2$ weakly bisimulates $\mathcal{A}_1$ if there exists a weak bisimulation $\mathcal{R}$ on the disjoint union $S_1 \uplus S_2$ such that $\bar{s}_1 \mathcal{R} \bar{s}_2$ and we say that the state t weakly bisimulates the state s if there exists a weak bisimulation $\mathcal{R}$ such that $s \mathcal{R} t$. We denote the coarsest weak bisimulation, called weak bisimilarity, by $\approx$.

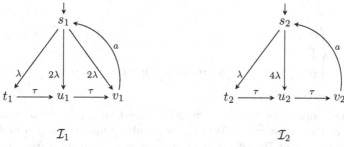

Fig. 13. Two *IMC*s with $L_i(x) = \emptyset$ for each state x except for $L_i(t_i) = \{t\}$, $i = 1, 2$, such that $\mathcal{I}_1 \approx \mathcal{I}_2$

Figure 13 shows two *IMC*s that are weak bisimilar. This is justified by the equivalence relation whose classes are $\mathcal{C}_s = \{s_1, s_2\}$, $\mathcal{C}_t = \{t_1, t_2\}$, and $\mathcal{C}_o = \{u_1, u_2, v_1, v_2\}$. The first two conditions about labeling and interactive transitions are trivial for all pairs of related states; in particular, classes $\mathcal{C}_t$ and $\mathcal{C}_o$ (or $\mathcal{C}_s$) cannot be merged since labels are different: for instance, $L(t_1) = \{t\} \neq \emptyset = L(u_1)$, so the first condition would be violated. Consider the classes $\mathcal{C}_o$ and $\mathcal{C}_s$: they have the same labeling (each state in them has label $\emptyset$) but they cannot be merged since for instance the state $u_1 \in \mathcal{C}_o$ enables an a weak transition reaching s_1 that cannot be matched by the state $s_2 \in \mathcal{C}_s$, so the second condition would not be satisfied. The third condition about rates is obvious as well for pairs of states in the classes $\mathcal{C}_t$ and $\mathcal{C}_o$ since none of their states enables a timed transition, so $\gamma(x, \mathcal{C}^\tau)$ is 0 for each $x \in \mathcal{C}_t \cup \mathcal{C}_o$ and $\mathcal{C} \in \{\mathcal{C}_s, \mathcal{C}_t, \mathcal{C}_o\}$. The only non-trivial case is the pair (s_1, s_2) (the symmetric case is analogous). The only weak transitions with label τ enabled by s_1 and s_2 are $s_1 \stackrel{\tau}{\Longrightarrow} s_1$ and $s_2 \stackrel{\tau}{\Longrightarrow} s_2$, since neither s_1 nor s_2 enables a transition with label τ; $s_1 \mathcal{R} s_2$ trivially holds. $\gamma(s_1, \mathcal{C}_s^\tau) = 0 = \gamma(s_2, \mathcal{C}_s^\tau)$ since $\mathcal{C}_s^\tau = \mathcal{C}_s$ and there is no timed transition reaching $\mathcal{C}_s$; $\gamma(s_1, \mathcal{C}_t^\tau) = \lambda = \gamma(s_2, \mathcal{C}_t^\tau)$ since $\mathcal{C}_t^\tau = \mathcal{C}_t$ and both s_1 and s_2 have a single timed transition with rate λ reaching $\mathcal{C}_t$; finally, $\gamma(s_1, \mathcal{C}_o^\tau) = 5\lambda = \gamma(s_2, \mathcal{C}_o^\tau)$ since $\mathcal{C}_o^\tau = \mathcal{C}_o \cup \mathcal{C}_t$.

Probabilistic Automata. Before introducing the weak (combined) transition for probabilistic automata, we need some preliminary definition.

A *scheduler* for a *PA* $\mathcal{P}$ is a function $\sigma \colon \mathit{frags}^*(\mathcal{P}) \to \mathrm{SubDisc}(\to)$ such that for each $\alpha \in \mathit{frags}^*(\mathcal{P})$, $\sigma(\alpha) \in \mathrm{SubDisc}(\{\, tr \in \to \mid src(tr) = last(\alpha)\,\})$. Given a scheduler σ and a finite execution fragment α, the distribution $\sigma(\alpha)$ describes how transitions are chosen to move on from $last(\alpha)$. We say that a scheduler σ is a *Dirac scheduler* if for each $\alpha \in \mathit{frags}^*(\mathcal{P})$, $\sigma(\alpha)$ is a Dirac distribution and we say that σ is a *determinate scheduler* if for each $\alpha, \alpha' \in \mathit{frags}^*(\mathcal{P})$, if $trace(\alpha) = trace(\alpha')$ and $last(\alpha) = last(\alpha')$, then $\dot\sigma(\alpha) = \sigma(\alpha')$. A scheduler σ and a state s induce a probability distribution $\mu_{\sigma,s}$ over execution fragments as follows. The basic measurable events are the cones of finite execution fragments, where the cone of α, denoted by C_α, is the set $\{\, \alpha' \in \mathit{frags}(\mathcal{P}) \mid \alpha \leqslant \alpha'\,\}$. The probability $\mu_{\sigma,s}$ of a cone C_α is defined recursively as follows:

$$
\mu_{\sigma,s}(C_\alpha) = \begin{cases} 0 & \text{if } \alpha = t \text{ for a state } t \neq s, \\ 1 & \text{if } \alpha = s, \\ \mu_{\sigma,s}(C_{\alpha'}) \cdot \sum_{tr \in \xrightarrow{a}} \sigma(\alpha')(tr) \cdot \mu_{tr}(t) & \text{if } \alpha = \alpha' at. \end{cases}
$$

Standard measure theoretical arguments ensure that $\mu_{\sigma,s}$ extends uniquely to the σ-field generated by cones. We call the resulting measure $\mu_{\sigma,s}$ a *probabilistic execution fragment* of $\mathcal{P}$ and we say that it is generated by σ from s. Given a finite execution fragment α, we define $\mu_{\sigma,s}(\alpha)$ as $\mu_{\sigma,s}(\alpha) = \mu_{\sigma,s}(C_\alpha) \cdot \sigma(\alpha)(\bot)$, where $\sigma(\alpha)(\bot)$ is the probability of terminating the computation after α has occurred.

We say that there is a *weak combined transition* from $s \in S$ to $\mu \in \mathrm{Disc}(S)$ labelled by $a \in \Sigma$, denoted by $s \xRightarrow{a}_C \mu$, if there exists a scheduler σ such that the following holds for the induced probabilistic execution fragment $\mu_{\sigma,s}$:

1. $\mu_{\sigma,s}(\mathit{frags}^*(\mathcal{P})) = 1$;

2. for each $\alpha \in \mathit{frags}^*(\mathcal{P})$, if $\mu_{\sigma,s}(\alpha) > 0$ then $trace(\alpha) = trace(a)$;

3. for each state t, $\mu_{\sigma,s}(\{\, \alpha \in \mathit{frags}^*(\mathcal{P}) \mid last(\alpha) = t\,\}) = \mu(t)$.

In this case, we say that the weak combined transition $s \xRightarrow{a}_C \mu$ is induced by σ. When σ is a Dirac scheduler, then we say that it induces a weak transition from $s \in S$ to $\mu \in \mathrm{Disc}(S)$ labelled by $a \in \Sigma$, denoted by $s \xRightarrow{a} \mu$.

Albeit the definition of weak (combined) transitions is somewhat intricate, this definition is just the obvious extension of weak transitions on labelled transition systems to the setting with probabilities. See [48] for more details on weak combined transitions.

As an example of weak combined transition, consider the *PA* in Figure 14. We now show that there exists a scheduler inducing the weak combined transition $s \xRightarrow{a}_C \mu$ where $\mu = \{(\square, \frac{4}{18}), (\pmb{\bullet}, \frac{7}{18}), (\pmb{\bullet}, \frac{7}{18})\}$. Let μ_s be $\{(t, \frac{1}{3}), (u, \frac{1}{3}), (v, \frac{1}{3})\}$

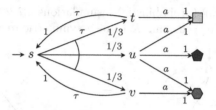

Fig. 14. A probabilistic automaton

and consider the scheduler σ defined as follows:

$$
\sigma(\alpha) = \begin{cases}
\delta_{s \xrightarrow{\tau} \mu_s} & \text{if } last(\alpha) = s, \\
\{(t \xrightarrow{\tau} \delta_s, \frac{1}{2}), (t \xrightarrow{a} \delta_\square, \frac{1}{2})\} & \text{if } \alpha = s\tau t, \\
\delta_{t \xrightarrow{a} \delta_\square} & \text{if } \alpha = s\tau t\tau s\tau t, \\
\delta_{u \xrightarrow{a} \delta_\pentagon} & \text{if } last(\alpha) = u, \\
\delta_{v \xrightarrow{a} \delta_\hexagon} & \text{if } last(\alpha) = v, \text{ and} \\
\delta_\perp & \text{otherwise.}
\end{cases}
$$

It is easy to show that indeed σ induces $s \xRightarrow{a}_C \mu$. For instance, consider the state $\square$; in fact, $\mu_{\sigma,s}(\{\alpha \in frags^*(\mathcal{P}) \mid last(\alpha) = \square\}) = \mu_{\sigma,s}(\{s\tau ta\square, s\tau t\tau s\tau ta\square\}) + \mu_{\sigma,s}(\{\alpha \in frags^*(\mathcal{P}) \mid last(\alpha) = \square\} \setminus \{s\tau ta\square, s\tau t\tau s\tau ta\square\}) = \mu_{\sigma,s}(s\tau ta\square) + \mu_{\sigma,s}(s\tau t\tau s\tau ta\square) + 0 = 1 \cdot 1 \cdot \frac{1}{3} \cdot \frac{1}{2} \cdot 1 \cdot 1 + 1 \cdot 1 \cdot \frac{1}{3} \cdot \frac{1}{2} \cdot 1 \cdot 1 \cdot \frac{1}{3} \cdot 1 \cdot 1 \cdot 1 = \frac{4}{18} = \mu(\square)$.

Note that σ is neither Dirac nor determinate; moreover it is not the only scheduler inducing $s \xRightarrow{a}_C \mu$: in fact, also the determinate scheduler σ' defined as follows induces $s \xRightarrow{a}_C \mu$.

$$
\sigma'(\alpha) = \begin{cases}
\delta_{s \xrightarrow{\tau} \mu_s} & \text{if } last(\alpha) = s, \\
\{(t \xrightarrow{\tau} \delta_s, \frac{3}{7}), (t \xrightarrow{a} \delta_\square, \frac{4}{7})\} & \text{if } last(\alpha) = t, \\
\delta_{u \xrightarrow{a} \delta_\pentagon} & \text{if } last(\alpha) = u, \\
\delta_{v \xrightarrow{a} \delta_\hexagon} & \text{if } last(\alpha) = v, \text{ and} \\
\delta_\perp & \text{otherwise.}
\end{cases}
$$

Definition 18 (Weak (probabilistic) simulation on PAs [6,9,43,51,56, 57]). *Let $\mathcal{P}$ be a PA. A relation $\mathcal{R}$ on S is a weak (probabilistic) simulation if, for each pair of states $s, t \in S$ such that $s \mathcal{R} t$,*

- *$L(s) = L(t)$ and*
- *if $s \xrightarrow{a} \mu_s$ for some probability distribution μ_s, then there exists μ_t such that $t \xRightarrow{a} \mu_t$ ($t \xRightarrow{a}_C \mu_t$) and $\mu_s \, \mathcal{L}(\mathcal{R}) \, \mu_t$.*

We say that the PA $\mathcal{P}_2$ weakly (probabilistically) simulates $\mathcal{P}_1$ if there exists a weak (probabilistic) simulation $\mathcal{R}$ on the disjoint union $S_1 \uplus S_2$ such that $\bar{s}_1 \, \mathcal{R} \, \bar{s}_2$ and we say that the state t weakly (probabilistically) simulates the state s if there exists a weak (probabilistic) simulation $\mathcal{R}$ such that $s \, \mathcal{R} \, t$. We denote the coarsest weak (probabilistic) simulation, called weak (probabilistic) similarity, by $\precsim$ ($\precsim_p$).

As usual, the weak (probabilistic) bisimulations [27, 29, 43, 51], denoted by $\approx$ ($\approx_p$), are obtained by requiring $\mathcal{R}$ to be a symmetric relation.

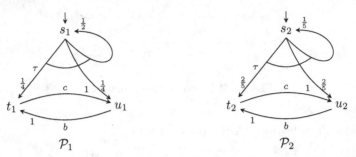

Fig. 15. Two *PAs* with $L_i(x_i) = \{x\}$ for each state x_i, $i = 1, 2$ such that $\mathcal{P}_1 \approx_p \mathcal{P}_2$

Figure 15 shows two *PAs* that are weak probabilistic bisimilar. The relation justifying $\mathcal{P}_1 \approx_p \mathcal{P}_2$ is the equivalence relation whose classes are $\{s_1, s_2\}$, $\{t_1, t_2\}$, and $\{u_1, u_2\}$. Checking the pairs in $\{t_1, t_2\}$ and $\{u_1, u_2\}$ is trivial, so consider for instance the pair (s_1, s_2) and the transition $s_1 \xrightarrow{\tau} \mu_1$ where $\mu_1 = \{(s_1, \frac{1}{2}), (t_1, \frac{1}{4}), (u_1, \frac{1}{4})\}$. s_2 can match such transition via the weak combined transition $s_2 \Longrightarrow_C \mu$ where $\mu = \{(s_2, \frac{1}{2}), (t_2, \frac{1}{4}), (u_2, \frac{1}{4})\}$ induced by the scheduler σ defined as

$$\sigma(\alpha) = \begin{cases} \{(s_2 \xrightarrow{\tau} \mu_2, \frac{5}{8}), (\bot, \frac{3}{8})\} & \text{if } \alpha = s_2 \text{ and} \\ \delta_\bot & \text{otherwise,} \end{cases}$$

where $\mu_2 = \{(s_2, \frac{1}{5}), (t_2, \frac{2}{5}), (u_2, \frac{2}{5})\}$. The transition $s_2 \xrightarrow{\tau} \mu_2$ can be matched by s_1 via the weak combined transition $s_1 \Longrightarrow_C \mu'$ where μ' is the distribution $\{(s_1, \frac{1}{5}), (t_1, \frac{2}{5}), (u_1, \frac{2}{5})\}$, transition that is induced by the scheduler σ' defined as

$$\sigma'(\alpha) = \begin{cases} \delta_{s_1 \xrightarrow{\tau} \mu_1} & \text{if } \alpha = s_1 \text{ or } \alpha = s_1 \tau s_1, \\ \{(s_1 \xrightarrow{\tau} \mu_1, \frac{2}{5}), (\bot, \frac{3}{5})\} & \text{if } \alpha = s_1 \tau s_1 \tau s_1, \text{ and} \\ \delta_\bot & \text{otherwise.} \end{cases}$$

4.4   Markov Automata

Finally, we discuss simulations and bisimulations for Markov Automata. For the strong (probabilistic) simulations and bisimulations, they are just the merge of the corresponding definitions for *PAs* and *CTMCs*, where the timed transitions are considered only if the states do not enable internal transitions, as happens for *IMCs*.

For the weak (probabilistic) simulations and bisimulations, the approach is quite different from the previous definitions since they relate distributions instead of states. This makes the definition quite involved and out of the scope of this survey, also by considering that the exponential decision algorithms [20, 46] for these bisimulations just make use of the algorithm for *PA* weak combined transitions we will see in Section 6.2 as a black box.

We refer the interested reader to [20, 46] for the technical details and theoretical considerations that allow to define a state-based bisimulation that is equivalent to the distribution-based one as defined in [17, 22, 23].

5   Networks and Maximum Flow Problem

Given a set V, we say that (V, E) is a *directed graph* with vertices V and edges E if $E \subseteq V \times V$. A network $\mathcal{N}$ is a tuple $(V, E, \triangle, \blacktriangledown, c)$ where (V, E) is a finite directed graph, $\triangle$ and $\blacktriangledown$ are distinguished vertices called *source* and *sink*, respectively, and $c \colon E \to \mathbb{R}^{\geq 0} \cup \{\infty\}$ is a total function called edge *capacity function*. The capacity function, however, can be generalized to all pairs of vertexes by defining $c(u, v) = 0$ for each $(u, v) \notin E$.

Definition 19 (Flow [1, 12]). *A flow f on $\mathcal{N}$ is a function $f \colon V \times V \to \mathbb{R}$ such that:*

1. $f(u, v) \leq c(u, v)$ *for each* $(u, v) \in V \times V$ capacity constraints
2. $f(u, v) = -f(v, u)$ *for each* $(u, v) \in V \times V$ antisymmetry constraint
3. $f(V, v) = 0$ *for each* $v \in V \setminus \{\triangle, \blacktriangledown\}$ conservation rule

The value of a flow function is computed as $f(\triangle, V)$, also denoted by $|f|$. A flow of maximum value is called a *maximum flow*.

5.1   Computing the Maximum Flow

A *preflow* [1] is a function $f \colon V \times V \to \mathbb{R}$ that satisfies the first two conditions above and the following relaxation of the last condition: $f(V, v) \geq 0$ for each $v \in V \setminus \{\triangle\}$.

For each vertex v, its *excess* $e(v)$ is defined by $f(V, v)$. A vertex $v \in V \setminus \{\triangle, \blacktriangledown\}$ is called *active* if $e(v) > 0$. It is easy to check that when no vertex $v \in V \setminus \{\triangle, \blacktriangledown\}$ is active, the preflow function is actually a flow function. A pair (u, v) is said to be a *residual edge* of f if $f(u, v) < c(u, v)$. We denote the set of residual edges with regard to f by E_f. Corresponding to each residual edge (u, v) we define the *residual capacity* $c_f(u, v)$ as $c(u, v) - f(u, v)$. We say that the edge (u, v) is *saturated* if it is not a residual edge. A *valid distance function* d (also known as *valid labeling* [25]) is a function $d \colon V \to \mathbb{N} \cup \{\infty\}$ such that $d(\triangle) = |V|$, $d(\blacktriangledown) = 0$, and $d(u) \leq d(v) + 1$ for each residual edge (u, v). A residual edge (u, v) is called *admissible* if $d(u) = d(v) + 1$.

The maximal flow can be computed by means of preflow as follows: the algorithm initializes the preflow f by defining $f(u, v) = 0$ for each $(u, v) \in V \times V$ except for $f(\triangle, v) = c(\triangle, v)$ for each $v \in V$. The distance function d has initial values $d(\triangle) = |V|$ and $d(v) = 0$ for each other vertex $v \in V$. In order to maintain the validity of the preflow f and of the distance function d, the algorithm looks for active vertices in the network. If there exists an active vertex v and a residual edge (v, u) that is admissible, then, we *push* through (v, u) the amount of flow $\chi = \min\{e(v), c_f(v, u)\}$. This is done by increasing $f(v, u)$ (and decreasing $f(u, v)$) by χ and similarly, the excesses of v and u are updated by setting $e(v) = e(v) - \chi$ and $e(u) = e(u) + \chi$. If v is active but there is no admissible edge leaving it, the algorithm *relabels* v by defining $d(v) = \min\{d(u) + 1 \mid (v, u) \in E_f\}$. Pushing and relabeling are repeated until all vertexes are not active. It can be proved that the resulting ultimate flow f is a maximum flow [1, 25]. The generic preflow-push algorithm terminates after $\mathcal{O}(n^2 m)$ iterations where n and m are the number of nodes and the number of arcs of the network G, respectively.

5.2   Relation between Lifting and Maximum Flow

As we have seen in Section 2, the lifting $\mathcal{L}(\mathcal{R}) \subseteq \mathrm{Disc}(X) \times \mathrm{Disc}(X)$ of a relation $\mathcal{R} \subseteq X \times X$ has two different characterizations: via weighting functions and via $\mathcal{R}$-closure. It can be indeed characterized also via the maximum flow in a network as follows. First, we construct the network induced by the relation $\mathcal{R}$ and the two (sub-)probability distributions ρ_1 and ρ_2 and then we compute the maximum flow for such network. Given a set X, let $\overline{X}$ be the set $\overline{X} = \{\overline{x} \mid x \in X\}$.

Definition 20 ([4, 57]). *Let* $\rho_1, \rho_2 \in \mathrm{Disc}(X)$ *and* $\mathcal{R} \subseteq X \times X$. *The induced network* $\mathcal{N}(\mathcal{R}, \rho_1, \rho_2) = (V, E, \triangle, \blacktriangledown, c)$ *is defined by*

- $V = X \cup \overline{X} \cup \{\triangle, \blacktriangledown\}$,
- $E = \{(x, \overline{y}) \mid (x, y) \in \mathcal{R}\} \cup \{(\triangle, x) \mid x \in X\} \cup \{(\overline{y}, \blacktriangledown) \mid y \in X\}$, *and*
- $c(\triangle, x) = \rho_1(x)$, $c(\overline{y}, \blacktriangledown) = \rho_2(y)$, *and* $c(x, \overline{y}) = 1$ *for all* $x, y \in X$.

As shown in [55], $\rho_1 \ \mathcal{L}(\mathcal{R}) \ \rho_2$ if and only if the maximum flow of the induced network $\mathcal{N}(\mathcal{R}, \rho_1, \rho_2)$ is 1.

It is worthwhile to note that for each $x \notin \mathrm{Supp}(\rho_1)$, $c(\triangle, x) = 0$ (and similarly, for each $y \notin \mathrm{Supp}(\rho_2)$, $c(y, \blacktriangledown) = 0$), thus the flow along the edge $(\triangle, x)$ is always 0. Therefore the induced network can be equivalently simplified as follows: let $\rho_1, \rho_2 \in \mathrm{Disc}(X)$ and $\mathcal{R} \subseteq X \times X$. The induced network $\mathcal{N}(\mathcal{R}, \rho_1, \rho_2) = (V, E, \triangle, \blacktriangledown, c)$ is defined by

- $V = \mathrm{Supp}(\rho_1) \cup \overline{\mathrm{Supp}(\rho_2)} \cup \{\triangle, \blacktriangledown\}$,
- $E = \{(x, \overline{y}) \mid (x, y) \in \mathcal{R}, x \in \mathrm{Supp}(\rho_1), y \in \mathrm{Supp}(\rho_2)\} \cup \{(\triangle, x) \mid x \in \mathrm{Supp}(\rho_1)\} \cup \{(\overline{y}, \blacktriangledown) \mid y \in \mathrm{Supp}(\rho_2)\}$, *and*
- $c(\triangle, x) = \rho_1(x)$, $c(\overline{y}, \blacktriangledown) = \rho_2(y)$, *and* $c(x, \overline{y}) = 1$ for all $x, y \in X$.

In the remaining of the paper we use the more appropriate definition of induced network without further mentioning which one we are considering.

6   The Algorithms

We now consider the algorithms and their complexity that are used to decide the simulations and the bisimulations introduced in Section 4. In the following, we denote by $s \xrightarrow{a}_{\preccurlyeq} \rho$ the matching transition involved in the step condition of the relation $\preccurlyeq$. For instance, when $\preccurlyeq$ is $\lesssim$, then $s \xrightarrow{a}_{\preccurlyeq} \rho$ stands for $s \xrightarrow{a} \rho$ while when $\preccurlyeq$ is $\lesssim_p$, then $s \xrightarrow{a}_{\preccurlyeq} \rho$ stands for $s \xRightarrow{a}_C \rho$.

$\text{SIM}(\preccurlyeq, \mathcal{A})$
1. $i \leftarrow 0$; $\mathcal{R}_i \leftarrow \{ (s_1, s_2) \in S \times S \mid L(s_1) = L(s_2) \}$;
2. **repeat**
3. $\mathcal{R}_{i+1} \leftarrow \mathcal{R}_i$;
4.   **for all** $s_1 \xrightarrow{a} \rho_1 \in s_1 \longrightarrow$ **do**
5.   **for all** $s_2 \in S$ such that $s_1 \mathcal{R}_i s_2$ **do**
6.   **if** there does not exist $s_2 \xrightarrow{a}_{\preccurlyeq} \rho_2$ satisfying the step condition
7. $\mathcal{R}_{i+1} \leftarrow \mathcal{R}_{i+1} \setminus \{(s_1, s_2)\}$;
8. $i \leftarrow i + 1$;
9. **until** $\mathcal{R}_i = \mathcal{R}_{i-1}$;
10. **return** $\mathcal{R}_i$;

Fig. 16. Algorithm for computing simulation

6.1 The General Algorithms for Simulations and Bisimulations

Simulation Algorithm. Figure 16 depicts the algorithm SIM, proposed for instance in [4], that computes the simulation $\preccurlyeq$ for the automaton $\mathcal{A}$. The procedure begins with the initial relation $\mathcal{R}_0 = \{ (s_1, s_2) \in S \times S \mid L(s_1) = L(s_2) \}$ which is coarser than $\preccurlyeq$. In each iteration i of the main loop, the relation $\mathcal{R}_i$, initialized with $\mathcal{R}_{i-1}$, is refined by deleting each pair (s_1, s_2) such that s_2 is not able to exhibit the transition $s_2 \xrightarrow{a}_{\preccurlyeq} \rho_2$ such that $\rho_1 \; \mathcal{L}(\mathcal{R}_{i-1}) \; \rho_2$ required by the step condition of $\preccurlyeq$. The main loop terminates when all pairs of $\mathcal{R}_i$ satisfy the step condition, that is, when $\mathcal{R}_i = \mathcal{R}_{i-1}$. The resulting similarity $\preccurlyeq$ is then $\mathcal{R}_i$.

It is immediate to see that the core of this algorithm is the check for the existence of the step condition for s_1 and s_2, and that this is also the main source of the complexity of the algorithm. In fact, if we denote by N the size of the automaton, i.e., $N = \max\{|S|, |\rightarrow|\}$, it is easy to derive that the complexity of the algorithm is $\mathcal{O}(N^4 \cdot C)$ where C is the complexity of deciding the existence of the matching transition $s_2 \xrightarrow{a}_{\preccurlyeq} \rho_2$.

Bisimulation Algorithm. In order to compute the bisimulation $\asymp$ for the automaton $\mathcal{A}$ we can follow the standard partition refinement approach [13,20,29, 35,41,43] depicted in Figure 17: the procedure BISIM takes as parameter the bisimulation $\asymp$ and the automaton $\mathcal{A}$ and iteratively constructs the set $S/\asymp$, the set of equivalence classes of states S under $\asymp$, starting with the partitioning $\mathcal{W} = \{ (s_1, s_2) \in S \times S \mid L(s_1) = L(s_2) \}$ and refining it until $\mathcal{W}$ satisfies the definition of $\asymp$ and thus the resulting partitioning is the coarsest one, i.e., the algorithm computes $\asymp$. In the refinement phase, the algorithm checks for each partition whether all pairs respect the step condition; if a pair (s_1, s_2) fails, then such partition is split in two partitions: one containing s_1 and all states satisfying the step condition with respect to transitions from s, the other containing the remaining states, including s_2. On termination, the partitioning $\mathcal{W}$ is the bisimilarity $\asymp$.

$$\boxed{\text{Bisim}(\precsim, \mathcal{A})}$$

1. $\mathcal{W} \leftarrow \{\, (s_1, s_2) \in S \times S \mid L(s_1) = L(s_2) \,\}$;
2. **repeat**
3. $\mathcal{W}' \leftarrow \mathcal{W}$;
4. $(\mathcal{C}, a, \rho) \leftarrow \text{FindSplit}(\mathcal{A}, \mathcal{W})$;
5. $\mathcal{W} \leftarrow \text{Refine}(\mathcal{C}, a, \rho)$;
6. **until** $\mathcal{W} \leftarrow \mathcal{W}'$
7. **return** $\mathcal{W}$

$$\boxed{\text{FindSplit}(\mathcal{A}, \mathcal{W})}$$

1. **for all** $(s_1, a, \rho_1) \in \,\rightarrow\,$ **do**
2.   **for all** $s_2 \in [s_1]_\mathcal{W}$ **do**
3.   **if** there does not exist $s_2 \xrightarrow{a}_\precsim \rho_2$ satisfying the step condition
4.   **return** $([s_1]_\mathcal{W}, a, \rho_1)$
5. **return** $(\emptyset, \tau, \delta_\perp)$

Fig. 17. Algorithm for computing bisimulation

As happens for Sim, it is immediate to see that the core of this algorithm is the check whether the step condition for s_1 and s_2 holds, and that this is also the main source of the complexity of the algorithm. In fact, as done for the simulation procedure, if we denote by N the size of the automaton, i.e., $N = \max\{|S|, |\rightarrow|\}$, it is easy to derive that the complexity of the algorithm is $\mathcal{O}(N^3 \cdot C)$ where C is the complexity of deciding the existence of the matching transition $s_2 \xrightarrow{a}_\precsim \rho_2$.

6.2   The Specialized Algorithms

Strong Simulation. We now present an improved algorithm [57] for the strong simulation on *sDTMCs*, *DTMCs*, and *CTMCs*, respectively, based on the properties of the network flow setting.

Let $\mathcal{A}$ be either a *sDTMC* or a *DTMC*. Since for *sDTMCs* and *DTMCs* every state s enables a single transition, checking the existence of the matching transition $s_2 \xrightarrow{\tau}_\precsim \mu_2$ reduces to check whether $\mu_1 \, \mathcal{L}(\mathcal{R}_i) \, \mu_2$ where $\mu_1 = \mathbf{P}(s_1, \cdot)$ and $\mu_2 = \mathbf{P}(s_2, \cdot)$. As we have seen in Section 5.2, $\mu_1 \, \mathcal{L}(\mathcal{R}_i) \, \mu_2$ is equivalent to check whether the induced network $\mathcal{N}(\mathcal{R}_i, \mu_1, \mu_2)$ has 1 as maximum flow. Since finding the maximum flow has complexity $\mathcal{O}(\frac{N^3}{\log N})$, thus the resulting complexity of Sim$(\precsim, \mathcal{A})$ is $\mathcal{O}(\frac{N^7}{\log N})$ [4, 57].

Improved Algorithm for sDTMCs. Consider the Sim$(\precsim, \mathcal{A})$ procedure and a pair of states $(s_1, s_2) \in \mathcal{R}_1$. Let $\mu_1 = \mathbf{P}(s_1, \cdot)$ and $\mu_2 = \mathbf{P}(s_2, \cdot)$ and suppose that (s_1, s_2) belongs to $\mathcal{R}_1, \ldots, \mathcal{R}_k$ during the whole of the iterations $i = 1, \ldots, k$ until the pair either violates the step condition with respect to $\mathcal{R}_k$ or the algorithm terminates after iteration k. This means that the maximum flow algorithm is used k times for this pair. As a matter of fact, the induced networks $\mathcal{N}(\mathcal{R}_i, \mu_1, \mu_2)$

built in successive iterations are very similar, and may often be the same across iterations. From iteration to iteration, in fact, they differ only for the removal of some edge $(t_1, \overline{t_2})$ induced by $\mathcal{R}_i \leftarrow \mathcal{R}_i \setminus \{(t_1, t_2)\}$ but this does not change the network when $t_1 \notin \text{Supp}(\mu_1)$ or $t_2 \notin \text{Supp}(\mu_2)$. This observation, inspired by [24], is the key point of [57] for improving the basic algorithm. In fact, the authors reuse the previous computed maximum flow in the sense that whatever happens to the network is good: if the network $\mathcal{N}(\mathcal{R}_i, \mu_1, \mu_2)$ is equal to $\mathcal{N}(\mathcal{R}_{i-1}, \mu_1, \mu_2)$, then the maximum flow is the same as the one in the previous iteration. On the other hand, if the two networks are different, then the preflow algorithm can be adapted to compute the new maximum flow using the previous maximum flow and distance function as a starting point.

$\text{SMF}_{init}(i, \mathcal{R}_i, \mu_1, \mu_2)$
1. Initialize the network $\mathcal{N}(\mathcal{R}_i, \mu_1, \mu_2)$;
2. Apply the preflow algorithm to compute the maximum flow for $\mathcal{N}(\mathcal{R}_i, \mu_1, \mu_2)$;
3. **return** $(\|f_i\| = 1, \mathcal{N}(\mathcal{R}_i, \mu_1, \mu_2), f_i, d_i)$;

$\text{SMF}(i, \mathcal{N}(\mathcal{R}_{i-1}, \mu_1, \mu_2), f_{i-1}, d_{i-1}, D_{i-1})$
1. $\mathcal{N}(\mathcal{R}_i, \mu_1, \mu_2) \leftarrow \mathcal{N}(\mathcal{R}_{i-1} \setminus D_{i-1}, \mu_1, \mu_2);\ f_i \leftarrow f_{i-1};\ d_i \leftarrow d_{i-1};$
2. **for all** $(v_1, v_2) \in D_{i-1}$ **do**
3. $f_i(\overline{v_2}, \blacktriangledown) \leftarrow f_i(\overline{v_2}, \blacktriangledown) - f_i(v_1, \overline{v_2});$
4. $f_i(v_1, v_2) \leftarrow 0;$
5. Apply the preflow algorithm initialized with f_i and d_i to compute the maximum flow for $\mathcal{N}(\mathcal{R}_i, \mu_1, \mu_2)$;
6. **return** $(\|f_i\| = 1, \mathcal{N}(\mathcal{R}_i, \mu_1, \mu_2), f_i, d_i)$;

Fig. 18. Algorithm for computing a sequence of maximum flows

To explain this approach in more detail, consider the network $\mathcal{N}(\mathcal{R}_1, \mu_1, \mu_2)$ and let $D_1, \ldots, D_k$ be pairwise disjoint subsets of $\mathcal{R}_1$ that correspond to the pairs deleted from $\mathcal{R}_1$ in iteration i, i.e., $\mathcal{R}_{i+1} = \mathcal{R}_i \setminus D_i$ for $1 \leq i \leq k$. Let $f_i^{(s_1,s_2)}$ be the flow and $d_i^{(s_1,s_2)}$ the distance function of the network $\mathcal{N}(\mathcal{R}_i, \mu_1, \mu_2)$ where $0 \leq i \leq k$, respectively. The algorithm for updating the sequences of maximum flows and distances of the network $\mathcal{N}(\mathcal{R}_i, \mu_1, \mu_2)$ where $1 \leq i \leq k$ is depicted in Figure 18 and it works as follows: starting from the network $\mathcal{N}(\mathcal{R}_{i-1} \setminus D_{i-1}, \mu_1, \mu_2)$ with flow f_{i-1} and distance d_{i-1}, it computes for each pair $(v_1, v_2) \in D_{i-1}$ the flow $f_i(\overline{v_2}, \blacktriangledown)$ by decreasing the previous value $f_{i-1}(\overline{v_2}, \blacktriangledown)$ by the value of the flow $f_{i-1}(v_1, \overline{v_2})$ and then forces $f_i(v_1, v_2)$ to be 0. Then it calls the preflow algorithm initialized with the updated flow and distance function to compute the maximum flow for the new network and returns the network, the updated flow and distance functions, and a boolean representing whether the maximum flow is 1.

The $\text{SMF}(i, \mathcal{N}(\mathcal{R}_{i-1}, \mu_1, \mu_2), f_{i-1}, d_{i-1}, D_{i-1})$ algorithm is the building block for the improved algorithm $\text{SIM}_{sDTMC}(\lesssim, \mathcal{S})$ that computes the strong simulation on sDTMCs, depicted in Figure 19.

$$\text{SIM}_{sDTMC}(\precsim, sDTMC)$$

---

1. $\mathcal{R}_1 \leftarrow \{ (s_1, s_2) \in S \times S \mid L(s_1) = L(s_2) \}; i \leftarrow 1;$
2. $\mathcal{R}_2 \leftarrow \emptyset;$
3. **for all** $(s_1, s_2) \in \mathcal{R}_1$ **do**
4. $Listener(s_1, s_2) \leftarrow \{ (u_1, u_2) \mid u_1 \in pre(s_1), u_2 \in pre(s_2), L(u_1) = L(u_2) \};$
5. $(match, \mathcal{N}(\mathcal{R}_1, \mu_1, \mu_2), f_1^{(s_1, s_2)}, d_1^{(s_1, s_2)}) \leftarrow \text{SMF}_{init}(1, \mathcal{R}_1, \mu_1, \mu_2)$
6.   **if** $match$
7. $\mathcal{R}_2 \leftarrow \mathcal{R}_2 \cup \{(s_1, s_2)\};$
8. **while** $\mathcal{R}_{i+1} \neq \mathcal{R}_i$ **do**
9. $i \leftarrow i + 1; \mathcal{R}_{i+1} \leftarrow \emptyset; D_{i-1} \leftarrow \mathcal{R}_{i-1} \setminus \mathcal{R}_i;$
10.   **for all** $(s_1, s_2) \in \mathcal{R}_i$ **do**
11. $D_{i-1}^{(s_1, s_2)} \leftarrow \emptyset;$
12.   **for all** $(s_1, s_2) \in D_{i-1}, (u_1, u_2) \in Listener(s_1, s_2) \cap \mathcal{R}_{i-1}$ **do**
13. $D_{i-1}^{(u_1, u_2)} \leftarrow D_{i-1}^{(u_1, u_2)} \cup \{(s_1, s_2)\};$
14.   **for all** $(s_1, s_2) \in \mathcal{R}_i$ **do**
15. $(match, \mathcal{N}(\mathcal{R}_i, \mu_1, \mu_2), f_i^{(s_1, s_2)}, d_i^{(s_1, s_2)})$
            $\leftarrow \text{SMF}(i, \mathcal{N}(\mathcal{R}_{i-1}, \mu_1, \mu_2), f_{i-1}^{(s_1, s_2)}, d_{i-1}^{(s_1, s_2)}, D_{i-1}^{(s_1, s_2)});$
16.   **if** $match$
17. $\mathcal{R}_{i+1} \leftarrow \mathcal{R}_{i+1} \cup \{(s_1, s_2)\};$
18. **return** $\mathcal{R}_i;$

---

where $\mu_1 = \mathbf{P}(s_1, \cdot)$ and $\mu_2 = \mathbf{P}(s_2, \cdot)$

Fig. 19. Improved algorithm for deciding strong simulation for $sDTMCs$

The first part of the algorithm, from line 1 to 7 is essentially the same as in $\text{SIM}(\precsim, sDTMC)$, except for line 4 where the set

$$Listener(s_1, s_2) = \{ (u_1, u_2) \mid u_1 \in pre(s_1), u_2 \in pre(s_2), L(u_1) = L(u_2) \}$$

is computed for the remainder of the procedure. In particular, this set contains all pairs (u_1, u_2) such that $(s_1, \overline{s_2})$ is an edge of $\mathcal{N}(\mathcal{R}_0, \mathbf{P}(u_1, \cdot), \mathbf{P}(u_2, \cdot))$.

In each iteration i of the loop at lines 8–17, the procedure generates $\mathcal{R}_{i+1}$ from $\mathcal{R}_i$ by performing several steps: first, with the loop at line 12, it collects in $D_{i-1}^{(u_1, u_2)}$ the edges that should be removed from $\mathcal{N}(\mathcal{R}_{i-1}, \mathbf{P}(u_1, \cdot), \mathbf{P}(u_2, \cdot))$. Then, at line 15, the algorithm SMF builds the maximum flow by using information from the previous iteration $i - 1$. Basically, $\mathcal{N}(\mathcal{R}_{i-1}, \mu_1, \mu_2)$, $f_{i-1}^{(s_1, s_2)}$, and $d_{i-1}^{(s_1, s_2)}$ are updated according to the set $D_{i-1}^{(s_1, s_2)}$; this generates the new maximum flow $f_i^{(s_1, s_2)}$ for the network $\mathcal{N}(\mathcal{R}_i, \mu_1, \mu_2)$ and if such flow is 1 (i.e., $match$ is true), then (s_1, s_2) is added to $\mathcal{R}_{i+1}$ and survives this iteration. Eventually the while loop terminates and the last candidate simulation $\mathcal{R}_i$ is actually a strong bisimilarity.

The correctness and time complexity of this algorithm is stated in [57, Theorem 4.5 and 4.6], respectively. In particular, the time complexity is $\mathcal{O}(m^2 \cdot N + N^2)$, where $m = \sum_{s \in S} |post(s)|$, that is significantly smaller than $\mathcal{O}(\frac{N^7}{\log N})$ of the general algorithm $\text{SIM}(\precsim, sDTMC)$.

Strong Simulation for DTMCs and CTMCs. We now take into account *DTMCs* and *CTMCs*. Since every *DTMC* $\mathcal{D}$ is a *sDTMC*, we can use directly the algorithm $\mathrm{SIM}_{sDTMC}(\precsim, \mathcal{D})$. For *CTMCs*, we have to take care of the rate condition; this is easily obtained by replacing the assignment to $\mathcal{R}_1$ at the line 1 of the algorithm with

$$\mathcal{R}_1 \leftarrow \{ (s_1, s_2) \in S \times S \mid L(s_1) = L(s_2), \mathbf{R}(s_1, S) \leq \mathbf{R}(s_2, S) \}$$

It is immediate to see that this change does not increase the complexity of the algorithm; in particular, the time complexity may be reduced, since there may be fewer pairs satisfying the rate condition as well.

Strong Probabilistic Simulation and Bisimulation. Strong probabilistic simulation and bisimulation are defined only for *PAs* and *CTPAs* since they are the only models that exhibit internal nondeterminism, thus they allow to combine transitions with the same label (and the same rate, for *CTPAs*). Checking the step condition thus requires to find such combined transition. One possibility is to check, for every possible combined transition, whether it satisfies the step condition; however this approach is not practical since given two transitions, there are uncountable many different convex combinations of them. The other possibility is to check whether there exists a choice for the coefficients of the convex combination by solving a linear programming problem encoding convex combination and lifting [57].

For a *PA* $\mathcal{P}$, the *LP* problem relative to relation $\mathcal{R}$, transition $s_1 \xrightarrow{a} \mu$ and state s_2 is:

$$\sum_{i=1}^{k} c_i = 1$$
$$0 \leq c_i \leq 1 \qquad\qquad \text{for } 1 \leq i \leq k$$
$$0 \leq f_{u,v} \leq 1 \qquad\qquad \text{for each } (u, v) \in \mathcal{R}_{\perp}$$
$$\mu(s) = \sum_{t \in \mathcal{R}_{\perp}(s)} f_{s,t} \qquad \text{for each } s \in S_{\perp}$$
$$\sum_{s \in \mathcal{R}_{\perp}^{-1}(t)} f_{s,t} = \sum_{i=1}^{k} c_i \rho_i(t) \qquad \text{for each } t \in S_{\perp}$$

where $\{\rho_1, \ldots, \rho_k\} = \{ \rho \mid (s_2, a, \rho) \in \rightarrow \}$.

For a *CTPA* $\mathcal{CP}$, the *LP* problem relative to relation $\mathcal{R}$, transition $s_1 \xrightarrow{a} r$ and state s_2 is similar:

$$\sum_{i=1}^{k} c_i = 1$$
$$0 \leq c_i \leq 1 \qquad\qquad \text{for } 1 \leq i \leq k$$
$$0 \leq f_{u,v} \leq 1 \qquad\qquad \text{for each } (u, v) \in \mathcal{R}_{\perp}$$
$$r(s) = r(S) \cdot \sum_{t \in \mathcal{R}_{\perp}(s)} f_{s,t} \qquad \text{for each } s \in S_{\perp}$$
$$E \cdot \sum_{s \in \mathcal{R}_{\perp}^{-1}(t)} f_{s,t} = \sum_{i=1}^{k} c_i r_i(t) \qquad \text{for each } t \in S_{\perp}$$

for some $E \in \{ r'(S) \mid (s_2, a, r') \in \mathbf{R} \}$, $E \geq r(S)$ where $\{r_1, \ldots, r_k\} = \{ r' \mid (s_2, a, r') \in \mathbf{R}, r'(S) = E \}$.

The complexity of the $\mathrm{SIM}(\precsim_p, \mathcal{A})$ and $\mathrm{BISIM}(\sim_p, \mathcal{A})$ algorithms is then polynomial and directly depends on the polynomial complexity [53] of solving the above *LP* problems, each one with at most $\mathcal{O}(N^2)$ constraints.

It is worthwhile to note that by combining a preflow approach, as the one adopted for SMF, and abstract interpretation techniques, the complexity can be reduced to $\mathcal{O}(N^3)$ for simulation and $\mathcal{O}(N^2 \cdot \log N)$ for bisimulation [15].

Weak Simulation and Bisimulation for *DTMC*s and *CTMC*s. Now, we focus our attention to weak simulations. As it was the case for strong simulations, the core of the algorithm is to check the step condition with respect to the current relation $\mathcal{R}$. Based on the definition of weak simulation, for fixed characteristic functions γ_i for $i = 1, 2$, maximum flow algorithms can be used in order to check condition (2) in definition 15. In order to improve this check, we can make use of the parametric maximum flow algorithm in order to determine whether functions γ_i exist, with the aid of *breakpoints*, as we will see in the following.

As shown in [57], checking the step condition of the weak simulation for the pair of states (s_1, s_2) for both *DTMC*s and *CTMC*s is equivalent to finding the parameter ψ that makes a parametric network valid. In particular, the considered parametric network is $\mathcal{N}_\psi(\mathcal{R}, \mu_1, \mu_2)$ that is defined as $\mathcal{N}(\mathcal{R}, \mu_1, \psi \cdot \mu_2)$; this means that $\mathcal{N}(\mathcal{R}, \mu_1, \psi \cdot \mu_2)$ is the network $\mathcal{N}(\mathcal{R}, \mu_1, \mu_2)$ where the capacities for the edges leading to the sink are $c(\bar{t}, \blacktriangledown) = \psi \cdot \mu_2(t)$. The network $\mathcal{N}_\psi(\mathcal{R}, \mu_1, \mu_2)$ is valid if there exists a flow f that saturates all edges $(\triangle, u_1)$ and $(\overline{u_2}, \blacktriangledown)$ where u_1 belongs to the set $MU_1 = post(s_1) \setminus PV_1$ with $PV_1 = post(s_1) \cap \mathcal{R}^{-1}(s_2)$ and u_2 belongs to the set $MU_2 = post(s_2) \setminus PV_2$ with $PV_2 = post(s_2) \cap \mathcal{R}(s_1)$.

Sets MU_i and PV_i are strictly related to the sets U_i and V_i used for the strong simulation algorithm. Indeed, MU_i stands for "must be in U_i" while PV_i stands for "potentially in V_i". Functions γ_i are extended as expected by $\gamma_i(u) = 1$ for $u \in MU_i$, $i \in \{1, 2\}$.

If we fix $\psi \in \mathbb{R}^{\geq 0}$, then checking whether $\mathcal{N}_\psi(\mathcal{R}, \mu_1, \mu_2)$ is valid reduces to verify the feasibility of a flow problem (f has to saturate edges to MU_1 and from MU_2); this can be done by applying a simple transformation to the graph (in time $\mathcal{O}(|MU_1| + |MU_2|)$), solving the maximum flow problem for the transformed graph, and checking whether the flow saturates all edges from the new source [1]. So now the problem is to find a good ψ that makes $\mathcal{N}_\psi(\mathcal{R}, \mu_1, \mu_2)$ valid, but there are uncountably many of such ψ we may check for. However, the candidates that really matter are finite, not uncountably many, and are called *breakpoints*. In particular, breakpoints can be identified by solving one more parametric maximum flow problem: Let $\kappa(\psi)$ be the *minimum cut capacity function* for the parameter ψ, that is, the capacity of a minimum cut of $\mathcal{N}_\psi(\mathcal{R}, \mu_1, \mu_2)$ as a function of ψ. Based on the Max flow Min cut theorem [1], the capacity of a minimum cut equals the value of a maximum flow. On the other hand, if the edge capacities in the network are linear functions of ψ, $\kappa(\psi)$ is a piecewise linear concave function with at most $|V| - 2$ breakpoints [24]. In particular, $|V| - 1$ or fewer line segments of the graph of $\kappa(\psi)$ are equivalent to $|V| - 1$ or fewer distinct minimal cuts. For some ψ^*, the capacity of a minimum cut gives an equation that leads to a line segment to the function $\kappa(\psi)$ at $\psi = \psi^*$. Furthermore, this line segment attaches the two points $(\psi_1, \kappa(\psi_1))$ and $(\psi_2, \kappa(\psi_2))$, where ψ_1, ψ_2 are the nearest breakpoints to the left and right,

$\text{STEPCONDITION}_{\lesssim}(\mathcal{D}, \mathcal{R}, s_1, s_2)$

1. **if** $post(s_1) \subseteq \mathcal{R}^{-1}(s_2)$
2.   **return true**
3. **if** $post(s_2) \subseteq \mathcal{R}(s_1)$
4. $U_1 \leftarrow \{\, s_1' \in post(s_1) \mid s_1' \notin \mathcal{R}^{-1}(s_2) \,\}$
5.   **return** $\forall u_1 \in U_1 . \exists s \in post(reach(s_2)) \cap \mathcal{R}(s_1) . s \in \mathcal{R}(u_1)$
6. Compute all of the breakpoints $\psi_1 < \psi_2 < \cdots < \psi_j$ of $\mathcal{N}_\psi(\mathcal{R}, \mu_1, \mu_2)$
7. **return** $\exists i \in \{1, \ldots, j\} . \psi_i$ is valid for $\mathcal{N}_{\psi_j}(\mathcal{R}, \mu_1, \mu_2)$

where $\mu_1 = \mathbf{P}(s_1, \cdot)$ and $\mu_2 = \mathbf{P}(s_2, \cdot)$

Fig. 20. Algorithm to check whether s_2 weakly simulates s_1 with respect to $\mathcal{R}$

respectively. Therefore, as it would be expected, it is enough to examine only the breakpoints of $\mathcal{N}_\psi(\mathcal{R}, \mu_1, \mu_2)$: there exists a valid ψ for $\mathcal{N}_\psi(\mathcal{R}, \mu_1, \mu_2)$ if and only if one of the breakpoints of $\mathcal{N}_\psi(\mathcal{R}, \mu_1, \mu_2)$ is valid.

For a fixed breakpoint, it is adequate to solve one feasible flow problem to check if it is valid. In the network $\mathcal{N}_\psi(\mathcal{R}, \mu_1, \mu_2)$ the capacities of the edges going to the sink are increasing functions of the real-valued parameter ψ. If $\mathcal{N}_\psi(\mathcal{R}, \mu_1, \mu_2)$ is reversed, a parametric network that satisfies the conditions in [24] can be derived: the capacities emanating from $\triangle$ are non-decreasing functions of ψ. Therefore, the *breakpoint algorithm* [24] can be applied to compute the breakpoints of $\mathcal{N}_\psi(\mathcal{R}, \mu_1, \mu_2)$.

The Algorithm for DTMCs. We are now able to provide the decision algorithm for the *DTMC* weak simulation: we just consider the $\text{SIM}(\lesssim, \mathcal{D})$ where the step condition is verified by invoking the algorithm in Figure 20 that, given two states s_1 and s_2, it actually computes whether $s_1 \lesssim s_2$.

By using this approach, the resulting complexity of the algorithm that computes the weak simulation for *DTMCs* is $\mathcal{O}(N^5)$. This complexity can be improved in practice by exploiting the network $\mathcal{N}_\psi(\mathcal{R}, \mu_1, \mu_2)$ whenever it can be parted into sub-networks. We refer the reader interested in this approach to [57, Section 5.1.4]

An Algorithm for CTMCs. The algorithm for computing weak simulation on *CTMCs* is very close to the one for *DTMCs* since the only difference is the last requirement of the step condition: "$K_1 \cdot \mathbf{R}(s_1, S) \leq K_2 \cdot \mathbf{R}(s_2, S)$" instead of "for $u_1 \in U_1$ there exist an execution fragment $s_2 t_1 \ldots t_n u_2$ with positive probability such that $n \in \mathbb{N}$, $s_1 \mathcal{R} t_j$ for $0 < j \leq n$, and $u_1 \mathcal{R} u_2$".

This makes the algorithm for $\mathcal{C}$ simpler: if $K_1 > 0$ and $K_2 = 0$, then $s_1 \not\mathcal{R} s_2$ for the rate condition. Therefore, the reachability condition does not need to be checked and the lines 3–5 of the algorithm $\text{STEPCONDITION}_{\lesssim}(\mathcal{D}, \mathcal{R}, s_1, s_2)$ can be omitted. In general, the rate condition can be verified by checking the validity of the network $\mathcal{N}_\psi(\mathcal{R}, \mu_1, \mu_2)$ induced in the embedded *DTMC* $emb(\mathcal{C})$. In particular, the step condition holds if and only if there exists $\psi \leq \frac{\mathbf{R}(s_2, S)}{\mathbf{R}(s_1, S)}$ such that ψ is valid for $\mathcal{N}_\psi(\mathcal{R}, \mu_1, \mu_2)$. This means that we can replace the returned value

$$\exists i \in \{1, \ldots, j\}.\psi_i \text{ is valid for } \mathcal{N}_{\psi_j}(\mathcal{R}, \mu_1, \mu_2)$$

of line 7 of $\text{STEPCONDITION}_{\lesssim}(\mathcal{D}, \mathcal{R}, s_1, s_2)$ with

$$\exists i \in \{1, \ldots, j\}.\psi_i \leq \frac{\mathbf{R}(s_2, S)}{\mathbf{R}(s_1, S)} \wedge \psi_i \text{ is valid for } \mathcal{N}_{\psi_j}(\mathcal{R}, \mu_1, \mu_2).$$

These improvements do not change the worst case complexity of the algorithm, but they improve it in practice, in particular when merged with the improved algorithm for DTMCs.

Weak Probabilistic Simulation and Bisimulation for PAs. To complete the survey on the simulations and bisimulations defined on PAs, we consider the weak probabilistic (bi)simulation and the weak (bi)simulation. The latter relation is a restriction of the former where the step condition for the pair (s, t) requires that t matches the challenging transition proposed by s via a weak transition instead of a weak combined transition. By using the PAs proposed by [16], it is possible to show that both weak simulation and bisimulation are not transitive, so we omit them. On the contrary, both weak probabilistic simulation and bisimulation are transitive [47] and they can be used whenever we want to abstract away from the internal computation of a probabilistic automaton.

The decidability of weak probabilistic bisimulation has been stated in [13] and it is based on the standard partition refinement approach. The complexity of such algorithm is exponential in the number of transitions and only recently it has been improved to polynomial [29]. Indeed, [29] reduces the complexity to polynomial by constructing a flow network enriched with side constraints that admits a valid flow if and only if there exists a determinate scheduler that induces the desired weak combined transition.

With some inspiration from network flow problems, authors of [29, 30] were able to see a transition $t \stackrel{a}{\Longrightarrow}_C \mu_t$ of the PA $\mathcal{P}$ as a *flow* where the initial probability mass δ_t flows and splits along internal transitions (and exactly one transition with label a for each stream when $a \neq \tau$) according to the transition target distributions and the scheduler resolution of the nondeterminism. The resulting flow problem is then translated into an LP extended with *balancing constraints* that encode the need to respect transition probability distributions. To describe the structure of the LP problem, we first recall the original definition of the network graph corresponding to a weak combined transition.

Given a PA $\mathcal{P} = (S, \bar{s}, \Sigma, \rightarrow)$ and a relation $\mathcal{R} \subseteq S \times S$, for $a \in \mathrm{E}$, the network $G(t, a, \mu, \mathcal{R}) = (V, E, \triangle, \blacktriangledown, c)$ has the set of vertices $V = \{\triangle, \blacktriangledown\} \cup S \cup S^{tr} \cup S_a \cup S_a^{tr} \cup S_{\mathcal{R}}$ where

$$S^{tr} = \{ v^{tr} \mid tr = v \stackrel{b}{\longrightarrow} \rho \in \rightarrow, b \in \{a, \tau\} \},$$
$$S_a = \{ v_a \mid v \in S \},$$
$$S_a^{tr} = \{ v_a^{tr} \mid v^{tr} \in S^{tr} \}, \text{ and}$$
$$S_{\mathcal{R}} = \{ s_{\mathcal{R}} \mid s \in S \}$$

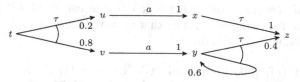

Fig. 21. A probabilistic automaton

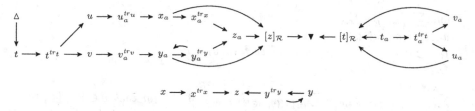

Fig. 22. The network graph $G(t, a, \delta_z, \mathcal{R})$ for the automaton $\mathcal{A}$

and the set of edges

$$E = \{(\triangle, t)\} \cup \{ (v_a, u_{\mathcal{R}}), (u_{\mathcal{R}}, \blacktriangledown) \mid u \in S, v \mathrel{\mathcal{R}} u \}$$
$$\cup \{ (v, v^{tr}), (v^{tr}, v'), (v_a, v_a^{tr}), (v_a^{tr}, v_a') \mid tr = v \xrightarrow{\tau} \rho \in \to, v' \in \mathrm{Supp}(\rho) \}$$
$$\cup \{ (v, v_a^{tr}), (v_a^{tr}, v_a') \mid tr = v \xrightarrow{a} \rho \in \to, v' \in \mathrm{Supp}(\rho) \}.$$

For each $(u, v) \in E$, $c(u, v) = \infty$. When $a \in \mathrm{H}$, the definition is similar: $V = \{\triangle, \blacktriangledown\} \cup S \cup S^{tr} \cup S_{\mathcal{R}}$ and $E = \{(\triangle, t)\} \cup \{ (v, u_{\mathcal{R}}), (u_{\mathcal{R}}, \blacktriangledown) \mid u \in S, v \mathrel{\mathcal{R}} u \} \cup \{ (v, v^{tr}), (v^{tr}, v') \mid tr = v \xrightarrow{\tau} \rho \in \to, v' \in \mathrm{Supp}(\rho) \}$. As a concrete example to illustrate the construction of the flow network, consider the probabilistic automaton $\mathcal{P}$ given in Figure 21. The flow network $G(t, a, \delta_z, \mathcal{R})$ is depicted in Figure 22, where $\mathcal{R}$ is the equivalence relation inducing classes $\{t, u, v\}$ and $\{x, y, z\}$.

As pointed out in [29], the fact that the network admits a flow that respects the probability distribution μ_t does not imply the existence of a corresponding weak combined transition, because the flow may not respect probability ratios. Moreover, in order to define a flow problem, we need to define the capacity for each arc. There are several possibilities for doing this: the first possibility is to use as capacity for the arc (v^{tr}, u) corresponding to the transition $tr = s \xrightarrow{\tau} \rho$ with $u \in \mathrm{Supp}(\rho)$ the probability $\rho(u)$; the capacity of the remaining arcs is 1. As we will see, such capacity in general is not suitable for arcs that are part of cycles. Another possibility is to use as capacity the value $\frac{1}{1-\rho(u)}$ for arcs of the kind (v^{tr}, u), $\max\{ \frac{1}{1-\rho(u)} \mid u \in \mathrm{Supp}(\mu) \}$ for the arc (v, v^{tr}), and 1 for other arcs; in this case such capacity is suitable for the arcs involved in cycles, but still it does not force to respect probability ratios. Finally, arcs have infinite capacity; this is the simplest choice that has been adopted in [29]. Therefore, the network is converted into a linear programming problem for which the feasibility is shown to be equivalent to the existence of the desired weak combined transition.

The idea is to convert the flow network into the canonical *LP* problem and then to add the balancing constraints that force the "flow" to split according to transition probability distributions.

Definition 21 (cf. [30, Definition 1]). *Given a PA* $\mathcal{P}$, $\mathcal{R} \subseteq S \times S$, $\mu \in$ Disc(S), *and* $t \in S$, *for* $a \in \mathbf{E}$ *we define the* $t \overset{a}{\Longrightarrow}_C \diamond \mathcal{L}(\mathcal{R})\, \mu$ *LP problem associated to the network graph* $(V, E) = G(t, a, \mu, \mathcal{R})$ *as follows:*

$$\max \sum\nolimits_{(x,y) \in E} -f_{x,y}$$

under constraints

$$
\begin{aligned}
&f_{u,v} \geq 0 && \text{for each } (u,v) \in E \\
&f_{\triangle,t} = 1 && \\
&f_{v_{\mathcal{R}},\blacktriangledown} = \mu(v) && \text{for each } v \in S_{\mathcal{R}} \\
&\sum\nolimits_{u \in \{\, x \mid (x,v) \in E \,\}} f_{u,v} - \sum\nolimits_{u \in \{\, y \mid (v,y) \in E \,\}} f_{v,u} = 0 && \text{for each } v \in V \setminus \{\triangle, \blacktriangledown\} \\
&f_{v^{tr},v'} - \rho(v') f_{v,v^{tr}} = 0 && \text{for each } tr = v \overset{\tau}{\to} \rho \in \to \text{ and } v' \in \mathrm{Supp}(\rho) \\
&f_{v^{tr}_a,v'_a} - \rho(v') f_{v_a,v^{tr}_a} = 0 && \text{for each } tr = v \overset{\tau}{\to} \rho \in \to \text{ and } v' \in \mathrm{Supp}(\rho) \\
&f_{v^{tr}_a,v'_a} - \rho(v') f_{v,v^{tr}_a} = 0 && \text{for each } tr = v \overset{a}{\to} \rho \in \to \text{ and } v' \in \mathrm{Supp}(\rho)
\end{aligned}
$$

When a is τ, the *LP* problem $t \overset{\tau}{\Longrightarrow}_C \diamond \mathcal{L}(\mathcal{R})\, \mu$ associated to $G(t, \tau, \mu, \mathcal{R})$ is defined as above without the last two groups of constraints.

Since it is possible to solve a linear programming problem in polynomial time [53], so it is to find a feasible solution for $t \overset{a}{\Longrightarrow}_C \diamond \mathcal{L}(\mathcal{R})\, \mu$ (cf. [29, Theorem 7]), hence computing the weak probabilistic similarity and bisimilarity for probabilistic automata is polynomial as well. A comprehensive efficiency analysis about deciding weak probabilistic bisimulation on *PAs* is presented in [27].

6.3   The Algorithms for Mixed Time Models Relations

We do not present explicitly the algorithms for the simulations and bisimulations defined on *IMCs* and *MAs*: the former just makes use of the algorithms for *CTMCs* and graph visiting (as a depth first search) [28], while the latter just takes the *LP* problem for finding a weak transition in a *PA* as a blackbox [20,46].

7   Conclusion

In this survey we have presented several discrete and continuous time systems with external and/or internal nondeterminism and investigated the models of *CTMCs*, *DTMCs*, *PAs*, and *CTPAs*, and discussed *IMCs* and *MAs*. For these models, we have recalled the behavioral relations they are equipped with like simulations and bisimulations and we have described the corresponding decision algorithms. These procedures follow the standard refinement approach and they improve their complexity by using algorithms for optimization and flow network problems. We omitted some of the technical details but provided extensive references to the original literature.

Acknowledgements. The authors would like to thank Holger Hermanns for his invaluable assistance to motivate and improve the results of the paper. We would also like to thank the three anonymous reviewers for their many constructive comments and suggestions on the manuscript. This work has been supported by the DFG/NWO Bilateral Research Programme ROCKS, by the DFG as part of the SFB/TR 14 "Automatic Verification and Analysis of Complex Systems" (AVACS), and by the European Union Seventh Framework Programme under grant agreement no. 295261 (MEALS) and 318490 (SENSATION).

References

[1] Ahuja, R.K., Magnanti, T.J., Orlin, J.B.: Network Flows: Theory, Algorithms, and Applications. Prentice Hall (1993)

[2] Andova, S., Willemse, T.A.C.: Branching bisimulation for probabilistic systems: Characteristics and decidability. TCS 356(3), 325–355 (2006)

[3] Aziz, A., Sanwal, K., Singhal, V., Brayton, R.K.: Model-checking continuous-time Markov chains. ACM Transactions on Computational Logic 1(1), 162–170 (2000)

[4] Baier, C., Engelen, B., Majster-Cederbaum, M.: Deciding bisimilarity and similarity for probabilistic processes. J. Computer and Systems Science 60(1), 187–231 (2000)

[5] Baier, C., Haverkort, B.R., Hermanns, H., Katoen, J.P.: Model-checking algorithms for continuous-time Markov chains. IEEE Transactions on Software Engineering 29(6), 524–541 (2003)

[6] Baier, C., Hermanns, H.: Weak bisimulation for fully probabilistic processes. In: Grumberg, O. (ed.) CAV 1997. LNCS, vol. 1254, pp. 119–130. Springer, Heidelberg (1997)

[7] Baier, C., Hermanns, H., Katoen, J.P., Haverkort, B.R.: Efficient computation of time-bounded reachability probabilities in uniform continuous-time Markov decision processes. TCS 345(1), 2–26 (2005)

[8] Baier, C., Katoen, J.P.: Principles of Model Checking. The MIT Press (2008)

[9] Baier, C., Katoen, J.P., Hermanns, H., Wolf, V.: Comparative branching-time semantics for Markov chains. I&C 200(2), 149–214 (2005)

[10] Bellman, R.: A Markovian decision process. Indiana University Mathematics Journal 6, 679–684 (1957)

[11] Bertsekas, D.P.: Dynamic Programming and Optimal Control. Athena Scientific (2005)

[12] Bertsimas, D., Tsitsiklis, J.N.: Introduction to Linear Optimization. Athena Scientific (1997)

[13] Cattani, S., Segala, R.: Decision algorithms for probabilistic bisimulation. In: Brim, L., Jančar, P., Křetínský, M., Kučera, A. (eds.) CONCUR 2002. LNCS, vol. 2421, pp. 371–385. Springer, Heidelberg (2002)

[14] Clarke, E.M., Grumberg, O., Long, D.E.: Model checking and abstraction. ACM Transactions on Programming Languages and Systems 16(5), 1512–1542 (1994)

[15] Crafa, S., Ranzato, F.: Probabilistic bisimulation and simulation algorithms by abstract interpretation. In: Aceto, L., Henzinger, M., Sgall, J. (eds.) ICALP 2011, Part II. LNCS, vol. 6756, pp. 295–306. Springer, Heidelberg (2011)

[16] Deng, Y.: Axiomatisations and Types for Probabilistic and Mobile Processes. Ph.D. thesis, École des Mines de Paris (2005)

[17] Deng, Y., Hennessy, M.: On the semantics of Markov automata. I&C 222, 139–168 (2012)

[18] Desharnais, J.: Labelled Markov Processes. Ph.D. thesis, McGill University (1999)

[19] Eisentraut, C., Hermanns, H., Katoen, J.-P., Zhang, L.: A semantics for every GSPN. In: Colom, J.-M., Desel, J. (eds.) PETRI NETS 2013. LNCS, vol. 7927, pp. 90–109. Springer, Heidelberg (2013)

[20] Eisentraut, C., Hermanns, H., Krämer, J., Turrini, A., Zhang, L.: Deciding bisimilarities on distributions. In: Joshi, K., Siegle, M., Stoelinga, M., D'Argenio, P.R. (eds.) QEST 2013. LNCS, vol. 8054, pp. 72–88. Springer, Heidelberg (2013)

[21] Eisentraut, C., Hermanns, H., Schuster, J., Turrini, A., Zhang, L.: The quest for minimal quotients for probabilistic automata. In: Piterman, N., Smolka, S.A. (eds.) TACAS 2013. LNCS, vol. 7795, pp. 16–31. Springer, Heidelberg (2013)

[22] Eisentraut, C., Hermanns, H., Zhang, L.: Concurrency and composition in a stochastic world. In: Gastin, P., Laroussinie, F. (eds.) CONCUR 2010. LNCS, vol. 6269, pp. 21–39. Springer, Heidelberg (2010)

[23] Eisentraut, C., Hermanns, H., Zhang, L.: On probabilistic automata in continuous time. In: LICS, pp. 342–351 (2010)

[24] Gallo, G., Grigoriadis, M.D., Tarjan, R.E.: A fast parametric maximum flow algorithm and applications. SIAM J. Comp. 18(1), 30–55 (1989)

[25] Goldberg, A.V., Tarjan, R.E.: A new approach to the maximum-flow problem. J. ACM 35(4), 921–940 (1988)

[26] Hansson, H., Jonsson, B.: A logic for reasoning about time and reliability. Formal Aspects of Computing 6(5), 512–535 (1994)

[27] Hashemi, V., Hermanns, H., Turrini, A.: On the efficiency of deciding probabilistic automata weak bisimulation. ECEASST 66 (2013)

[28] Hermanns, H.: Interactive Markov Chains. LNCS, vol. 2428. Springer, Heidelberg (2002)

[29] Hermanns, H., Turrini, A.: Deciding probabilistic automata weak bisimulation in polynomial time. In: FSTTCS, pp. 435–447 (2012)

[30] Hermanns, H., Turrini, A.: Cost preserving bisimulations for probabilistic automata. In: D'Argenio, P.R., Melgratti, H. (eds.) CONCUR 2013 – Concurrency Theory. LNCS, vol. 8052, pp. 349–363. Springer, Heidelberg (2013)

[31] Howard, R.A.: Dynamic Programming and Markov Processes. John Wiley and Sons, Inc. (1960)

[32] Howard, R.A.: Dynamic Probabilistic Systems: Semi-Markov and Decision Processes, vol. II. Dover Publications (2007)

[33] Jansen, D.N., Song, L., Zhang, L.: Revisiting weak simulation for substochastic Markov chains. In: Joshi, K., Siegle, M., Stoelinga, M., D'Argenio, P.R. (eds.) QEST 2013. LNCS, vol. 8054, pp. 209–224. Springer, Heidelberg (2013)

[34] Jonsson, B., Larsen, K.G.: Specification and refinement of probabilistic processes. In: LICS, pp. 266–277 (1991)

[35] Kanellakis, P.C., Smolka, S.A.: CCS expressions, finite state processes, and three problems of equivalence. I&C 86(1), 43–68 (1990)

[36] Katoen, J.-P., Kemna, T., Zapreev, I., Jansen, D.N.: Bisimulation minimisation mostly speeds up probabilistic model checking. In: Grumberg, O., Huth, M. (eds.) TACAS 2007. LNCS, vol. 4424, pp. 87–101. Springer, Heidelberg (2007)

[37] Knast, R.: Continuous-time probabilistic automata. Information and Control 15(4), 335–352 (1969)

[38] Larsen, K.G., Skou, A.: Bisimulation through probabilistic testing (preliminary report). In: POPL, pp. 344–352 (1989)

[39] Milner, R.: Communication and Concurrency. Prentice-Hall International, Englewood Cleiffs (1989)

[40] Milner, R.: Communicating and Mobile Systems: the π-calculus. Cambridge University Press (1999)
[41] Paige, R., Tarjan, R.E.: Three partition refinement algorithms. SIAM J. on Computing 16(6), 973–989 (1987)
[42] Peterson, M.: An Introduction to Decision Theory. Cambridge University Press (2009)
[43] Philippou, A., Lee, I., Sokolsky, O.: Weak bisimulation for probabilistic systems. In: Palamidessi, C. (ed.) CONCUR 2000. LNCS, vol. 1877, pp. 334–349. Springer, Heidelberg (2000)
[44] Puterman, M.L.: Markov Decision Processes: Discrete Stochastic Dynamic Programming. Wiley Series in Probability and Statistics, vol. (594). John Wiley & Sons, Inc. (2005)
[45] Sack, J., Zhang, L.: A general framework for probabilistic characterizing formulae. In: Kuncak, V., Rybalchenko, A. (eds.) VMCAI 2012. LNCS, vol. 7148, pp. 396–411. Springer, Heidelberg (2012)
[46] Schuster, J., Siegle, M.: Markov automata: Deciding weak bisimulation by means of "non-naïvely" vanishing states. I&C (to appear, 2014), http://dx.doi.org/10.1016/j.ic.2014.02.001
[47] Segala, R.: Modeling and Verification of Randomized Distributed Real-Time Systems. Ph.D. thesis, MIT (1995)
[48] Segala, R.: Probability and nondeterminism in operational models of concurrency. In: Baier, C., Hermanns, H. (eds.) CONCUR 2006. LNCS, vol. 4137, pp. 64–78. Springer, Heidelberg (2006)
[49] Segala, R., Lynch, N.: Probabilistic simulations for probabilistic processes. In: Jonsson, B., Parrow, J. (eds.) CONCUR 1994. LNCS, vol. 836, pp. 481–496. Springer, Heidelberg (1994)
[50] Segala, R., Lynch, N.A.: Probabilistic simulations for probabilistic processes. Nordic J. Computing 2(2), 250–273 (1995)
[51] Segala, R., Turrini, A.: Comparative analysis of bisimulation relations on alternating and non-alternating probabilistic models. In: QEST, pp. 44–53 (2005)
[52] Stewart, W.J.: Introduction to the Numerical Solution of Markov Chains. Princeton University Press (1994)
[53] Todd, M.J.: The many facets of linear programming. Mathematical Programming 91(3), 417–436 (2002)
[54] Wolovick, N., Johr, S.: A characterization of meaningful schedulers for continuous-time Markov decision processes. In: Asarin, E., Bouyer, P. (eds.) FORMATS 2006. LNCS, vol. 4202, pp. 352–367. Springer, Heidelberg (2006)
[55] Zhang, L.: Decision Algorithm for Probabilistic Simulations. Ph.D. thesis, Saarland University (2008)
[56] Zhang, L., Hermanns, H.: Deciding simulations on probabilistic automata. In: Namjoshi, K.S., Yoneda, T., Higashino, T., Okamura, Y. (eds.) ATVA 2007. LNCS, vol. 4762, pp. 207–222. Springer, Heidelberg (2007)
[57] Zhang, L., Hermanns, H., Eisenbrand, F., Jansen, D.N.: Flow faster: Efficient decision algorithms for probabilistic simulations. In: Grumberg, O., Huth, M. (eds.) TACAS 2007. LNCS, vol. 4424, pp. 155–169. Springer, Heidelberg (2007)

Markov Reward Models and Markov Decision Processes in Discrete and Continuous Time: Performance Evaluation and Optimization

Alexander Gouberman and Markus Siegle

Department of Computer Science
Universität der Bundeswehr, München, Germany
{alexander.gouberman,markus.siegle}@unibw.de

Abstract. State-based systems with discrete or continuous time are often modelled with the help of Markov chains. In order to specify performance measures for such systems, one can define a reward structure over the Markov chain, leading to the Markov Reward Model (MRM) formalism. Typical examples of performance measures that can be defined in this way are time-based measures (e.g. mean time to failure), average energy consumption, monetary cost (e.g. for repair, maintenance) or even combinations of such measures. These measures can also be regarded as target objects for system optimization. For that reason, an MRM can be enhanced with an additional control structure, leading to the formalism of Markov Decision Processes (MDP).

In this tutorial, we first introduce the MRM formalism with different types of reward structures and explain how these can be combined to a performance measure for the system model. We provide running examples which show how some of the above mentioned performance measures can be employed. Building on this, we extend to the MDP formalism and introduce the concept of a policy. The global optimization task (over the huge policy space) can be reduced to a greedy local optimization by exploiting the non-linear Bellman equations. We review several dynamic programming algorithms which can be used in order to solve the Bellman equations exactly. Moreover, we consider Markovian models in discrete and continuous time and study value-preserving transformations between them. We accompany the technical sections by applying the presented optimization algorithms to the example performance models.

1 Introduction

State-based systems with stochastic behavior and discrete or continuous time are often modelled with the help of Markov chains. Their efficient evaluation and optimization is an important research topic. There is a wide range of application areas for such kind of models, coming especially from the field of Operations Research, e.g. economics [4,23,31] and health care [14,36], Artificial Intelligence, e.g. robotics, planning and automated control [11,40] and Computer Science [2,6, 12,13,21,25,34]. In order to specify performance and dependability measures for

A. Remke and M. Stoelinga (Eds.): ROCKS Autumn School 2012, LNCS 8453, pp. 156–241, 2014.

such systems, one can define a reward structure over the Markov chain, leading to the Markov Reward Model (MRM) formalism. Typical examples of performance measures that can be defined in this way are time-based measures (e.g. mean time to failure), average energy consumption, monetary cost (e.g. for repair, maintenance) or even combinations of such measures. These measures can also be regarded as target objects for system optimization. For that reason, an MRM can be enhanced with an additional control structure, leading to the formalism of Markov Decision Processes (MDP) [20]. There is a huge number of optimization algorithms for MDPs in the literature, based on dynamic programming and linear programming [7, 8, 17, 33] – all of them rely on the Bellman optimality principle [5].

In many applications, the optimization criteria are a trade-off between several competing goals, e.g. minimization of running cost and maximization of profit at the same time. For sure, in these kinds of trade-off models, it is important to establish an optimal policy which in most cases is not intuitive. However, there are also examples of target functions with no trade-off character (e.g. pure lifetime maximization [16]) which can also lead to counterintuitive optimal policies. Therefore, using MDPs for optimization of stochastic systems should not be neglected, even if a heuristically established policy seems to be optimal.

In order to build up the necessary theoretical background in this introductory tutorial, we first introduce in Sect. 2 the discrete-time MRM formalism with finite state space and define different types of reward measures typically used in performance evaluation, such as total reward, discounted reward and average reward. In contrast to the majority of literature, we follow a different approach to deduce the definition of the discounted reward through a special memoryless horizon-expected reward. We discuss properties of these measures and create a fundamental link between them, which is based on the Laurent series expansion of the discounted reward (and involves the deviation matrix for Markov chains). We derive systems of linear equations used for evaluation of the reward measures and provide a running example based on a simple queueing model, in order to show how these performance measures can be employed.

Building on this, in Sect. 3 we introduce the MDP formalism and the concept of a policy. The global optimization task (over the huge policy space) can be reduced to a greedy local optimization by exploiting the set of non-linear Bellman equations. We review some of the basic dynamic programming algorithms (policy iteration and value iteration) which can be used in order to solve the Bellman equations. As a running example, we extend the queueing model from the preceding section with a control structure and compute optimal policies with respect to several performance measures.

From Sect. 4 on we switch to the continuous-time setting and present the CT-MRM formalism with a reward structure consisting of the following two different types: impulse rewards which measure discrete events (i.e. transitions) and rate rewards which measure continuous time activities. In analogy to the discrete-time case, we discuss the performance measures given by the total, horizon-expected, discounted and average reward measures. In order to be able to evaluate these measures, we present model transformations that can be used for discretizing the CTMRM to a DTMRM by embedding or uniformization [24]. As a third transformation, we define the continuization which integrates the discrete impulse rewards into a continuous-time rate, such that the whole CTMRM possesses only rate rewards. We further study the soundness of these transformations, i.e. the preservation of the aforementioned performance measures. Similar to DTMRMs, the discounted reward measure can be expanded into a Laurent series which once again shows the intrinsic structure between the measures. We accompany Sect. 4 with a small wireless sensor network model.

In Sect. 5 we finally are able to define the continuous-time MDP formalism which extends CTMRMs with a control structure, as for discrete-time MDPs. With all the knowledge collected in the preceding sections, the optimization algorithms for CTMDPs can be performed by MDP algorithms through time discretization. For evaluation of the average reward measure, we reveal a slightly different version of policy iteration [17], which can be used for the continuization transformation. As an example, we define a bridge circuit CTMDP model and optimize typical time-based dependability measures like mean time to failure and the availability of the system.

Figure 1.1 shows some dependencies between the sections in this tutorial in form of a roadmap. One can read the tutorial in a linear fashion from beginning to end, but if one wants to focus on specific topics it is also possible to skip certain sections. For instance, readers may wish to concentrate on Markov Reward Models in either discrete or continuous time (Sects. 2 and 4), while neglecting the optimization aspect. Alternatively, they may be interested in the discrete time setting only, ignoring the continuous time case, which would mean to read only Sects. 2 (on DTMRMs) and 3 (on MDPs). Furthermore, if the reader is interested in the discounted reward measure, then he may skip the average reward measure in every subsection.

Readers are assumed to be familiar with basic calculus, linear algebra and probability theory which should suffice to follow most explanations and derivations. For those wishing to gain insight into the deeper mathematical structure of MRMs and MDPs, additional theorems and proofs are provided, some of which require more involved concepts such as measure theory, Fubini's theorem or Laurent series. For improved readability, long proofs are moved to the Appendix in Sect. A.

The material presented in this tutorial paper has been covered previously by several authors, notably in the books of Puterman [33], Bertsekas [7, 8], Bertsekas/Tsitsiklis [10] and Guo/Hernandez-Lerma [17]. However, the present paper offers its own new points of view: Apart from dealing also with non-standard measures, such as horizon-expected reward measures and the unified treatment of rate rewards and impulse rewards through the concept of continuization, the paper puts an emphasis on transformations between the different model classes by embedding and uniformization. The paper ultimately anwers the interesting question of which measures are preserved by those transformations.

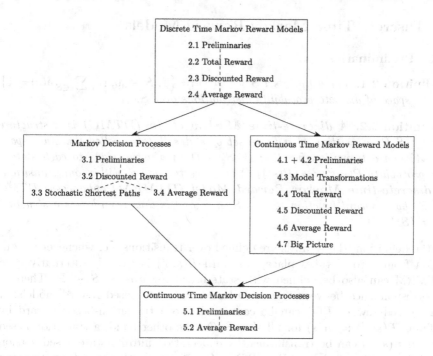

Fig. 1.1. Roadmap of the tutorial with dependencies between the sections and subsections. As an example, Sect. 3.3 on stochastic shortest paths, needs Sect. 3.2 and therefore also Sect. 2.3 but not Sect. 2.4. The Big Picture in Sect. 4.7 is one of the main goals of this tutorial and also necessary for the section on CTMDPs.

Symbols

B^A	set of functions $f : A \to B$
2^A	power set of A
$\mathcal{D}(S)$	set of probability distributions over S
P	probability measure, probability transition matrix
P_s	probability measure assuming initial state s
$\mathbb{E}_s$	expectation from initial state s
$\mathbb{1}_A(x)$	indicator function, $\mathbb{1}_A(x) = 1$ if $x \in A$ and 0 otherwise
$\delta_{s,s'}$	Kronecker-δ, $\delta_{s,s'} = 1$ if $s = s'$ and 0 otherwise
I	identity matrix
$\mathbf{1}$	column vector consisting of ones in each entry
$\bar{p}$	$\bar{p} = 1 - p$
$\ker(A)$	kernel of a matrix A
γ	discount factor, $0 < \gamma < 1$
α	discount rate, $\alpha > 0$
V	value function
g	average reward
h	bias

2   Discrete Time Markov Reward Models

2.1   Preliminaries

Definition 2.1. *For a finite set S let $\mathcal{D}(S) := \left\{ \delta \colon S \to [0,1] \mid \sum_{s \in S} \delta(s) = 1 \right\}$ be the space of discrete probability distributions over S.*

Definition 2.2. *A **discrete-time Markov chain (DTMC)** is a structure $\mathcal{M} = (S, P)$ consisting of a finite set of states S (also called the state space of $\mathcal{M}$) and a transition function $P \colon S \to \mathcal{D}(S)$ which assigns to each state s the probability $P(s, s') := (P(s))(s')$ to move to state s' within one transition. A **discrete-time Markov Reward Model (DTMRM)** enhances a DTMC (S, P) by a reward function $R \colon S \times S \to \mathbb{R}$ and is thus regarded as a structure $\mathcal{M} = (S, P, R)$.*

In the definition, the rewards are defined over transitions, i.e. whenever a transition from s to s' takes place, a reward $R(s, s')$ is gained. Alternatively, a DTMRM can also be defined with state-based rewards $R \colon S \to \mathbb{R}$. There is a correspondence between transition-based and state-based reward models: A state-based reward $R(s)$ can be converted into a transition-based reward by defining $R(s, s') := R(s)$ for all $s' \in S$. On the other hand, a transition-based reward $R(s, s')$ can be transformed by expectation into its state-based version by defining $R(s) := \sum_{s' \in S} R(s, s') P(s, s')$. Of course, this transformation can not be inverted, but as we will see, the reward measures that we consider do not differ. Note that for state-based rewards there are two canonical but totally different possibilities to define the point in time, when such a reward can be gained: either when a transition into the state or out of the state is performed. This corresponds to the difference in the point of view for "arrivals" and "departures" of jobs in queueing systems. When working with transition-based rewards $R(s, s')$ as in Definition 2.2, then such a confusion does not occur since $R(s, s')$ is gained in state s after transition to s' and thus its expected value $R(s)$ corresponds to the "departure" point of view. For our purposes we will mix both representations and if we write R for a reward then it is assumed to be interpreted in a context-dependent way as either the state-based or the transition-based version.

Each bijective representation

$$\varphi \colon S \to \{1, 2, \dots, n\}, \ n := |S| \tag{2.1}$$

of the state space as natural numbered indices allows to regard the rewards and the transition probabilities as real-valued vectors in $\mathbb{R}^n$ respectively matrices in $\mathbb{R}^{n \times n}$. We indirectly take such a representation φ, especially when we talk about P and R in vector notation. In this case the transition function $P \colon S \to \mathcal{D}(S)$ can be regarded as a **stochastic** matrix, i.e. $P\mathbf{1} = \mathbf{1}$, where $\mathbf{1} = (1, \dots, 1)^T$ is the column vector consisting of all ones.

Example 2.1 (Queueing system). As a running example, consider a system consisting of a queue of capacity k and a service unit which can be idle, busy or

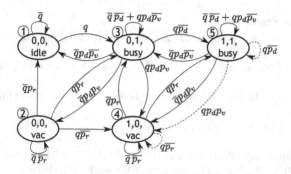

Fig. 2.1. Queueing model with queue capacity $k = 1$ and a service unit which can be idle, busy or on vacation. State $(0, 1, \text{busy})$ represents 0 jobs in the queue, 1 job in service and server is busy. The dashed transitions represent the event that an incoming job is discarded. The parameter values are $q = 0.25$, $p_d = 0.5$, $p_v = 0.1$ and $p_r = 0.25$. Overlined probabilities are defined as $\overline{p} := 1 - p$.

on vacation (Fig. 2.1) [38]. A job arrives with probability $q = 0.25$ and gets enqueued if the queue is not full. If the server is idle and there is some job waiting in the queue, the server immediately gets busy and processes the job. At the end of each unit of service time, the server is done with the job with probability $p_d = 0.5$ and can either go idle (and possibly getting the next job) or the server needs vacation with probability $p_v = 0.1$. From the vacation mode the server returns with probability $p_r = 0.25$. Figure 2.1 shows the model for the case of queue capacity $k = 1$, where transitions are split into regular transitions (solid lines) and transitions indicating that an incoming job gets discarded (dashed lines). The reward model is as follows: For each accomplished service a reward of $R_{\text{acc}} = \$100$ is gained, regardless of whether the server moves to idle or to vacation. However, the loss of a job during arrival causes a cost of $C_{\text{loss}} = -\$1000$. Therefore we consider the following state-based reward structures:

$$
R_{\text{profit}} = \begin{pmatrix} 0 \\ 0 \\ R_{\text{acc}}p_d \\ 0 \\ R_{\text{acc}}p_d \end{pmatrix}, \ R_{\text{cost}} = \begin{pmatrix} 0 \\ 0 \\ 0 \\ C_{\text{loss}}q\overline{p_r} \\ C_{\text{loss}}\left(q\overline{p_d} + qp_dp_v\right) \end{pmatrix}, \ R_{\text{total}} = R_{\text{profit}} + R_{\text{cost}},
$$

$$(2.2)$$

where $\overline{p} := 1 - p$ for some probability p. $\qquad\square$

There are several ways how the rewards gained for each taken transition (respectively for a visited state) can contribute to certain measures of interest. In typical applications of performance evaluation, rewards are accumulated over some time period or a kind of averaging over rewards is established. In order to be able to define these reward measures formally, we need to provide some basic knowledge on the stochastic state process induced by the DTMC part of a DTMRM.

2.1.1 Sample Space

For a DTMC $\mathcal{M} = (S, P)$ define Ω as the set of infinite paths, i.e.

$$\Omega := \big\{(s_0, s_1, s_2, \dots) \in S^{\mathbb{N}} \mid P(s_{i-1}, s_i) > 0 \text{ for all } i \geq 1\big\} \qquad (2.3)$$

and let $\mathcal{B}(\Omega)$ be the Borel σ-algebra over Ω generated by the cylinder sets

$$C(s_0, s_1, \dots, s_N) := \{\omega \in \Omega \mid \omega_i = s_i \; \forall i \leq N\}.$$

Each $s \in S$ induces a probability space $(\Omega, \mathcal{B}(\Omega), P_s)$ with a probability distribution $P_s \colon \mathcal{B}(\Omega) \to \mathbb{R}$ over paths, such that for each cylinder set $C(s_0, \dots, s_N) \in \mathcal{B}(\Omega)$

$$P_s\left(C(s_0, \dots, s_N)\right) = \delta_{s_0, s} P(s_0, s_1) P(s_1, s_2) \dots P(s_{N-1}, s_N),$$

where δ is the Kronecker-δ, i.e. $\delta_{s, s'} = 1$ if $s = s'$ and 0 otherwise.

Definition 2.3. *The DTMC $\mathcal{M}$ induces the stochastic **state process** $(X_n)_{n \in \mathbb{N}}$ over Ω which is a sequence of S-valued random variables such that $X_n(\omega) := s_n$ for $\omega = (s_0, s_1, \dots) \in \Omega$.*

Note that

$$P_s(X_n = s') = \sum_{s_1, \dots, s_{n-1}} P(s, s_1) P(s_1, s_2) \dots P(s_{n-2}, s_{n-1}) P(s_{n-1}, s') = P^n(s, s'),$$

where $\sum_{s_1, \dots, s_{n-1}}$ denotes summation over all tuples $(s_1, \dots, s_{n-1}) \in S^{n-1}$. The process X_n also fulfills the **Markov property** (or **memorylessness**): For all $s, s', s_0, s_1, \dots s_{n-1} \in S$ it holds that

$$P_{s_0}(X_{n+1} = s' \mid X_1 = s_1, \dots, X_{n-1} = s_{n-1}, X_n = s) = P(s, s'), \qquad (2.4)$$

i.e. if the process is in state s at the current point in time n, then the probability to be in state s' after the next transition does not depend on the history of the process consisting of the initial state $X_0 = s_0$ and the traversed states $X_1 = s_1, \dots, X_{n-1} = s_{n-1}$ up to time $n - 1$.

We denote the expectation operator over $(\Omega, \mathcal{B}(\Omega), P_s)$ as $\mathbb{E}_s$. For a function $f \colon S^{n+1} \to \mathbb{R}$ it holds

$$\mathbb{E}_s\left[f(X_0, X_1, \dots, X_n)\right] = \sum_{s_1, \dots, s_n} f(s, s_1 \dots, s_n) P(s, s_1) P(s_1, s_2) \dots P(s_{n-1}, s_n).$$

In vector representation we often write $\mathbb{E}[Y] := (\mathbb{E}_s[Y])_{s \in S}$ for the vector consisting of expectations of a real-valued random variable Y.

2.1.2 State Classification

In the following, we briefly outline the usual taxonomy regarding the classification of states for a discrete-time Markov chain $\mathcal{M} = (S, P)$. The state process X_n induced by $\mathcal{M}$ allows to classify the states S with respect to their recurrence and reachability behavior. If $X_0 = s$ is the initial state then the random variable $M_s := \inf\{n \geq 1 \mid X_n = s\} \in \mathbb{N} \cup \{\infty\}$ is the first point in time when the process X_n returns to s. If along a path $\omega \in \Omega$ the process never returns to s then $M_s(\omega) = \inf \emptyset = \infty$. If there is a positive probability to never come back to s, i.e. $P_s(M_s = \infty) > 0$ then the state s is called **transient**. Otherwise, if $P_s(M_s < \infty) = 1$ then s is **recurrent**. We denote S^t as the set of transient states and S^r as the set of recurrent states. A state s' is **reachable** from s (denoted by $s \to s'$), if there exists $n \in \mathbb{N}$ with $P_s(X_n = s') > 0$. The notion of reachability induces the (communcation) equivalence relation

$$s \leftrightarrow s' \quad \Leftrightarrow \quad s \to s' \text{ and } s' \to s.$$

This relation further partitions the set of recurrent states S^r into the equivalence classes S_i^r, $i = 1, \ldots, k$ such that the whole state space S can be written as the disjoint union $S = \bigcup_{i=1}^k S_i^r \cup S^t$. Each of these equivalence classes S_i^r is called a **closed recurrent class**, since (by the communication relation) for every $s \in S_i^r$ there are no transitions out of this class, i.e. $P(s, s') = 0$ for all $s' \in S \backslash S_i^r$. For this reason each S_i^r is a minimal closed subset of S, i.e. there is no proper nonempty subset of S_i^r which is closed. In case a closed recurrent class consists only of one state s, then s is called **absorbing**. The DTMC $\mathcal{M}$ is **unichain** if there is only one recurrent class ($k = 1$) and if in addition $S^t = \emptyset$ then $\mathcal{M}$ is **irreducible**. A DTMC that is not unichain will be called **multichain**. The queueing system in Example 2.1 is irreducible, since every state is reachable from every other state. For discrete-time Markov chains there are some peculiarities regarding the long-run behavior of the Markov chain as $n \to \infty$. If $X_0 = s$ is the initial state and the limit $\rho_s(s') := \lim_{n \to \infty} P_s(X_n = s')$ exists for all s', then $\rho_s \in \mathcal{D}(S)$ is called the **limiting distribution** from s. In general this limit does not need to exist, since the sequence $P_s(X_n = s') = P^n(s, s')$ might have oscillations between distinct accumulation points. This fact is related to the periodicity of a state: From state s the points in time of possible returns to s are given by the set $R_s := \{n \geq 1 \mid P_s(X_n = s) > 0\}$. If all $n \in R_s$ are multiples of some natural number $d \geq 2$ (i.e. $R_s \subseteq \{kd \mid k \in \mathbb{N}\}$) then the state s is called **periodic** and the **periodicity** of s is the largest such integer d. Otherwise s is called **aperiodic** and the periodicity of s is set to 1. The periodicity is a class property and means that for every closed recurrent class S_i^r and for all $s, s' \in S_i^r$ the periodicity of s and s' is the same. A Markov chain which is irreducible and aperiodic is often called **ergodic** in the literature. As an example, a two-state Markov chain with transition probability matrix

$$P = \begin{pmatrix} 0 & 1 \\ 1 & 0 \end{pmatrix}$$

is irreducible but not ergodic, since it is periodic with periodicity 2.

One can show that a recurrent state s in a DTMC with finite state space is aperiodic if and only if for all $s' \in S$ the sequence $P^n(s, s')$ converges. Therefore, the limiting distribution ρ_s exists (for s recurrent) if and only if s is aperiodic. In this case $\rho_s(s') = \sum_t \rho_s(t) P(t, s')$ for all s', which is written in vector notation by $\rho_s = \rho_s P$. This equation is often interpreted as the invariance (or stationarity) condition: If $\rho_s(s')$ is the probability to find the system in state s' at some point in time, then this probability remains unchanged after the system performs a transition. In general, there can be several distributions $\rho \in \mathcal{D}(S)$ with the invariance property $\rho = \rho P$ and any such distribution ρ is called a **stationary distribution**. It holds that the set of all stationary distributions forms a simplex in $\mathbb{R}^S$ and the number of vertices of this simplex is exactly the number of recurrent classes k. Therefore, in a unichain model $\mathcal{M}$ there is only one stationary distribution ρ and if $\mathcal{M}$ is irreducible then $\rho(s) > 0$ for all $s \in S$. Since a limiting distribution is stationary it further holds that if $\mathcal{M}$ is unichain and aperiodic (or even ergodic), then for all initial states s the limiting distribution ρ_s exists and $\rho_s = \rho$ is the unique stationary distribution and thus independent of s.

We will draw on the stationary distributions (and also the periodic behavior) of a Markov chain in Sect. 2.4, where we will outline the average reward analysis. In order to make the intuition on the average reward clear, we will also work with the splitting $S = \bigcup_{i=1}^{k} S_i^r \cup S^t$ into closed recurrent classes and transient states. The computation of this splitting can be performed by the Fox-Landi state classification algorithm [15]. It finds a representation φ of S (see (2.1)) such that P can be written as

$$P = \begin{pmatrix} P_1 & 0 & 0 & \dots & 0 & 0 \\ 0 & P_2 & 0 & \dots & 0 & 0 \\ \vdots & \vdots & \vdots & \ddots & \vdots & 0 \\ 0 & 0 & 0 & \dots & P_k & 0 \\ \widetilde{P}_1 & \widetilde{P}_2 & \widetilde{P}_3 & \dots & \widetilde{P}_k & \widetilde{P}_{k+1} \end{pmatrix} \qquad (2.5)$$

where $P_i \in \mathbb{R}^{r_i \times r_i}$, $\widetilde{P}_i \in \mathbb{R}^{t \times r_i}$, $\widetilde{P}_{k+1} \in \mathbb{R}^{t \times t}$ with $r_i := |S_i^r|$ for $i = 1, \dots, k$ and $t := |S^t|$. The matrix P_i represents the transition probabilities within the i-th recurrent class S_i^r and $\widetilde{P}_i$ the transition probabilities from transient states S^t into S_i^r if $i = 1, \dots, k$, respectively transitions within S^t for $i = k + 1$. Every closed recurrent class S_i^r can be seen as a DTMC $\mathcal{M}^i = (S_i^r, P_i)$ by omitting incoming transitions from transient states. It holds that $\mathcal{M}^i$ is irreducible and thus has a unique stationary distribution $\rho_i \in \mathcal{D}(S^{r_i})$ with $\rho_i(s) > 0$ for all $s \in S^{r_i}$. This distribution ρ_i can be extended to a distribution $\rho_i \in \mathcal{D}(S)$ on S by setting $\rho_i(s) := 0$ for all $s \in S \setminus S_i^r$. Note that ρ_i is also stationary on $\mathcal{M}$. Since transient states S^t of $\mathcal{M}$ are left forever with probability 1 (into the recurrent classes), every stationary distribution $\rho \in \mathcal{D}(S)$ fulfills that $\rho(s) = 0$ for all $s \in S^t$. Thus, an arbitrary stationary distribution $\rho \in \mathcal{D}(S)$ is a convex combination of all the ρ_i, i.e. $\rho(s) = \sum_{i=1}^{k} a_i \rho_i(s)$ with $a_i \geq 0$ and $\sum_{i=1}^{k} a_i = 1$. (This forms the k-dimensional simplex with vertices ρ_i as mentioned above.)

2.1.3   Reward Measure

We now consider a DTMRM $\mathcal{M} = (S, P, R)$ and want to describe in the following sections several ways to accumulate the rewards $R(s, s')$ along paths $\omega = (s_0, s_1, \dots) \in \Omega$. As an example, for a fixed $N \in \mathbb{N} \cup \{\infty\}$ (also called the *horizon length*) we can accumulate the rewards for the first N transitions by simple summation: the reward gained for the i-th transition is $R(s_{i-1}, s_i)$ and is summed up to $\sum_{i=1}^{N} R(s_{i-1}, s_i)$, which is regarded as the value of the path ω for the first N transitions. The following definition introduces the notion of a value for state-based models, with which we will be concerned in this tutorial.

Definition 2.4. *Consider a state-based model $\mathcal{M}$ with state space S and a real vector space $\mathcal{V}$. A **reward measure** $\mathcal{R}$ is an evaluation of the model $\mathcal{M}$ that maps $\mathcal{M}$ with an optional set of parameters to the **value** $V \in \mathcal{V}$ of the model. If $\mathcal{V}$ is a vector space of functions over S, i.e. $\mathcal{V} = \mathbb{R}^S = \{V : S \to \mathbb{R}\}$, then a value $V \in \mathcal{V}$ is also called a **value function** of the model.*

Note that we consider in this definition an arbitrary state-based model, which can have discrete time (e.g. DTMRM or MDP, cf. Sect. 3) or continuous time (CTMRM or CTMDP, cf. Sects. 4 and 5). We will mainly consider vector spaces $\mathcal{V}$ which consist of real-valued functions. Beside value functions $V \in \mathbb{R}^S$ which map every state $s \in S$ to a real value $V(s) \in \mathbb{R}$, we will also consider value functions that are time-dependent. For example, if T denotes a set of time values then $\mathcal{V} = \mathbb{R}^{S \times T}$ consists of value functions $V : S \times T \to \mathbb{R}$ such that $V(s, t) \in \mathbb{R}$ is the real value of state $s \in S$ at the point in time $t \in T$. In typical applications T is a discrete set for discrete-time models (e.g. $T = \mathbb{N}$ or $T = \{0, 1, \ldots, N\}$), or T is an interval for continuous-time models (e.g. $T = [0, \infty)$ or $T = [0, T_{\max}]$). The difference between the notion of a reward measure $\mathcal{R}$ and its value function V is that a reward measure can be seen as a measure type which needs additional parameters in order to be able to formally define its value function V. Examples for such parameters are the horizon length N, a discount factor γ (in Sect. 2.3) or a discount rate α (in Sect. 4.5). If clear from the context, we use the notions reward measure and value (function) interchangeably.

2.2   Total Reward Measure

We now define the finite-horizon and infinite-horizon total reward measures which formalize the accumulation procedure along paths by summation as mentioned in the motivation of Definition 2.4. The finite-horizon reward measure is used as a basis upon which all the following reward measures will be defined.

Definition 2.5. *Let $\mathcal{M}$ be a DTMRM with state process $(X_n)_{n \in \mathbb{N}}$ and $N < \infty$ a fixed finite horizon length. We define the **finite-horizon total value function** $V_N : S \to \mathbb{R}$ by*

$$V_N(s) := \mathbb{E}_s \left[\sum_{i=1}^{N} R(X_{i-1}, X_i) \right]. \tag{2.6}$$

*If for all states s the sequence $\mathbb{E}_s \left[\sum_{i=1}^{N} |R(X_{i-1}, X_i)| \right]$ converges with $N \to \infty$, we define the **(infinite-horizon) total value function** as*

$$V_\infty(s) := \lim_{N \to \infty} V_N(s).$$

In general $V_N(s)$ does not need to converge as $N \to \infty$. For example, if all rewards for every recurrent state are strictly positive, then accumulation of positive values diverges to ∞. Even worse, if the rewards have different signs then their accumulation can also oscillate. In order not to be concerned with such oscillations, we impose as a stronger condition the absolute convergence for the infinite-horizon case as in the definition.

As next we want to provide a method which helps to evaluate the total reward measure for the finite and infinite horizon cases. The proof of the following theorem can be found in the Appendix (page 234).

Theorem 2.1 (Evaluation of the Total Reward Measure).

(i) The finite-horizon total value $V_N(s)$ can be computed iteratively through

$$V_N(s) = R(s) + \sum_{s' \in S} P(s, s')V_{N-1}(s'),$$

where $V_0(s) := 0$ for all $s \in S$.

(ii) If the infinite-horizon total value $V_\infty(s)$ exists, then it solves the system of linear equations

$$V_\infty(s) = R(s) + \sum_{s' \in S} P(s, s')V_\infty(s'). \tag{2.7}$$

We formulate the evaluation of the total value function in vector notation:

$$V_N = R + PV_{N-1} = \sum_{i=1}^{N} P^{i-1}R. \tag{2.8}$$

For the infinite-horizon total value function it holds

$$V_\infty = R + PV_\infty \quad \text{respectively} \quad (I - P)V_\infty = R. \tag{2.9}$$

Note that Theorem 2.1 states that if V_∞ exists, then it solves (2.9). On the other hand the system of equations $(I - P)X = R$ with the variable X may have several solutions, since P is stochastic and thus the rank of $I - P$ is not full. The next proposition shows a necessary and sufficient condition for the existence of V_∞ in terms of the reward function R. Furthermore, it follows that if V_∞ exists then $V_\infty(s) = 0$ on all recurrent states s and V_∞ is also the unique solution to $(I - P)X = R$ with the property that $X(s) = 0$ for all recurrent states s. A proof (for aperiodic Markov chains) can be found in the Appendix on page 235.

Proposition 2.1. For a DTMRM (S, P, R) let $S = \bigcup_{i=1}^{k} S_i^r \cup S^t$ be the partitioning of S into k closed recurrent classes S_i^r and transient states S^t. The infinite-horizon total value function V_∞ exists if and only if for all $i = 1, \ldots, k$ and for all $s, s' \in S_i^r$ it holds that

$$R(s, s') = 0.$$

Example 2.2. Let us go back to the queueing model introduced in Example 2.1. The finite-horizon total value function for the first 30 transitions and reward function $R := R_{\text{total}}$ is shown in Fig. 2.2 for the initial state $s_{\text{init}} = (0, 0, \text{idle})$.

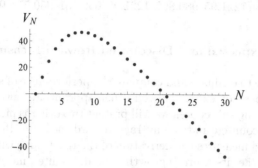

Fig. 2.2. Finite-horizon total value function with horizon length $N = 30$ for the queueing model in Example 2.1 and initial state $s_{\text{init}} = (0, 0, \text{idle})$

As one can see, at the beginning the jobs need some time to fill the system (i.e. both the queue and the server) and thus the expected accumulated reward increases. But after some time steps the high penalty of $C_{\text{loss}} = -\$1000$ for discarding a job outweighs the accumulation of the relatively small reward $R_{\text{acc}} = \$100$ for accomplishing a job and the total value decreases. The infinite-horizon total value does not exist in this model, since $V_N(s_{\text{init}})$ diverges to $-\infty$. However, in case the total reward up to the first loss of a job is of interest, one can introduce an auxiliary absorbing state loss with reward 0, which represents that an incoming job has been discarded (Fig. 2.3).

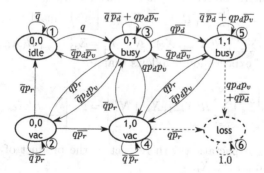

Fig. 2.3. Queueing model enhanced with an auxiliary absorbing state 'loss' representing the loss of an incoming job due to a full queue

Since the single recurrent state loss in this DTMRM has reward 0, the total value function exists and fulfills (2.7) (respectively (2.9)) with $R(s) := R_{\text{profit}}(s)$ for

$s \neq \text{loss}$ (see (2.2)) and $R(\text{loss}) := 0$. Note that $I - P$ has rank 5 since P defines an absorbing state. From $R(\text{loss}) = 0$ it follows that the total value function V_∞ is the unique solution for (2.7) with the constraint $V_\infty(\text{loss}) = 0$ and is given by

$$V_\infty \approx (1221.95, 980.892, 1221.95, 659.481, 950.514, 0)^T .$$ □

2.3   Horizon-Expected and Discounted Reward Measure

In order to be able to evaluate and compare the performance of systems in which the total value does not exist, we need other appropriate reward measures. In this and the following subsection, we will present two other typically used reward measures: the discounted and the average reward measure. Roughly speaking, the average reward measures the derivation of the total value with respect to the horizon length N, i.e. its average growth. The discounted measure can be used if the horizon length for the system is finite but a priori unknown and can be assumed as being random (and memoryless). In order to define the discounted reward measure, we first introduce the more general horizon-expected reward measure.

Definition 2.6. *Let $\mathcal{M} = (S, P, R)$ be a DTMRM and consider a random horizon length N for $\mathcal{M}$, i.e. N is a random variable over $\mathbb{N}$ that is independent of the state process X_n of $\mathcal{M}$. Let $V_{(N)}$ denote the random finite-horizon total value function that takes values in $\{V_n \in \mathbb{R}^S \mid n \in \mathbb{N}\}$. Define the **horizon-expected value function** by*

$$V(s) := \mathbb{E}\left[V_{(N)}(s)\right],$$

if the expectation exists for all $s \in S$, i.e. $|V_{(N)}(s)|$ has finite expectation.

In order to be formally correct, the random variable $V_{(N)}(s)$ is the conditional expectation $V_{(N)}(s) = \mathbb{E}_s\left[\sum_{i=1}^N R(X_{i-1}, X_i) \mid N\right]$ and thus if $N = n$ then $V_{(N)}(s)$ takes the value $\mathbb{E}_s\left[\sum_{i=1}^N R(X_{i-1}, X_i) \mid N = n\right] = \mathbb{E}_s[\sum_{i=1}^n R(X_{i-1}, X_i)] = V_n(s)$. By the law of total expectation it follows that

$$V(s) = \mathbb{E}\left[\mathbb{E}_s\left[\sum_{i=1}^N R(X_{i-1}, X_i) \mid N\right]\right] = \mathbb{E}_s\left[\sum_{i=1}^N R(X_{i-1}, X_i)\right],$$

i.e. $V(s)$ is a joint expectation with respect to the product of the probability measures of N and all the X_i.

The following lemma presents a natural sufficient condition that ensures the existence of the horizon-expected value function.

Lemma 2.1. *If the random horizon length N has finite expectation $\mathbb{E}[N] < \infty$ then $V(s)$ exists.*

Proof. Since the state space is finite there exists $C \in \mathbb{R}$ such that $|R(s, s')| \leq C$ $\forall s, s' \in S$. Therefore

$$|V_n(s)| \leq \mathbb{E}_s \left[\sum_{i=1}^{n} |R(X_{i-1}, X_i)| \right] \leq n \cdot C$$

and thus

$$\mathbb{E}\left[|V_{(N)}(s)|\right] = \sum_{n=0}^{\infty} |V_n(s)| \cdot P(N = n) \leq C \cdot \mathbb{E}[N] < \infty. \qquad \square$$

In many applications the horizon length is considered to be memoryless, i.e. $P(N > n + m \mid N > m) = P(N > n)$ and is therefore geometrically distributed. This fact motivates the following definition.

Definition 2.7. *For $\gamma \in (0, 1)$ let N be geometrically distributed with parameter $1 - \gamma$, i.e. $P(N = n) = \gamma^{n-1}(1 - \gamma)$ for $n = 1, 2, \ldots$. In this case the horizon-expected value function is called **discounted value function** with **discount factor** γ (or just γ-discounted value function) and is denoted by V^γ.*

As for the total value function in Sect. 2.2 we can explicitly compute the discounted value function:

Theorem 2.2 (Evaluation of the Discounted Reward Measure). *For a discount factor $\gamma \in (0, 1)$ it holds that*

$$V^\gamma(s) = \lim_{n \to \infty} \mathbb{E}_s \left[\sum_{i=1}^{n} \gamma^{i-1} R(X_{i-1}, X_i) \right]. \tag{2.10}$$

Furthermore, V^γ is the unique solution to the system of linear equations

$$V^\gamma(s) = R(s) + \gamma \sum_{s' \in S} P(s, s') V^\gamma(s') \tag{2.11}$$

which is written in vector notation as

$$(I - \gamma P) V^\gamma = R.$$

Proof. Let N be geometrically distributed with parameter γ. By conditional expectation we get

$$V^\gamma(s) = \mathbb{E}\left[\mathbb{E}_s \left[\sum_{i=1}^{N} R(X_{i-1}, X_i) \mid N \right] \right] = \sum_{n=1}^{\infty} \mathbb{E}_s \left[\sum_{i=1}^{n} R(X_{i-1}, X_i) \right] P(N = n)$$

$$= (1 - \gamma) \sum_{i=1}^{\infty} \mathbb{E}_s \left[R(X_{i-1}, X_i) \right] \sum_{n=i}^{\infty} \gamma^{n-1} = \sum_{i=1}^{\infty} \mathbb{E}_s \left[R(X_{i-1}, X_i) \right] \gamma^{i-1},$$

which gives (2.10). The derivation of the linear equations (2.11) is completely analogous to the total case by comparing (2.10) to (2.6) (see proof of Theorem 2.1). Since $I - \gamma P$ has full rank for $\gamma \in (0, 1)$ the solution is unique. $\qquad \square$

Equation (2.10) yields also another characterization of the discounted reward measure. Along a path $\omega = (s_0, s_1, s_2, \dots)$ the accumulated discounted reward is $R(s_0, s_1) + \gamma R(s_1, s_2) + \gamma^2 R(s_2, s_3) + \dots$. The future rewards $R(s_i, s_{i+1})$ are reduced by the factor $\gamma^i < 1$ in order to express some kind of uncertainty about the future reward value (e.g. induced by inflation). In many applications, the discounted reward measure is used with a high discount factor close to 1 which still avoids possible divergence of the infinite-horizon total value. Qualitatively speaking, for high $\gamma < 1$ the sequence γ^i decreases for the first few points in time i slowly, but exponentially to 0. If we assume that the rewards $R(s)$ are close to each other for each state s, then the rewards accumulated within the first few time steps approximately give the discounted value.

Remark 2.1. Note that the discounted value function can be equivalently characterized as an infinite-horizon total value function by adding an absorbing and reward-free final state *abs* to the state space S such that *abs* is reachable from any other state with probability $1 - \gamma$ and any other transition probability is multiplied with γ (see Fig. 2.4). Since *abs* is eventually reached on every path within a finite number of transitions with probability 1 and has reward 0, it characterizes the end of the accumulation procedure. The extended transition probability matrix $P' \in \mathbb{R}^{(|S|+1) \times (|S|+1)}$ and the reward vector $R' \in \mathbb{R}^{|S|+1}$ are given by

$$P' = \begin{pmatrix} \gamma P & (1-\gamma)\mathbf{1} \\ 0 & 1 \end{pmatrix} \quad \text{and} \quad R' = \begin{pmatrix} R \\ 0 \end{pmatrix},$$

where $\mathbf{1} = (1, \dots, 1)^T$. Since *abs* is the single recurrent state and $R(abs) = 0$ it follows by Proposition 2.1 that V_∞ exists. Furthermore, it satisfies $(I - P')V_\infty = R'$ and because $V_\infty(abs) = 0$ it is also the unique solution with this property. On the other hand $(V^\gamma, 0)^T$ is also a solution and thus $V_\infty = (V^\gamma, 0)^T$.

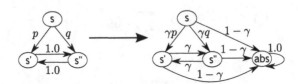

Fig. 2.4. Equivalent characterization of the γ-discounted reward measure as a total reward measure by extending the state space with an absorbing reward-free state

Example 2.3. As an example we analyze the queueing model from Example 2.1 with respect to the discounted reward measure. Figure 2.5 shows $V^\gamma(s)$ for the initial state $s_{\text{init}} = (0, 0, \text{idle})$ as a function of γ. As we see, for small values of γ the discounted values are positive, since the expected horizon length $\frac{1}{1-\gamma}$ is also small and thus incoming jobs have a chance to be processed and not discarded within that time. However for γ approaching 1, the expected horizon length gets larger and the accumulation of the negative reward $C_{\text{loss}} = -\$1000$ for discarding jobs in a full queue prevails.   □

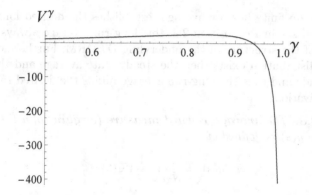

Fig. 2.5. Discounted reward $V^\gamma(s_{\text{init}})$ as a function of γ for the queueing model in Example 2.1 and initial state $s_{\text{init}} = (0, 0, \text{idle})$

2.4   Average Reward Measure

We now provide another important reward measure for the case that the horizon length is infinite (and not random as assumed in Sect. 2.3). We assume for this section that the reader is familiar with the concept of periodicity as presented in Sect. 2.1.2. If for a DTMRM $\mathcal{M} = (S, P, R)$ the infinite-horizon total value function V_∞ does not exist, then either $V_N(s)$ diverges to $\pm\infty$ for some state s or $V_N(s)$ is oscillating over time. The problem with this measure is that V_∞ infinitely often collects the rewards and sums them all up. Instead of building such a total accumulation, one can also measure the system by considering the gained rewards only per time step. As an example, consider an ergodic model $\mathcal{M}$ in the long run with limiting distribution $\rho := \rho_s$ given by $\rho_s(s') := \lim_{n\to\infty} P^n(s, s')$ for an arbitrary intial state s. Then in steady-state the system is rewarded at each time step with $\rho R = \sum_s \rho(s) R(s) \in \mathbb{R}$, i.e. an average of all the rewards $R(s)$ weighted by the probability $\rho(s)$ that the system occupies state s in steady-state. But averaging the rewards $R(s)$ shall not be restricted to only those models $\mathcal{M}$ for which a limiting distribution exists. First of all, the limiting distribution ρ_s can depend on the initial state s, if $\mathcal{M}$ has several closed recurrent classes. More important, the limiting distribution might even not exist, as one can see for the periodic model with

$$P = \begin{pmatrix} 0 & 1 \\ 1 & 0 \end{pmatrix} \quad \text{and} \quad R = \begin{pmatrix} 2 \\ 4 \end{pmatrix}.$$

However, in this case one would also expect an average reward of 3 for each time step and for every state, since P is irreducible and has the unique stationary distribution $\rho = (0.5, 0.5)$. This means that the average reward measure shall also be applicable to models for which the limiting distribution does not exist. Instead of computing an average per time step in steady-state, one can also think of calculating an average in the long-run by accumulating for each horizon length N the total value V_N and dividing it by the time N. The limit of the sequence of these finite horizon averages establishes the desired long-run average. As we will see in Proposition 2.2, this long-run average always converges, independent of the structure of the underlying DTMRM. Furthermore, in case the limiting distribution exists then the steady-state average and the long-run average coincide and thus the long-run average fulfills the desired requirements from the motivation.

Definition 2.8. *The **average reward measure** (or **gain**) of a DTMRM with value function $g(s)$ is defined by*

$$g(s) := \lim_{N \to \infty} \frac{1}{N} V_N(s)$$

if the limit exists.

In the following, we summarize well-known results from linear algebra which first of all directly imply that the average reward exists (at least for finite state spaces) and furthermore also allow us to provide methods for its evaluation.

Definition 2.9. *For a stochastic matrix P define the **limiting matrices** P^∞ and P^* as:*

$$P^\infty := \lim_{N \to \infty} P^N \quad \text{and} \quad P^* := \lim_{N \to \infty} \frac{1}{N} \sum_{i=1}^{N} P^{i-1},$$

if the limits exist. (P^* is the Cesàro limit of the sequence P^i.)

Suppose that P^* exists. Then the average reward can be computed by

$$g = P^* R, \tag{2.12}$$

since by (2.8) it holds that

$$g(s) = \lim_{N \to \infty} \frac{1}{N} V_N(s) = \lim_{N \to \infty} \frac{1}{N} \sum_{i=1}^{N} \left(P^{i-1} R \right)(s) = \left(P^* R \right)(s).$$

Note also that if P^∞ exists, then the i-th row in P^∞ (with $i = \varphi(s)$, see (2.1)) represents the limiting distribution ρ_i of the model, given that the initial state of the system is s. By the motivation from above it should also hold that $g(s) = \rho_i R$. The following proposition relates these two quantities to each other. We refer for proof to [33].

Proposition 2.2. *Consider a DTMC $\mathcal{M} = (S, P)$ with finite state space.*

(i) *The limiting matrix P^* exists.*
(ii) *If P^∞ exists, then $P^* = P^\infty$.*
(iii) *If P is aperiodic then P^∞ exists and if in addition P is unichain (or ergodic) with limiting distribution ρ then $\rho P = \rho$ and P^∞ has identical rows ρ, i.e.*

$$P^* = P^\infty = \mathbf{1}\rho,$$

where $\mathbf{1} = (1, \dots, 1)^T$ is the column vector consisting of all ones.

From Proposition 2.2(ii) it follows that the definition of the average reward corresponds to its motivation from above in the case that the limiting distribution ρ_i exists for all i. However, in the case of periodic DTMRMs (when the limiting distribution is not available), Proposition 2.2(i) ensures that at least P^* exists and due to (2.12) this is sufficient for the computation of the average reward.

Remark 2.2. The limiting matrix P^* satisfies the equalities

$$PP^* = P^*P = P^*P^* = P^*. \tag{2.13}$$

P^* can be computed by partitioning the state space $S = \cup_{i=1}^k S_i^r \cup S^t$ into closed recurrent classes S_i^r and transient states S^t which results in a representation of P as in (2.5). Let the row vector $\rho_i \in \mathbb{R}^{r_i}$ denote the unique stationary distribution of P_i, i.e. $\rho_i P_i = \rho_i$. Then

$$P^* = \begin{pmatrix} P_1^* & 0 & 0 & \dots & 0 & 0 \\ 0 & P_2^* & 0 & \dots & 0 & 0 \\ \vdots & \vdots & \vdots & \ddots & \vdots & 0 \\ 0 & 0 & 0 & \dots & P_k^* & 0 \\ \widetilde{P}_1^* & \widetilde{P}_2^* & \widetilde{P}_3^* & \dots & \widetilde{P}_k^* & 0 \end{pmatrix} \tag{2.14}$$

where $P_i^* = \mathbf{1}\rho_i$ has identical rows $\rho_i \in \mathbb{R}^{r_i}$ and $\widetilde{P}_i^* = (I - \widetilde{P}_{k+1})^{-1}\widetilde{P}_i P_i^*$ consists of trapping probabilities from transient states into the i-th recurrent class. It follows that the average reward g is constant on each recurrent class, i.e. $g(s) = g_i := \rho_i R_i$ for all $s \in S_i^r$ where $R_i \in \mathbb{R}^{r_i}$ is the vector of rewards on S_i^r. On transient states the average reward g is a weighted sum of all the g_i with weights given by the trapping probabilities.

We want to provide another method to evaluate the average reward measure, because it will be useful for the section on MDPs. This method relies on the key aspect of a Laurent series decomposition which also links together the three proposed measures total reward, discounted reward and average reward. Consider the discounted value $V^\gamma = (I - \gamma P)^{-1}R$ as a function of γ (cf. Theorem 2.2). If the total value function $V_\infty = (I - P)^{-1}R$ exists then V^γ converges to V_∞ as $\gamma \nearrow 1$. But what happens if V_∞ diverges to ∞ or $-\infty$? In this case V^γ has a pole singularity at $\gamma = 1$ and can be expanded into a Laurent series. Roughly speaking, a Laurent series generalizes the concept of a power series for (differentiable) functions f with poles, i.e. points c at the boundary of the domain of f with $\lim_{x \to c} f(x) = \pm\infty$. In such a case, f can be expanded in some neighborhood of c into a function of the form $\sum_{n=-N}^\infty a_n(x - c)^n$ for some $N \in \mathbb{N}$ and $a_n \in \mathbb{R}$, which is a sum of a rational function and a power series. In our case, since $\gamma \mapsto V^\gamma$ might have a pole at $\gamma = 1$, the Laurent series is of the form $V^\gamma = \sum_{n=-N}^\infty a_n(\gamma - 1)^n$ for γ close to 1. The coefficients a_n in this expansion are given in Theorem 2.3 in the sequel which can be deduced from the following Lemma. A proof for the lemma can be found in [33].

Lemma 2.2 (Laurent Series Expansion). *For a stochastic matrix P the matrix $(I - P + P^*)$ is invertible. Let*

$$H := (I - P + P^*)^{-1} - P^*.$$

There exists $\delta > 0$ such that for all $0 < \rho < \delta$ the Laurent series of the matrix-valued function $\rho \mapsto (\rho I + (I - P))^{-1}$ is given by

$$(\rho I + (I - P))^{-1} = \rho^{-1} P^* + \sum_{n=0}^{\infty} (-\rho)^n H^{n+1}.$$

Theorem 2.3 (Laurent Series of the Discounted Value Function). *Let $M = (S, P, R)$ be a DTMRM. For a discount factor $\gamma < 1$ close to 1 write $\gamma(\rho) := \frac{1}{1+\rho}$ where $\rho = \frac{1-\gamma}{\gamma} > 0$ and consider the discounted value $V^{\gamma(\rho)}$ as a function of ρ.*

(i) The Laurent series of $\rho \mapsto V^{\gamma(\rho)}$ at 0 is given (for small $\rho > 0$) by

$$V^{\gamma(\rho)} = (1 + \rho) \left(\rho^{-1} g + \sum_{n=0}^{\infty} (-\rho)^n H^{n+1} R \right), \tag{2.15}$$

where g is the average reward.

(ii) It holds that

$$V^{\gamma} = \frac{1}{1-\gamma} g + h + f(\gamma) \tag{2.16}$$

where $h := HR$ and f is some function with $\lim_{\gamma \nearrow 1} f(\gamma) = 0$. Furthermore,

$$g = \lim_{\gamma \nearrow 1} (1 - \gamma) V^{\gamma}.$$

Proof. (i) We apply the Laurent series from Lemma 2.2 as follows:

$$V^{\gamma} = (I - \gamma P)^{-1} R = (1 + \rho)(\rho I + (I - P))^{-1} R$$

$$= (1 + \rho) \left(\rho^{-1} P^* R + \sum_{n=0}^{\infty} (-\rho)^n H^{n+1} R \right)$$

and the claim follows from (2.12). Furthermore, by substituting $\gamma = (1 + \rho)^{-1}$

$$V^{\gamma} = \frac{1}{\gamma} \left(\frac{\gamma}{1-\gamma} g + h + \sum_{n=1}^{\infty} \left(\frac{\gamma-1}{\gamma} \right)^n H^{n+1} R \right) = \frac{1}{1-\gamma} g + h + f(\gamma)$$

where $f(\gamma) := \frac{1-\gamma}{\gamma} h + \sum_{n=1}^{\infty} \left(\frac{\gamma-1}{\gamma} \right)^n H^{n+1} R$ and $f(\gamma) \to 0$ when $\gamma \nearrow 1$ such that (ii) follows. $\qquad \square$

The vector h in (2.16) is called the **bias** for the DTMRM. We provide an equivalent characterization for h that allows a simpler interpretation of the term "bias" as some sort of deviation. If the reward function R is replaced by the average reward g, such that in every state s the average reward $g(s)$ is gained instead of $R(s)$, then the finite horizon total reward is given by $G_N(s) := \mathbb{E}_s \left[\sum_{i=1}^{N} g(X_i) \right]$, where X_i is the state process. In this case $\Delta_N := V_N - G_N$ describes the deviation in accumulation between the specified reward R and its corresponding average reward $g = P^*R$ within N steps. By (2.8) it holds that

$$\Delta_N = \sum_{n=0}^{N-1} P^n(R - g) = \sum_{n=0}^{N-1} (P^n - P^*)R.$$

Note that $P^n - P^* = (P - P^*)^n$ for all $n \geq 1$ which follows from (2.13) and $\sum_{k=0}^{n}(-1)^k \binom{n}{k} = 0$ applied on $(P - P^*)^n = \sum_{k=0}^{n} \binom{n}{k} P^{n-k}(-P^*)^k$. If we assume that Δ_N converges for any reward function R then $\sum_{n=0}^{\infty}(P - P^*)^n$ converges and it holds that $\sum_{n=0}^{\infty}(P - P^*)^n = (I - (P - P^*))^{-1}$. It follows

$$\lim_{N\to\infty} \Delta_N = \sum_{n=0}^{\infty}(P^n - P^*)R = \left((I - P^*) + \sum_{n=1}^{\infty}(P - P^*)^n \right) R =$$

$$\left(\sum_{n=0}^{\infty}(P - P^*)^n - P^* \right) R = \left((I - (P - P^*))^{-1} - P^* \right) R = HR = h.$$

Therefore, the bias $h(s)$ is exactly the long-term deviation between $V_N(s)$ and $G_N(s)$ as $N \to \infty$. This means that the value $h(s)$ is the excess in the accumulation of rewards beginning in state s until the system reaches its steady-state. Remember that g is constant on recurrent classes. Thus, for a recurrent state s it holds that $G_N(s) = \mathbb{E}_s\left[\sum_{i=1}^{N} g(X_i) \right] = g(s)N$ is linear in N and $g(s)N + h(s)$ is a linear asymptote for $V_N(s)$ as $N \to \infty$, i.e. $V_N(s) - (g(s)N + h(s)) \to 0$ (see Fig. 2.6). The matrix H is often called the **deviation matrix** for the DTMRM since it maps any reward function R to the corresponding long-term deviation represented by the bias $h = HR$.

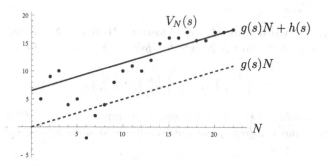

Fig. 2.6. Interpretation of the bias $h(s)$ as the limit of the deviation $V_N(s) - G_N(s)$ as $N \to \infty$. For a recurrent state s it holds that $G_N(s) = g(s)N$.

Another characterization for the bias can be given by considering Δ_N as a finite-horizon total value function for the average-corrected rewards $R - g$, i.e.

$\Delta_N(s) = \mathbb{E}_s \left[\sum_{i=1}^{N} (R(X_i) - g(X_i)) \right]$. For this reason the bias h is a kind of infinite-horizon total value function for the model $(S, P, R - g)$ [1].

In the above considerations we have assumed that Δ_N converges (for any reward function R), which is guaranteed if the DTMRM is aperiodic [33]. On the other hand, one can also construct simple periodic DTMRMs for which Δ_N is oscillating. For this reason, there is a similar interpretation of the bias h if the periodicity of P is averaged out (by the Cesàro limit). This is related to the distinction between the two limiting matrices P^∞ and P^* from Definition 2.9 and goes beyond the scope of this tutorial. In Sect. 4 we will introduce the average reward for continuous-time models (where periodicity is not a problem) and define the deviation matrix H by a continuous-time analogon of the discrete representation $H = \sum_{n=0}^{\infty} (P^n - P^*)$.

Remark 2.3. In the Laurent series expansion of V^γ respectively $V^{\gamma(\rho)}$ as in (2.15) the vector value $H^{n+1}R$ for $n \geq 0$ is often called the n-bias of V^γ and therefore the bias $h = HR$ is also called the 0-bias. We will see in Sect. 3.4 (and especially Remark 3.6) that these values play an important role in the optimization of MDPs with respect to the average reward and the n-bias measures.

We now provide some methods for computing the average reward based on the bias h of a DTMRM $\mathcal{M} = (S, P, R)$. If $\mathcal{M}$ is ergodic then from Proposition 2.2 it holds that $P^* = \mathbf{1}\rho$ and thus the average reward $g = P^*R = \mathbf{1}(\rho R)$ is constantly ρR for each state. In the general case (e.g. $\mathcal{M}$ is multichain or periodic), the following theorem shows how the bias h can be involved into the computation of g. The proof is based on the following equations that reveal some further connections between P^* and H:

$$P^* = I - (I - P)H \quad \text{and} \quad HP^* = P^*H = 0. \tag{2.17}$$

These equations can be deduced from the defining equation for H in Lemma 2.2 together with (2.13).

Theorem 2.4 (Evaluation of the Average Reward Measure). *The average reward g and the bias h satisfy the following system of linear equations:*

$$\begin{pmatrix} I - P & 0 \\ I & I - P \end{pmatrix} \begin{pmatrix} g \\ h \end{pmatrix} = \begin{pmatrix} 0 \\ R \end{pmatrix}. \tag{2.18}$$

Furthermore, a solution (u, v) to this equation implies that $u = P^*R = g$ is the average reward and v differs from the bias h up to some $w \in \ker(I - P)$, i.e. $v - w = h$.

---

[1] Note that in Definition 2.5 we restricted the existence of the infinite-horizon total value function to an absolute convergence of the finite-horizon total value function Δ_N. By Proposition 2.1 this is equivalent to the fact that the rewards are zero on recurrent states. For the average-corrected model this restriction is in general not satisfied. The reward function $R - g$ can take both positive and negative values on recurrent states which are balanced out by the average reward g such that Δ_N is converging (at least as a Cesàro limit).

Proof. We first show that g and h are solutions to (2.18). From $PP^* = P^*$ it follows that $Pg = PP^*R = P^*R = g$ and from (2.17) we have

$$(I - P)h = (I - P)HR = (I - P^*)R = R - g$$

and thus (g, h) is a solution to (2.18). Now for an arbitrary solution (u, v) to (2.18) it follows that

$$(I-P+P^*)u = (I-P)u+P^*u+P^*(I-P)v = (I-P)u+P^*(u+(I-P)v) = 0+P^*R.$$

Since $I - P + P^*$ is invertible by Lemma 2.2, we have

$$u = (I - P + P^*)^{-1}P^*R = \left((I - P + P^*)^{-1} - P^* + P^*\right)P^*R = HP^*R + P^*R.$$

From (2.17) it holds that $HP^* = 0$ and thus $u = P^*R = g$. Furthermore, since both h and v fulfill (2.18) it follows that

$$(I - P)v = R - g = (I - P)h$$

and thus $w := v - h \in \ker(I - P)$, such that $h = v - w$. $\qquad\square$

From (2.18) it holds that $h = (R - g) + Ph$, which reflects the motivation of the bias h as a total value function for the average-corrected model $(S, P, R - g)$. Furthermore, the theorem shows that the equation $h = (R - u) + Ph$ is only solvable for $u = g$. This means that there is only one choice for u in order to balance out the rewards R such that the finite-horizon total value function Δ_N for the model $(S, P, R - u)$ converges (as a Cesàro limit).

Remark 2.4. Assume that $S = \bigcup_{i=1}^{k} S_i^r \cup S^t$ is the splitting of the state space of $\mathcal{M}$ into closed recurrent classes S_i^r and transient states S^t.

(i) As we saw from (2.12) and (2.14) one can directly compute g from the splitting of S. Equation (2.18) shows that such a splitting is not really necessary. However, performing such a splitting (e.g. by the Fox-Landi algorithm [15]) for the computation of g by P^*R can be more efficient than simply solving (2.18) [33].

(ii) In order to compute g from (2.18) it is enough to compute the bias h up to $\ker(I - P)$. The dimension of $\ker(I - P)$ is exactly the number k of closed recurrent classes S_i^r. Hence, if the splitting of S is known, then $v(s)$ can be set to 0 for some arbitrary chosen state s in each recurrent class S_i^r (leading to a reduction in the number of equations in Theorem 2.4).

(iii) In order to determine the bias h, it is possible to extend (2.18) to

$$\begin{pmatrix} I - P & 0 & 0 \\ I & I - P & 0 \\ 0 & I & I - P \end{pmatrix} \begin{pmatrix} u \\ v \\ w \end{pmatrix} = \begin{pmatrix} 0 \\ R \\ 0 \end{pmatrix}. \tag{2.19}$$

It holds that if (u, v, w) is a solution to (2.19) then $u = g$ and $v = h$ [33]. In a similar manner, one can also establish a system of linear equations in order to compute the n-bias values (see Remark 2.3), i.e. the coefficients in the Laurent series of the discounted value function $V^{\gamma(\rho)}$.

The following corollary drastically simplifies the evaluation of the average reward for the case of unichain models $\mathcal{M} = (S, P, R)$. In this case, the state space can be split to $S = S^r \cup S^t$ and consists of only one closed recurrent class S^r.

Corollary 2.1. *For a unichain model (S, P, R) the average reward g is constant on S. More precisely, $g = g_0 1$ for some $g_0 \in \mathbb{R}$ and in order to compute g_0 one can either solve*

$$g_0 1 + (I - P)h = R \tag{2.20}$$

or compute $g_0 = \rho R$, where ρ is the unique stationary distribution of P.

The proof is obvious and we leave it as an exercise to the reader.

Note that (2.20) is a reduced version of (2.18) since $(I - P)g = 0$ for all constant g. Many models in applications are unichain or even ergodic (i.e. irreducible and aperiodic), thus the effort for the evaluation of the average reward is reduced by (2.20). If it is a priori not known if a model is unichain or multichain, then either a model classification algorithm can be applied (e.g. Fox-Landi [15]) or one can directly solve (2.18). In the context of MDPs an analogous classification into unichain and multichain MDPs is applicable. We will see in Sect. 3.4 that Theorem 3.8 describes an optimization algorithm, which builds upon (2.18). In case the MDP is unichain, this optimization algorithm can also be built upon the simpler equation (2.20), thus gaining in efficiency. However, the complexity for the necessary unichain classification is shown to be NP-hard [39]. We refer for more information on classification of MDPs to [19].

Example 2.4. We want to finish this section by showing an example with multi-chain structure based on the queuing model from Example 2.1. Assume a queue with capacity size k in which jobs are enqueued with probability q and a server with the processing states "idle" and "busy" (with no vacation state). Once again, an accomplished job is rewarded $R_{\mathrm{acc}} = \$100$ and a discarded job costs $C_{\mathrm{loss}} = -\$1000$. Additional to the processing behavior a server can also occupy one of the following modes: "normal", "intense" or "degraded" (see Fig. 2.7).

In normal mode the server accomplishes a job with probability $p_{d,n} = 0.5$. From every normal state the server degrades with probability $r_d = 0.01$. In this degraded mode jobs are accomplished with a lower probability $p_{d,d} = 0.25$. If the system is in normal mode, the queue is full and a job enters the system, then the server moves from the normal mode to the intense mode. This move can only happen, if the system does not degrade (as in state $(1, 1, \mathrm{normal}, \mathrm{busy})$). In the intense mode the processing probability increases to $p_{d,i} = 1.0$ but with the drawback that a job can be served not correctly with probability 0.1. A non-correctly served job behaves as if it would be lost, i.e. the job involves a cost of $C_{\mathrm{loss}} = -\$1000$. Being in intense mode or degraded mode, there is no way to change to any other mode. This means that both the intense mode and the degraded mode represent closed recurrent classes in the state space and thus the model is multichain.

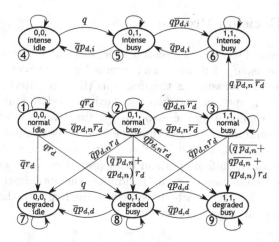

Fig. 2.7. Queueing model with queue capacity $k = 1$ and a service unit with processing behavior idle or busy and processing modes normal, intense or degraded. State $(0, 1, \text{normal}, \text{busy})$ represents 0 jobs in the queue, 1 job served, server is busy and in normal mode. The parameter values are $q = 0.25$, $p_{d,n} = 0.5$, $p_{d,i} = 1.0$ and $p_{d,d} = 0.25$ and $r_d = 0.01$. Probabilities for self-loops complete all outgoing probabilities to 1.0.

By solving (2.18) with P as in Fig. 2.7 and

$$
R = \begin{pmatrix}
0 \\
R_{\mathrm{acc}}p_{d,n} \\
R_{\mathrm{acc}}p_{d,n} \\
\hline
0 \\
R_{\mathrm{acc}}p_{d,i} \cdot 0.9 + C_{\mathrm{loss}}p_{d,i} \cdot 0.1 \\
R_{\mathrm{acc}}p_{d,i} \cdot 0.9 + C_{\mathrm{loss}}p_{d,i} \cdot 0.1 \\
\hline
0 \\
R_{\mathrm{acc}}p_{d,d} \\
R_{\mathrm{acc}}p_{d,d}
\end{pmatrix}
+
\begin{pmatrix}
0 \\
0 \\
C_{\mathrm{loss}}q\overline{p_{d,n}} \\
\hline
0 \\
0 \\
C_{\mathrm{loss}}q\overline{p_{d,i}} \\
\hline
0 \\
0 \\
C_{\mathrm{loss}}q\overline{p_{d,d}}
\end{pmatrix}
=
\begin{pmatrix}
0 \\
50 \\
-75 \\
\hline
0 \\
-10 \\
-10 \\
\hline
0 \\
25 \\
-162.5
\end{pmatrix}
$$

the average reward can be computed to

$$
g \approx (-25.8, -24.8, -19.8 \,|\, -2.5, -2.5, -2.5 \,|\, -50, -50, -50)^T .
$$

As we see the average reward is constant on recurrent classes, i.e. -2.5 in the intense mode and -50 in the degraded mode. For the transient states (represented by the normal mode) the average reward is a state-dependent convex combination of the average rewards for the recurrent classes, since the probability to leave the normal mode to one of the recurrent classes depends on the particular transient state. The bias h can be determined from (2.19) to

$$
h \approx (1853.3, 1811.3, 1213.9 \,|\, 2.5, -7.5, -17.5 \,|\, 363.6, 163.6, -436.364).
$$

This means that if $(0, 0, \text{normal}, \text{idle})$ is the initial state then the reward accumulation process of the finite-horizon total value function V_N follows the linear function $-25.8 \cdot N + 1853.3$ asymptotically as $N \to \infty$ (see Fig. 2.6). $\square$

3   Markov Decision Processes

This section is devoted to Markov Decision Processes with discrete time. Section 3.1 provides the necessary definitions and terminology, and Sect. 3.2 introduces the discounted reward measure as the first optimization criterion. We present its most important properties and two standard methods (value iteration and policy iteration) for computing the associated optimal value function. Depending on the reader's interest, one or both of the following two subsections may be skipped: Stochastic Shortest Path Problems, the topic of Sect. 3.3, are special MDPs together with the infinite-horizon total reward measure as optimization criterion. The final subsection (Sect. 3.4) addresses the optimization of the average reward measure for MDPs, where we involve the bias into the computation of the average-optimal value function.

3.1   Preliminaries

We first state the formal definition of an MDP and then describe its execution semantics. Roughly speaking, an MDP extends the purely stochastic behavior of a DTMRM by introducing actions, which can be used in order to control state transitions.

Definition 3.1. *A **discrete-time Markov Decision Process (MDP)** is a structure $\mathcal{M} = (S, Act, e, P, R)$, where S is the finite state space, $Act \neq \emptyset$ a finite set of actions, $e \colon S \to 2^{Act} \setminus \emptyset$ the action-enabling function, $P \colon S \times Act \to \mathcal{D}(S)$ an action-dependent transition function and $R \colon S \times Act \times S \to \mathbb{R}$ the action-dependent reward function. We denote $P(s, a, s') := (P(s, a))(s')$.*

From a state $s \in S$ an enabled action $a \in e(s)$ must be chosen which induces a probability distribution $P(s, a)$ over S to target states. If a transition to s' takes place then a reward $R(s, a, s')$ is gained and the process continues in s'. In analogy to DTMRMs we denote $R(s, a) := \sum_{s' \in S} P(s, a, s') R(s, a, s')$ as the expected reward that is gained when action a has been chosen and a transition from state s is performed. The mechanism which chooses an action in every state is called a policy. In the theory of MDPs there are several possibilities to define policies. In this tutorial, we restrict to the simplest type of policy.

Definition 3.2. *A **policy** is a function $\pi \colon S \to Act$ with $\pi(s) \in e(s)$ for all $s \in S$. Define $\Pi \subseteq Act^S$ as the set of all policies.*

A policy π of an MDP $\mathcal{M}$ resolves the non-deterministic choice between actions and thus reduces $\mathcal{M}$ into a DTMRM $\mathcal{M}^\pi := (S, P^\pi, R^\pi)$, where

$$P^\pi(s, s') := P(s, \pi(s), s') \quad \text{and} \quad R^\pi(s, s') := R(s, \pi(s), s').$$

Remark 3.1. In the literature one often finds more general definitions of policies in which the choice of an action a in state s does not only depend on the current state s but also

- on the history of both the state process and the previously chosen actions and
- can be randomized, i.e. the policy prescribes for each state s a probability distribution $\pi(s) \in \mathcal{D}(e(s))$ over all enabled actions and action a is chosen with probability $(\pi(s))(a)$.

The policy type as in Definition 3.2 is often referred to "stationary Markovian deterministic". Here, deterministic is in contrast to randomized and means that the policy assigns a fixed action instead of some probability distribution over actions. A policy is Markovian, if the choice of the action does not depend on the complete history but only on the current state and point in time of the decision. A Markovian policy is stationary, if it takes the same action a everytime it visits the same state s and is thus also independent of time. For simplicity we stick to the type of stationary Markovian deterministic policies as in Definition 3.2 since this is sufficient for the MDP optimization problems we discuss in this tutorial. The more general types of policies are required if e.g. the target function to be optimized is of a finite-horizon type or if additional constraints for optimization are added to the MDP model (see also Remark 3.3).

Example 3.1 (Queueing model). We consider the queueing model introduced in Example 2.4. Assume that the server can be either idle or busy and operate in normal or intense mode. Figure 3.1 shows an MDP for queue capacity size $k = 2$ and the two actions "keep" and "move", which enable swichting between the processing modes.

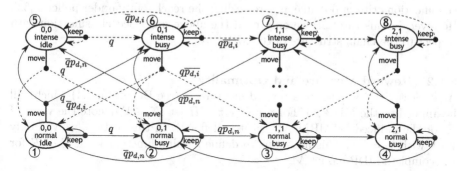

Fig. 3.1. An excerpt of the MDP queueing model with queue capacity $k = 2$, a service unit with processing behavior idle or busy and processing modes normal or intense. In each state the "keep" action keeps the current processing mode, while the "move" action changes the mode. State $(0, 1, \text{normal}, \text{busy})$ represents 0 jobs in the queue, 1 job served, server is busy and in normal mode. The parameter values are $q = 0.25$, $p_{d,n} = 0.5$ and $p_{d,i} = 1.0$. For better overview transitions from normal mode are bold, whereas transitions from intense mode are dashed.

A job enters the system with probability $q = 0.25$. If the queue is full then the system refuses the job, which causes a cost of $C_{\text{loss}} = -\$1000$. A normal operating server accomplishes the job with probability $p_{d,n} = 0.5$, whereas in intense mode the server succeeds with probability $p_{d,i} = 1.0$. If a job is accomplished the system is rewarded with $R_{acc} = \$100$. In contrast to the normal mode, in intense mode the system raises a higher operating cost of $C_{int} = -\$10$ per time step. Furthermore a change from normal to intense mode causes additionally $C_{move} = -\$50$. All together the rewards can be represented by

$$R^{keep} = \begin{pmatrix} 0 \\ R_{acc}p_{d,n} \\ R_{acc}p_{d,n} \\ \dfrac{R_{acc}p_{d,n} + C_{loss}q\overline{p_{d,n}}}{C_{int}} \\ R_{acc}p_{d,i} + C_{int} \\ R_{acc}p_{d,i} + C_{int} \\ R_{acc}p_{d,i} + C_{loss}q\overline{p_{d,i}} + C_{int} \end{pmatrix}, \quad R^{move} = R^{keep} + \begin{pmatrix} C_{move} \\ C_{move} \\ C_{move} \\ C_{move} \\ 0 \\ 0 \\ 0 \end{pmatrix}. \quad (3.1)$$

□

3.1.1 Classification of MDPs

The state classification introduced in Sect. 2.1.2 for DTMRMs also implies a classification of MDP models. An MDP $\mathcal{M}$ is called **unichain**, if for all policies $\pi \in \Pi$ the induced DTMRM $\mathcal{M}^\pi$ is unichain. Otherwise, if there is a policy π, for which $\mathcal{M}^\pi$ has at least two closed recurrent classes, then $\mathcal{M}$ is called **multichain**. As for DTMRMs, this classification will be mainly used for the analysis and optimization of the average reward in Sect. 3.4. There are also other classification criteria possible, e.g. regarding the reachability under policies [33]. However, in this tutorial, we only need the above described classification with respect to the chain structure of the MDP.

3.1.2 Reward Measure and Optimality

As in Sect. 2.1.3 we choose a reward measure $\mathcal{R}$ (see Definition 2.4) which will be applied to the MDP $\mathcal{M}$. For a policy $\pi \in \Pi$ let $V^\pi \in \mathcal{V}$ denote the value of $\mathcal{R}$ for the induced DTMRM $\mathcal{M}^\pi$ (if it is defined). In this section, we will only work with $\mathcal{V} = \mathbb{R}^S$. This allows us to define a value function $V^* \in \mathcal{V}$ of $\mathcal{R}$ for the complete MDP model $\mathcal{M}$.

Definition 3.3. *Let $\mathcal{M}$ be an MDP with reward measure $\mathcal{R}$. Define the (optimal) value function V^* of $\mathcal{M}$ by*

$$V^*(s) := \sup_{\pi \in \Pi} V^\pi(s), \quad (3.2)$$

*if $V^*(s)$ is finite for all s. A policy $\pi^* \in \Pi$ is called **optimal** if V^{π^*} is defined and*

$$\forall s \in S \; \forall \pi \in \Pi : \; V^{\pi^*}(s) \geq V^\pi(s). \quad (3.3)$$

Note that in the definition of V^* the supremum is taken in a state-dependent way. Furthermore, since the state and action spaces are finite, the policy space

Π is finite as well. Therefore, the supremum in the above definition is indeed a maximum. It follows that if for all $\pi \in \Pi$ the value $V^\pi(s)$ is defined and finite then $V^*(s)$ is also finite for all s.

In case $\mathcal{R}$ is the infinite-horizon total reward measure, we allow in contrast to Definition 2.5 the value $V^\pi(s)$ to converge improperly to $-\infty$. Taking the supremum in (3.2) doesn't care of this kind of convergence, if there is at least one policy providing a finite value $V^\pi(s)$. The same holds for (3.3) since here V^{π^*} has to exist in the sense of Definition 2.5 (and thus be finite).

Note further that through the definition of an optimal policy it is not clear if an optimal policy π^* exists, since π^* has to fulfill the inequality in (3.3) uniformly over all states. Definition 3.3 gives rise to the following natural question: Under which conditions does an optimal policy π^* exist and how is it related to the optimal value V^*? These questions will be answered in the following subsections.

3.2   Discounted Reward Measure

We first address the above mentioned questions in the case of the discounted reward measure with discount factor $\gamma \in (0,1)$, since this measure is analytically simpler to manage than the infinite-horizon or the average reward measure. For motivation, let us first provide some intuition on the optimization problem. Assume we have some value function $V : S \to \mathbb{R}$ and we want to check whether V is optimal or alternatively in what way can we modify V in order to approach the optimal value V^*. When in some state s one has a choice between enabled actions $a \in e(s)$, then for each of these actions one can perform a look-ahead step and compute $Q(s,a) := R(s,a) + \gamma \sum_{s' \in S} P(s,a,s')V(s')$. The value $Q(s,a)$ combines the reward $R(s,a)$ gained for the performed action and the expectation over the values $V(s')$ for the transition to the target state s' induced by action a. If now $Q(s,a') > V(s)$ for some $a' \in e(s)$ then clearly one should improve V by the updated value $V(s) := Q(s,a')$ or even better choose the best improving action and set $V(s) := \max_{a \in e(s)} Q(s,a)$. This update procedure can be formalized by considering the **Bellman operator** $\mathcal{T} : \mathbb{R}^S \to \mathbb{R}^S$ which assigns to each value function $V \in \mathbb{R}^S$ its update $\mathcal{T}V := \mathcal{T}(V) \in \mathbb{R}^S$ defined by

$$(\mathcal{T}V)(s) := \max_{a \in e(s)} \left\{ R(s,a) + \gamma \sum_{s' \in S} P(s,a,s')V(s') \right\}.$$

Note that $\mathcal{T}$ is a non-linear operator on the vector space $\mathbb{R}^S$, since it involves maximization over actions. If we proceed iteratively, a sequence of improving value functions V is generated and the hope is that this sequence convergences to the optimal value function V^*. In case V is already optimal, there should be no strict improvement anymore possible. This means that for every state s the value $V(s)$ is maximal among all updates $Q(s,a)$, $a \in e(s)$ on $V(s)$, i.e.

$$V(s) = \max_{a \in e(s)} \left\{ R(s,a) + \gamma \sum_{s' \in S} P(s,a,s')V(s') \right\}. \tag{3.4}$$

This non-linear fixed-point equation $V = \mathcal{T}V$ is also known as the **Bellman optimality equation** and we have to solve it, if we want to detemine V^*. The following Theorem 3.1 establishes results on existence and uniqueness of solutions to this equation. Furthermore, it also creates a connection between the optimal value function V^* and optimal policies π^*.

Theorem 3.1 (Existence Theorem). *Consider an MDP (S, Act, e, P, R) and the discounted reward measure with discount factor $\gamma \in (0, 1)$.*

(i) *There exists an optimal value $(V^\gamma)^*$ which is the unique fixed point of $\mathcal{T}$, i.e. the Bellman optimality equation holds:*

$$(V^\gamma)^* = \mathcal{T}(V^\gamma)^*. \tag{3.5}$$

(ii) *There exists an optimal policy π^* and it holds that $(V^\gamma)^{\pi^*} = (V^\gamma)^*$.*
(iii) *Every optimal policy π^* can be derived from the optimal value $(V^\gamma)^*$ by*

$$\pi^*(s) \in \operatorname*{argmax}_{a \in e(s)} \left\{ R(s, a) + \gamma \sum_{s' \in S} P(s, a, s') (V^\gamma)^* (s') \right\}.$$

The complete proof can be found in [33]. The key ingredient for this proof relies on the following lemma, which provides an insight into the analytical properties of the Bellman operator $\mathcal{T}$.

Lemma 3.1. *(i) $\mathcal{T}$ is monotonic, i.e. if $U(s) \leq V(s)$ for all $s \in S$ then $(\mathcal{T}U)(s) \leq (\mathcal{T}V)(s)$ for all s.*
(ii) $\mathcal{T}$ is a contraction with respect to the maximum norm $\|V\| := \max_{s \in S} |V(s)|$, i.e. there exists $q \in \mathbb{R}$ with $0 \leq q < 1$ such that

$$\|\mathcal{T}U - \mathcal{T}V\| \leq q\|U - V\|.$$

The constant q is called Lipschitz constant and one can choose $q := \gamma$.

Remark 3.2. Lemma 3.1(ii) allows to apply the Banach fixed point theorem on the contraction $\mathcal{T}$ which ensures existence and uniqueness of a fixed point V_{fix}. Furthermore the sequence

$$V_{n+1} := \mathcal{T}V_n \tag{3.6}$$

converges to V_{fix} for an arbitrary initial value function $V_0 \in \mathbb{R}^S$. From the monotonicity property of $\mathcal{T}$ it can be shown that the sequence in (3.6) also converges to the optimal value $(V^\gamma)^*$ and thus $V_{\text{fix}} = (V^\gamma)^*$.

Writing (3.5) in component-wise notation yields exactly the Bellman optimality equation (3.4) as motivated, for which $(V^\gamma)^*$ is the unique solution. Note that $(V^\gamma)^*$ is defined in (3.2) by a maximization over the whole policy space Π, i.e. for each state s all policies $\pi \in \Pi$ have to be considered in order to establish the supremum. In contrast, the Bellman optimality equation reduces this global optimization task into a local state-wise optimization over the enabled actions $a \in e(s)$ for every $s \in S$. Note also that from Theorem 3.1 one can deduce that

$\mathcal{T}((V^\gamma)^{\pi^*}) = (V^\gamma)^{\pi^*}$, such that an optimal policy π^* can be also considered as a "fixed-point" of the Bellman operator $\mathcal{T}$. However, an optimal policy does not need to be unique.

From the Bellman equation (3.4) several algorithms based on fixed-point iteration can be derived which can be used in order to compute the optimal policy together with its value function. The following value iteration algorithm is based on (3.6). Its proof can be found in the Appendix on page 236.

Theorem 3.2 (Value Iteration). *For an arbitrary initial value function $V_0 \in \mathbb{R}^S$ define the sequence of value functions*

$$V_{n+1}(s) := (\mathcal{T}V_n)(s) = \max_{a \in e(s)} \left\{ R(s,a) + \gamma \sum_{s' \in S} P(s,a,s')V_n(s') \right\}.$$

Then V_n converges to $(V^\gamma)^*$. As a termination criterion choose $\varepsilon > 0$ and continue iterating until $\|V_{n+1} - V_n\| < \frac{1-\gamma}{2\gamma}\varepsilon$ and let

$$\pi_\varepsilon(s) \in \operatorname*{argmax}_{a \in e(s)} \left\{ R(s,a) + \gamma \sum_{s' \in S} P(s,a,s')V_{n+1}(s') \right\}. \qquad (3.7)$$

Then $\|V^{\pi_\varepsilon} - (V^\gamma)^*\| < \varepsilon$.

The value iteration algorithm iterates on the vector space $\mathbb{R}^S$ of value functions. From an arbitrary value function an improving policy can be generated by (3.7). In contrast, the following policy iteration algorithm iterates on the policy space Π. From a policy π its value can be generated the other way round by solving a system of linear equations.

Theorem 3.3 (Policy Iteration). *Let $\pi_0 \in \Pi$ be an initial policy. Define the following iteration scheme.*

1. *Policy evaluation: Compute the value V^{π_n} of π_n by solving*

$$(I - \gamma P^{\pi_n})V^{\pi_n} = R^{\pi_n}$$

and define the set of improving actions

$$A_{n+1}(s) := \operatorname*{argmax}_{a \in e(s)} \left\{ R(s,a) + \gamma \sum_{s' \in S} P(s,a,s')V^{\pi_n}(s') \right\}.$$

Termination: If $\pi_n(s) \in A_{n+1}(s)$ for all s then π_n is an optimal policy.

2. *Policy improvement: Otherwise choose an improving policy π_{n+1} such that $\pi_{n+1}(s) \in A_{n+1}(s)$ for all $s \in S$.*

The sequence of values V^{π_n} is non-decreasing and policy iteration terminates within a finite number of iterations.

Proof. By definition of π_{n+1} it holds for all s that

$$V^{\pi_{n+1}}(s) = \max_{a \in e(s)} \left\{ R(s,a) + \gamma \sum_{s' \in S} P(s,a,s') V^{\pi_n}(s') \right\}$$

$$\geq R(s, \pi_n(s)) + \gamma \sum_{s' \in S} P(s, \pi_n(s), s') V^{\pi_n}(s') = V^{\pi_n}(s).$$

Since there are only finitely many policies and the values V^{π_n} are non-decreasing, policy iteration terminates in a finite number of iterations. Clearly, if $\pi_n(s) \in A_{n+1}(s)$ for all s then π_n is optimal, since

$$V^{\pi_n}(s) = \max_{a \in e(s)} \left\{ R(s,a) + \gamma \sum_{s' \in S} P(s,a,s') V^{\pi_n}(s') \right\} = (\mathcal{T} V^{\pi_n})(s).$$

The conclusion follows by Theorem 3.1.   □

Both presented algorithms value iteration and policy iteration create a converging sequence of value functions. For value iteration we mentioned in Remark 3.2 that the generated sequence $V_{n+1} = \mathcal{T} V_n$ converges to the fixed point of $\mathcal{T}$ which is also the global optimal value $(V^\gamma)^*$ of the MDP since $\mathcal{T}$ is monotonic. Same holds for the sequence V^{π_n} in policy iteration, since V^{π_n} is a fixed-point of $\mathcal{T}$ for some $n \in \mathbb{N}$ and thus $V^{\pi_n} = (V^\gamma)^*$. The convergence speed of these algorithms is in general very slow. Value iteration updates in every iteration step the value function V_n on every state. This means especially that states s that in the current iteration step do not contribute to a big improvement $|V_{n+1}(s) - V_n(s)|$ in their value will be completely updated like every other state. However, it can be shown that convergence in value iteration can also be guaranteed, if every state is updated infinitely often [37]. Thus, one could modify the order of updates to states regarding their importance or contribution in value improvement (asynchronous value iteration).

Policy iteration on the other hand computes at every iteration step the exact value V^{π_n} of the current considered policy π_n, by solving a system of linear equations. If π_n is not optimal, then after improvement to π_{n+1} the effort for the accurate computation of V^{π_n} is lost. Therefore, the algorithms value iteration and policy iteration just provide a foundation for potential algorithmic improvements. Examples for such improvements are relative value iteration, modified policy iteration or action elimination [33]. Of course, heuristics which use model-dependent meta-information can also be considered in order to provide a good initial value V_0 or initial policy π_0.

Note that MDP optimization underlies the curse of dimensionality: The explosion of the state space induces an even worse explosion of the policy space since $|\Pi| \in O\left(|Act|^{|S|}\right)$. There is a whole branch of Artificial and Computational Intelligence, which develops learning algorithms and approximation methods for large MDPs (e.g. reinforcement learning [37], evolutionary algorithms [29], heuristics [28] and approximate dynamic programming [8, 10, 26, 27, 32]).

Example 3.2. We now want to optimize the queueing model MDP from Example 3.1 by applying both algorithms value iteration and policy iteration. Table 3.1 shows a comparison between values for the initial state $s_{\text{init}} = (0,0,\text{normal},\text{idle})$ under the policies π^{normal} (respectively π^{intense}) which keeps normal (intense) mode or moves to normal (intense) mode and the discount-optimal policy π^* for $\gamma = 0.99$, i.e.

$$\pi^{\text{normal}} = \begin{pmatrix} \text{keep} \\ \text{keep} \\ \text{keep} \\ \text{keep} \\ \text{move} \\ \text{move} \\ \text{move} \\ \text{move} \end{pmatrix}, \quad \pi^{\text{intense}} = \begin{pmatrix} \text{move} \\ \text{move} \\ \text{move} \\ \text{move} \\ \text{keep} \\ \text{keep} \\ \text{keep} \\ \text{keep} \end{pmatrix}, \quad \pi^* = \begin{pmatrix} \text{keep} \\ \text{keep} \\ \text{keep} \\ \text{move} \\ \text{move} \\ \text{move} \\ \text{keep} \\ \text{keep} \end{pmatrix}.$$

Table 3.1. Discounted values with $\gamma = 0.99$ for different policies from initial state $s_{\text{init}} = (0,0,\text{normal},\text{idle})$.

policy	$\left(V^{0.99}\right)^{\pi}(s_{\text{init}})$
π^{normal}	1952.36
π^{intense}	1435.00
π^*	2220.95

The optimal policy was computed by policy iteration with initial policy π^{normal} and converged after 3 iterations. In each iteration a system of linear equations with $|S| = 8$ variables had to be solved and $\sum_{s \in S} |e(s)| = 16$ updates (sets of improving actions) computed. With standard algorithms like Gaussian elimination for solving the system of linear equations the worst-case complexity in each iteration is $O(|S|^3 + |Act||S|)$. Note that the number of all available policies is $|\Pi| = |S|^{|Act|} = 256$. Thus policy iteration is a huge gain in efficiency in contrast to the brute-force method, which computes the values V^{π} for every policy $\pi \in \Pi$ in order to establish the global maximum $V^*(s) = \max_{\pi \in \Pi} V^{\pi}(s)$.

The value iteration algorithm with initial value function constantly 0, $\varepsilon = 0.1$ and maximum norm $||.||$ converged after 1067 iterations to the value $V_{1067}^{0.99}(s_{\text{init}}) = 2220.90$ and the value-maximizing policy $\pi_{\varepsilon} = \pi^*$. In each iteration the values $V_n(s)$ for all states s have to be updated, thus the worst-case complexity for one iteration is $O(|Act||S|)$. Also note that value iteration already finds π^* after only 6 iterations with value $V_6^{0.99}(s_{\text{init}}) = 89.08$ – all the other remaining steps just solve the linear equation $V^{0.99} = R^{\pi^*} + 0.99 P^{\pi^*} V^{0.99}$ for π^* by the iterative procedure $V_{n+1}^{0.99} = R^{\pi^*} + 0.99 P^{\pi^*} V_n^{0.99}$. Nevertheless, value iteration in its presented form has to run until it terminates, in order to find a value function which can be guaranteed to be close to the optimal value (distance measured in maximum norm). Furthermore, it is a priori not known at which iteration step the optimization phase stops, i.e. the actions not improvable anymore. □

Remark 3.3. (i) One of the main theorems in Markov Decision Theory states
    (at least for the models we consider in this tutorial) that if one searches
    for an optimal value function within the broader space of randomized and
    history-dependent policies, then an optimal policy is still stationary Marko-
    vian deterministic, i.e. lies within Π (see also Remark 3.1). This is the reason
    why we stick in this introductory tutorial from beginning to the smaller pol-
    icy space Π.

(ii) An MDP problem can also be transformed to a linear programming problem,
    such that methods coming from the area of linear optimization can be applied
    to solve MDPs [33]. In its generality, a linear optimization formulation also
    allows to add further constraints to the set of linear constraints induced
    by the Bellman equation. An MDP model with an additional set of linear
    constraints is also known as a Constrained MDP [1]. It can be shown that in
    this case stationary Markovian deterministic policies $\pi \in \Pi$ can indeed be
    outperformed in their value by randomized and history-dependent policies.
    Also note that history-dependent policies can also be better then stationary
    Markovian deterministic policies in the finite-horizon case.

3.3   Stochastic Shortest Paths

We now address the problem of optimizing MDPs with respect to the infinite-
horizon total reward measure. In contrast to the discounted case, the existence
of an optimal value in general can not be guaranteed. In case it exists, it is
difficult to provide convergence criteria for dynamic programming algorithms.
Therefore, in literature one finds existence results and convergence criteria only
for special classes of MDPs with respect to this measure. In the following, we
show only one frequently used application for this type of measure.

Definition 3.4. *A **stochastic shortest path problem (SSP)** is an MDP
(S, Act, e, P, R) with an absorbing and reward-free state goal $\in S$, i.e. for all
policies $\pi \in \Pi$ it holds*

$$P^{\pi}(goal, goal) = 1 \quad and \quad R^{\pi}(goal) = 0.$$

Typically, in SSPs the goal is to minimize costs and not to maximize rewards
(see Definition 3.3). Of course, maximization of rewards can be transformed to
a minimization of costs, where costs are defined as negative rewards. Note that
we allow for rewards (and therefore also for costs) to have both a positive and a
negative sign. In order to be consistent with the rest of this tutorial, we stick in
the following to the maximization of rewards. For being able to provide results
on the existence of optimal solutions, we have to define the notion of a proper
policy.

Definition 3.5. *A policy π is called **proper** if there is $m \in \mathbb{N}$ such that under
π the goal state can be reached from every state with positive probability within
m steps, i.e.*

$$\exists m \in \mathbb{N} \; \forall s \in S : \; (P^{\pi})^m(s, goal) > 0.$$

*A policy is called **improper** if it is not proper.*

Define the Bellman operator $\mathcal{T}\colon \mathbb{R}^S \to \mathbb{R}^S$ by

$$(\mathcal{T}V)(s) := \max_{a \in e(s)} \left\{ R(s,a) + \sum_{s' \in S} P(s,a,s')V(s') \right\}.$$

In analogy to the discounted case it holds that $\mathcal{T}$ is monotonic (see Lemma 3.1). But the contraction property with respect to the maximum norm is in general not satisfied. However, Bertsekas and Tsitsiklis proved for typical SSPs the existence and uniqueness of optimal values and the existence of optimal policies [9, 10].

Theorem 3.4 (Existence Theorem). *Consider an SSP $\mathcal{M} = (S, Act, e, P, R)$ with infinite-horizon total reward measure. Further assume that there exists a proper policy $\pi_p \in \Pi$ and for every improper policy π_i there exists $s \in S$ such that $V_\infty^{\pi_i}(s) = -\infty$.*

(i) There exists an optimal value V_∞^* which is the unique fixed point of $\mathcal{T}$, i.e $\mathcal{T}V_\infty^* = V_\infty^*$.
(ii) There exists an optimal policy π^* and it holds that $V_\infty^{\pi^*} = V_\infty^*$.
(iii) Every optimal policy π^* can be derived from the optimal value V_∞^* by

$$\pi^*(s) \in \operatorname*{argmax}_{a \in e(s)} \left\{ R(s,a) + \sum_{s' \in S} P(s,a,s')V_\infty^*(s') \right\}.$$

The dynamic programming algorithms value iteration and policy iteration as presented for discounted MDPs (Theorems 3.2 and 3.3) can be applied for SSPs in an analogous way and are shown in Theorems 3.5 and 3.6. The most important difference is the termination criterion in value iteration, which in contrast to the discounted case uses a weighted maximum norm in order to measure the distance between the iterated values and the optimal value. The proofs for both theorems can be found in [8, 10].

Theorem 3.5 (Value Iteration). *Consider an SSP $\mathcal{M}$ with the assumptions from Theorem 3.4. Let $V_0(s)$ be an arbitrary value function with $V_0(goal) = 0$. Define the sequence*

$$V_{n+1}(s) := (\mathcal{T}V_n)(s) = \max_{a \in e(s)} \left\{ R(s,a) + \sum_{s' \in S} P(s,a,s')V_n(s') \right\}.$$

(i) V_n converges to V^*.
(ii) If every policy is proper, then there exists $\xi \in \mathbb{R}^S$ with $\xi(s) \geq 1$ for all $s \in S$, such that $\mathcal{T}$ is a contraction with respect to the ξ-weighted maximum norm $||.||_\xi$ defined by

$$||V||_\xi := \max_{s \in S} \frac{|V(s)|}{\xi(s)}.$$

As Lipschitz constant q for contraction of $\mathcal{T}$ choose $q := \max_{s \in S} \frac{\xi(s)-1}{\xi(s)}$. For a given $\varepsilon > 0$ stop value iteration when $||V_{n+1} - V_n||_\xi < \frac{1-q}{2q}\varepsilon$ and choose

$$\pi_\varepsilon(s) \in \operatorname*{argmax}_{a \in e(s)} \left\{ R(s,a) + \sum_{s' \in S} P(s,a,s')V_{n+1}(s') \right\}.$$

Then $||V^{\pi_\varepsilon} - V^*||_\xi < \varepsilon$.

In the following, we want to briefly outline, how $\xi(s)$ can be determined. Consider an arbitrary policy π. Since π is proper by assumption, the expected number of transitions $t(s)$ from s to *goal* is finite. If π is the single available policy, then take $\xi(s) := t(s)$. In case there are more policies available, it could be the case, that there exists another policy which enlarges for some state s the expected number of transitions towards *goal*. Thus $\xi(s)$ can be chosen as the maximal expected number of transitions to *goal* among all policies. In order to compute ξ exactly, a modified SSP can be considered: Define a reward $R(s, a)$ which acts as a counter for the number of transitions to *goal*, i.e. each state $s \neq goal$ is rewarded $R(s, a) := 1$ independent of $a \in e(s)$ and the *goal* state is rewarded 0. Choose ξ as the optimal solution to the induced Bellman equation:

$$\xi(s) = 1 + \max_{a \in e(s)} \left\{ \sum_{s' \in S} P(s, a, s')\xi(s') \right\} \text{ for } s \neq goal \quad \text{and} \quad \xi(goal) = 0. \quad (3.8)$$

Note that if we allow improper policies π_i in the termination check of Theorem 3.5 then by definition of π_i there is some state s_i from which *goal* is reached with probability 0. In this case the Bellman equation (3.8) is not solvable, since any solution ξ would imply that $\xi(s_i) = \infty$. The proof for the termination criterion in Theorem 3.5 is completely analogous to the proof of Theorem 3.2 (see Appendix, page 236). It holds that for every policy π the linear operator T^π defined by $T^\pi V = R^\pi + P^\pi V$ is also a contraction with respect to $||.||_\xi$ and Lipschitz constant q as defined in Theorem 3.5.

For completeness we also state the policy iteration algorithm which is directly transfered from the discounted case as in Theorem 3.3 by setting the discount factor $\gamma := 1$. We omit the proof since it is analogous.

Theorem 3.6 (Policy Iteration). *Let $\pi_0 \in \Pi$ an arbitrary initial policy. Define the following iteration scheme.*

1. *Policy evaluation: Compute the value V^{π_n} of π_n by solving*

$$(I - P^{\pi_n})V^{\pi_n} = R^{\pi_n} \quad \text{with} \quad V^{\pi_n}(goal) = 0$$

and define the set of improving actions

$$A_{n+1}(s) := \operatorname*{argmax}_{a \in e(s)} \left\{ R(s, a) + \sum_{s' \in S} P(s, a, s')V^{\pi_n}(s') \right\}.$$

Termination: If $\pi_n(s) \in A_{n+1}(s)$ for all s then π_n is an optimal policy.

2. *Policy improvement: Otherwise choose an improving policy $\pi_{n+1}(s)$ such that $\pi_{n+1}(s) \in A_{n+1}(s)$.*

The sequence of values V^{π_n} is non-decreasing and policy iteration terminates in a finite number of iterations.

Example 3.3. Coming back to our queueing model from Example 3.1, we are now interested in the following two SSP problems:

(i) $\mathcal{M}_1$: the total expected profit up to first loss and
(ii) $\mathcal{M}_2$: the total expected number of accomplished jobs up to first loss.

For both models we add to $\mathcal{M}$ from Fig. 3.1 a reward-free *goal* state and redirect from the states representing a full queue into *goal* the probability mass for the job loss event (i.e. $q\overline{p_{d,n}}$ respectively $q\overline{p_{d,i}}$). Since by Definition 3.1 every state must have at least one action, we add an artificial action *idle* for looping in *goal* with probability 1.0. For model $\mathcal{M}_1$ and $s \neq goal$ the rewards $R_1^{\text{keep}}(s)$ and $R_1^{\text{move}}(s)$ are given as in (3.1) with $C_{\text{loss}} = \$0$. For model $\mathcal{M}_2$ the rewards R_2^{move} and R_2^{keep} are independent of the action and equal to $1 \cdot p_{d,n}$ in normal mode and $1 \cdot p_{d,i}$ in intense mode. We set all the rewards for both models to 0 in state *goal* when taking action *idle*.

If we set the completion probability $p_{d,i} = 1.0$ in the intense mode, it is obvious that going to intense mode would be optimal, since the expected reward for accomplishing a job is greater than the running costs in intense mode. Furthermore, no jobs would be lost in intense mode and thus the total value would diverge to ∞ for all states. Comparing this fact to the assumptions of Theorem 3.4, it holds that $p_{d,i} = 1.0$ implies the existence of an improper policy (moving respectively keeping in intense mode) which does not diverge to $-\infty$. Therefore, we set in the following $p_{d,i} = 0.6$. Now, every policy is proper, since the probability to be absorbed in *goal* is positive for all states and all policies. The following optimal policies π_i^* for model $\mathcal{M}_i$ and values V_i^* were computed by policy iteration.

$$
\pi_1^* = \begin{pmatrix} \text{keep} \\ \text{keep} \\ \text{move} \\ \text{move} \\ \hline \text{keep} \\ \text{keep} \\ \text{keep} \\ \text{keep} \\ \hline \text{idle} \end{pmatrix}, V_1^* = \begin{pmatrix} 11447.5 \\ 11447.5 \\ 11047.5 \\ 8803.75 \\ \hline 11450.0 \\ 11490.0 \\ 11170.0 \\ 9230.0 \\ \hline 0.0 \end{pmatrix}, \quad \pi_2^* = \begin{pmatrix} \text{move} \\ \text{move} \\ \text{move} \\ \text{move} \\ \hline \text{keep} \\ \text{keep} \\ \text{keep} \\ \text{keep} \\ \hline \text{idle} \end{pmatrix}, V_2^* = \begin{pmatrix} 193.5 \\ 193.25 \\ 186.13 \\ 148.06 \\ \hline 193.5 \\ 193.5 \\ 187.5 \\ 154.5 \\ \hline 0.0 \end{pmatrix}
$$

Note also that if value iteration is applied, then the maximal expected number $\xi(s)$ of transitions from s to *goal* is given as the optimal solution of the SSP in (3.8) by

$$
\xi = (790.0, 785.0, 752.5, 596.25, 790.0, 786.0, 758.0, 622.0, 0.0).
$$

Therefore, the Lipschitz constant $q = 0.9988$ can be chosen, which is very high and makes value iteration in the ξ-weighted maximum norm terminating very late. Thus, the termination criterion in the ξ-weighted norm is a theoretical guarantee, but in general not applicable in practice.   □

We finish the section on the infinite-horizon total reward measure with an outlook to other typical total value problems discussed in [33].

Remark 3.4. Beside SSPs one also often considers the following model types:

(i) **Positive models**: For each state s there exists $a \in e(s)$ with $R(s, a) \geq 0$ and for all policies π the value $V^\pi(s) < \infty$ (this assumption can be relaxed).
(ii) **Negative models**: For each state s all rewards $R(s, a) \leq 0$ and there exists a policy π with $V^\pi(s) > -\infty$ for all s.

In a positive model the goal is to maximize the accumulation of positive rewards towards ∞. The model assumptions make this possible, since in every state there is at least one non-negative reward available and the accumulation does not diverge to ∞. In contrast, the goal in a negative model is to maximize negative rewards, i.e. to minimize costs towards 0. The value iteration algorithm for both models is the same as in Theorem 3.5, but without convergence criteria. The policy iteration algorithm however differs from the policy iteration for SSPs. In a positive model, the initial policy π_0 has to be chosen suitable with $R^{\pi_0}(s) \geq 0$. Furthermore, in the policy evaluation phase the solutions to the linear equation $V = R^\pi + P^\pi V$ for a policy π spans up a whole subspace of $\mathbb{R}^S$. In this subspace a minimal solution $V_{\min} \geq 0$ has to be chosen in order to perform the policy improvement phase on $V_{\min}$. In contrast, in a negative model, the initial policy π_0 has to fulfill $V^{\pi_0} > -\infty$ and in the policy evaluation phase the maximal negative solution has to be computed. Puterman shows in [33] that for both model types value iteration converges for $V_0 = 0$, but convergence of policy iteration is only assured for positive models.

3.4 Average Reward Measure

We now come to the final part of the MDP section, which is devoted to the optimization of the average reward measure. As a reminder, let us consider for the moment a discrete-time Markov Reward Model $\mathcal{M} = (S, P, R)$. From Theorem 2.4 we know that if g is the average reward of $\mathcal{M}$ then $g = Pg$. On the other hand, the average reward cannot be uniquely determined by this single equation. Any solution u to $u = Pu$ defines a further linear equation $u + (I - P)v = R$ in v. If this additional equation is solvable, then and only then $u = g$ is the average reward. In this case the bias h is one of the possible solutions to the second equation (and unique modulo $\ker(I - P)$). We write both of these equations in the fixed-point form

$$g = Pg \quad \text{and} \quad h = (R + Ph) - g. \tag{3.9}$$

Also remember, that if $\mathcal{M}$ is unichain, then by Corollary 2.1 the equation $g = Pg$ can be simplified to "g is constant".

We are concerned in this section with the optimization of the average reward for an MDP $\mathcal{M} = (S, Act, e, P, R)$. Surely, every policy $\pi \in \Pi$ of $\mathcal{M}$ induces a DTMRM $\mathcal{M}^\pi = (S, P^\pi, R^\pi)$ and we can compute for each π the average reward

g^π as described above. By Definition 3.3 the optimal average value function is $g^*(s) = \sup_{\pi \in \Pi} g^\pi(s)$ and in case an optimal policy π^* exists then $g^{\pi^*}(s) \geq g^\pi(s)$ for all s and all π. In analogy to the previous sections, it is possible to establish Bellman equations which reduce the global optimization task over all policies $\pi \in \Pi$ to a local state-wise search over actions $e(s)$ for all states s. Since the formal derivation of these Bellman equations is slightly involved, we just state them and refer for proof to [33]. Define for each of the two linear fixed-point equations in (3.9) the Bellman operators $\mathcal{T}_{av} : \mathbb{R}^S \to \mathbb{R}^S$ and $\mathcal{T}^g_{bias} : \mathbb{R}^S \to \mathbb{R}^S$ (parametrized by $g \in \mathbb{R}^S$) as follows:

$$(\mathcal{T}_{av}g)(s) := \max_{a \in e(s)} \left\{ \sum_{s' \in S} P(s, a, s')g(s') \right\} \tag{3.10}$$

$$(\mathcal{T}^g_{bias}h)(s) := \max_{a \in e^g(s)} \left\{ R(s, a) + \sum_{s' \in S} P(s, a, s')h(s') \right\} - g(s) \tag{3.11}$$

$$\text{where } e^g(s) := \left\{ a \in e(s) \mid g(s) = \sum_{s' \in S} P(s, a, s')g(s') \right\}.$$

The corresponding Bellman optimality equations for the average reward are just the fixed-point equations of these operators and read as

$$g(s) = \max_{a \in e(s)} \left\{ \sum_{s' \in S} P(s, a, s')g(s') \right\} \tag{3.12}$$

$$h(s) = \max_{a \in e^g(s)} \left\{ R(s, a) + \sum_{s' \in S} P(s, a, s')h(s') \right\} - g(s). \tag{3.13}$$

Equations (3.12) and (3.13) are referred to as the first and the second optimality equation. In order to provide an intuition for these equations, assume for the moment that there is an optimal policy π^* with $g^{\pi^*} = g^*$ and moreover that the MDP $\mathcal{M}$ is unichain. In this case, the average reward $g^\pi(s)$ is a constant function for any policy π and thus the first optimality equation does not yield any further restriction since it is satisfied for every constant function g. Only the second equation takes the reward values $R(s, a)$ into account that are needed in order determine their average. For each policy π the average reward g^π and the bias h^π satisfy

$$h^\pi(s) = (R^\pi(s) - g^\pi(s)) + \sum_{s' \in S} P^\pi(s, s')h^\pi(s'),$$

i.e. the bias h^π is the total value function for the DTMRM with average-corrected rewards $R^\pi(s) - g^\pi(s)$. Since this holds especially for $\pi = \pi^*$, the second optimality equation can be seen as a Bellman equation for maximizing the total value function for the MDP model with rewards $R(s, a) - g^*(s)$. In other words, if π is an arbitrary policy then the DTMRM with rewards $R(s, \pi(s)) - g^*(s)$

has a total value function if and only if $g^*(s)$ is the average reward for the DTMRM (S, P^π, R^π) and this holds especially for $\pi = \pi^*$. In case $\mathcal{M}$ is multichain, then $g^{\pi^*}(s)$ is only constant on recurrent classes of $\mathcal{M}^{\pi^*}$, whereas if s is transient then $g^{\pi^*}(s)$ is a weighted sum over all those average rewards on recurrent classes. This means that g^{π^*} has to fulfill $g^{\pi^*} = P^{\pi^*} g^{\pi^*}$ in addition and thus g^{π^*} is a solution to the first optimality equation. Since both Bellman equations are nested and have to be satisfied simultaneously, it is possible to reduce the set of actions $e(s)$ in the second equation to the maximizing actions $e^g(s) = \mathrm{argmax}_{a \in e(s)} \left\{ \sum_{s' \in S} P(s, a, s') g(s') \right\}$ for a solution g of the first equation. Note that in case of unichain models it holds that $e^g(s) = e(s)$ for all s.

The following theorem formalizes the explanations in the motivation above and connects the Bellman equations to the optimal value and optimal policies. We refer for proof to [33].

Theorem 3.7 (Existence Theorem). *Consider an MDP $\mathcal{M} = (S, Act, e, P, R)$ with average reward measure.*

(i) *The average optimal value function g^* is a solution to (3.12), i.e $g^* = \mathcal{T}_{av} g^*$. For $g = g^*$ there exists a solution h to (3.13), i.e. $h = \mathcal{T}^{g^*}_{bias} h$. If g and h are solutions to (3.12) and (3.13) then $g = g^*$.*

(ii) *There exists an optimal policy π^* and it holds that $g^{\pi^*} = g^*$.*

(iii) *For any solution h to (3.13) with $g = g^*$ an optimal policy π^* can be derived from*

$$\pi^*(s) \in \underset{a \in e^{g^*}(s)}{\mathrm{argmax}} \left\{ R(s, a) + \sum_{s' \in S} P(s, a, s') h(s') \right\}.$$

As a special case of part (iii) in this theorem it holds that if for a policy π the average reward g^π and the bias h^π solve the Bellman optimality equations, then π is optimal. In contrast to the discounted and total reward cases, the converse does not hold. This means that if a policy π is optimal then g^π and h^π are not necessary solutions to the Bellman equations.

The following policy iteration algorithm, can be applied in order to compute the optimal average reward g^* as well as an optimal policy π^*. A proof can be found in [33].

Theorem 3.8 (Policy Iteration). *Let $\pi_0 \in \Pi$ be an initial policy. Define the following iteration scheme:*

1. **Policy evaluation:** *Compute a solution $(g^{\pi_n}, h^{\pi_n}, w)^T$ to*

$$\begin{pmatrix} I - P^{\pi_n} & 0 & 0 \\ I & I - P^{\pi_n} & 0 \\ 0 & I & I - P^{\pi_n} \end{pmatrix} \begin{pmatrix} g^{\pi_n} \\ h^{\pi_n} \\ w \end{pmatrix} = \begin{pmatrix} 0 \\ R^{\pi_n} \\ 0 \end{pmatrix}. \tag{3.14}$$

Define

$$G_{n+1}(s) := \underset{a \in e(s)}{\mathrm{argmax}} \left\{ \sum_{s' \in S} P(s, a, s') g^{\pi_n}(s') \right\}.$$

2. **Policy improvement:** If $\pi_n(s) \notin G_{n+1}(s)$ for some s then choose an improving policy π_{n+1} with $\pi_{n+1}(s) \in G_{n+1}(s)$ and go to the policy evaluation phase. Otherwise if $\pi_n(s) \in G_{n+1}(s)$ for all s then define

$$H_{n+1}(s) := \operatorname*{argmax}_{a \in e^{g^{\pi_n}}(s)} \left\{ R(s,a) + \sum_{s' \in S} P(s,a,s') h^{\pi_n}(s') \right\}.$$

If $\pi_n(s) \notin H_{n+1}(s)$ for some s then choose an improving policy π_{n+1} such that $\pi_{n+1}(s) \in H_{n+1}(s)$ and go to the policy evaluation phase.
Termination: If $\pi_n(s) \in H_{n+1}(s)$ for all s then π_n is an optimal policy.

The values g^{π_n} are non-decreasing and policy iteration terminates in a finite number of iterations with an optimal policy π_n and optimal average value g^{π_n}.

Note that (3.14) corresponds to (2.19) instead of (2.18) for the following reason: For each policy π (2.18) provides a unique solution g^π to the average reward, but in general the bias h^π cannot be uniquely determined. Policy iteration can be assured to converge if the bias h^{π_n} is computed for each iterated policy π_n [33]. As described in Remark 2.4(iii) this can be done by solving (2.19), i.e. the equation $v + (I - P^{\pi_n})w = 0$ in addition to (2.18) for which $v = h^{\pi_n}$ is the unique solution. There are also other possibilities to assure convergence of policy iteration by solving only (2.18) and fixing a scheme that chooses a solution v to $g + (I - P)v = R$ in order to prevent cycles in policy iteration (see Remark 2.4(ii)).

Before showing the application of policy iteration on our queueing model running example, we first state the following remark regarding some algorithmic aspects.

Remark 3.5. (i) During policy iteration the action set $e^{g^{\pi_n}(s)}$ can be replaced by the whole action set $e(s)$ – this leads to the so-called modified optimality equations. The convergence and the optimality of the solution in policy iteration are not influenced by this replacement.

(ii) In the policy improvement phase, there are two jumps to the policy evaluation phase, which represent two nested cycles of evaluation and improvement phases. First, a policy π_n has to be found, which solves the first optimality equation. Then in a nested step, π_n is tested on the second optimality equation. If π_n can be improved by a better policy π_{n+1} with actions from H_{n+1} then π_{n+1} has to be sent back to the first evaluation and improvement cycle until it again solves the first optimality equation, and so on.

(iii) As already mentioned in the introducing motivation, if it is a priori known that the MDP is unichain, i.e. for all policies there is only one closed recurrent class of states, then the optimal average reward is constant and the first optimality equation is automatically satisfied (see Corollary 2.1). This reduces the complexity of policy iteration, since only the second optimality equation has to be considered for optimization.

(iv) We skip the value iteration algorithm in this tutorial since it is exactly the same as for the discounted case (Theorem 3.2) with $\gamma := 1$. It can be proven that the sequence $V_{n+1} - V_n$ converges to the optimal average reward g^*, if for every (optimal) policy the transition matrix is aperiodic [33]. The aperiodicity constraint is not a restriction, since every periodic DTMRM can be made aperiodic, by inserting self-loops with strictly positive probability for every state. (The reward function has to be transformed accordingly.) However, [33] presents a termination criterion for value iteration only for models with $g^*(s)$ constant for all s (e.g. unichain models).

Example 3.4. Consider the queueing MDP model from Example 3.1. We want to compute the optimal average value function for the queueing model with parameters $q = 0.25$, $p_{d,n} = 0.5$ and $p_{d,i} = 1.0$ and the reward structure as specified in (3.1). Note that the model is multichain, since the policy that takes the action keep in every state induces a DTMRM with two recurrent classes. Policy iteration converges after three iterations (with initial policy π_0 which keeps in normal mode or moves to it from intense mode) and results in the following optimal policy π^*, optimal average value $g^* = g^{\pi^*}$ and bias h^{π^*}:

$$\pi^* = \begin{pmatrix} \text{keep} \\ \text{keep} \\ \text{keep} \\ \hline \text{move} \\ \text{move} \\ \text{move} \\ \text{keep} \\ \text{keep} \end{pmatrix}, \quad g^{\pi^*} = \begin{pmatrix} 22.7 \\ 22.7 \\ 22.7 \\ 22.7 \\ \hline 22.7 \\ 22.7 \\ 22.7 \\ 22.7 \end{pmatrix}, \quad h^{\pi^*} = \begin{pmatrix} -49.4 \\ 41.3 \\ 95.3 \\ 38.7 \\ -59.4 \\ 40.6 \\ 130.3 \\ 220.0 \end{pmatrix}.$$

Thus, the optimal policy induces a DTMRM that is unichain with constant optimal average reward 22.7. The finite-horizon total value function $V_N^{\pi^*}$ from state $(0, 0, \text{normal}, \text{idle})$ increases asymptotically with $22.7 \cdot N - 49.4$ as $N \to \infty$. $\quad\square$

We conclude the MDP section by a further remark, which presents a short outlook on other optimization criteria that are applicable for MDPs.

Remark 3.6. The average reward considers reward accumulation in the long-run. Therefore, it is not very sensitive in the selection between policies with the same average reward: If two policies have the same long-run average reward but different short-run rewards, then one would prefer among all policies with the same average reward such a policy that also maximizes the short-run reward accumulation. This idea leads to the so-called bias (or more general *n-bias* or *n-discount*) optimization criteria, which belongs to the class of **sensitive discount optimality** criteria. In more detail, policies $\pi \in \Pi$ are compared regarding their Laurent series expansions (2.15)

$$(V^\gamma)^\pi = a_{-1} g^\pi + a_0 H^\pi R^\pi + a_1 (H^\pi)^2 R^\pi + \dots,$$

where a_i are constants (depending only on γ) and $H^\pi R^\pi = h^\pi$ is the bias, which represents the excess in reward accumulation up to steady-state. Now if a subset of policies $\Pi^*_{-1} \subseteq \Pi$ maximize g^π then this subset can be further refined to a reduced subset $\Pi^*_0 \subseteq \Pi^*_{-1}$ by maximizing the bias h^π for $\pi \in \Pi^*_{-1}$ in addition to the average reward g^π. If Π^*_0 still consists of more then one policy, then one can proceed iteratively and compare the higher order n-bias terms sequentially. Note that the n-bias reward measure specifies the vector space $\mathcal{V}$ of value functions as $\mathcal{V} = \left\{ V : S \to \mathbb{R}^{n+1} \right\}$ and values can be compared in Definition 3.3 by lexicographic order.

The most sensitive optimization criterion is the **Blackwell optimality criterion**, which selects a policy π^*, such that the entire Laurent series expansion (i.e. the complete discounted value) is maximal among the discounted values for all policies and for all discount factors high enough, i.e.

$$\exists \gamma^* \in [0,1) \; \forall \gamma \in [\gamma^*, 1) \; \forall \pi \in \Pi : \; (V^\gamma)^{\pi^*} \geq (V^\gamma)^\pi .$$

It can be shown, that Blackwell optimal policies exist and a Blackwell optimal policy is n-bias optimal for all n [33]. Furthermore, such a policy can be computed by proceeding with the policy space reduction as described above until some Π^*_n consists of only one policy, which is Blackwell optimal (or if $n \geq |S| - 2$ then Π^*_n is a set of Blackwell optimal policies).

4   Continuous Time Markov Reward Models

In both DTMRMs and MDPs time is discrete, i.e. it proceeds in a step by step manner. However, this is unrealistic for many applications – one rather wishes to work with a continuous notion of time. Therefore, in this and the following section, we study continuous-time models. Since we stick to the principle of memorylessness, it will turn out that the state sojourn times follow an exponential distribution (as opposed to the geometrically distributed sojourn times in the discrete-time setting).

4.1   Preliminaries

Definition 4.1. *A **continuous-time Markov chain (CTMC)** is a structure $\mathcal{M} = (S, Q)$ with finite state space S and generator function $Q \colon S \times S \to \mathbb{R}$ such that $Q(s, s') \geq 0$ for $s' \neq s$ and $\sum_{s' \in S} Q(s, s') = 0$ for all s. A **continuous-time Markov Reward Model (CTMRM)** is a structure (S, Q, i, r) which enhances a CTMC (S, Q) by a reward structure consisting of an impulse reward function $i \colon S \times S \to \mathbb{R}$ for transitions with $i(s, s) = 0$ for all $s \in S$ and a rate reward function $r \colon S \to \mathbb{R}$ for states.*

From state $s \in S$ each quantity $Q(s, s')$ with $s' \neq s$ defines an event which occurs after a random amount of time $\tau_{s,s'} \in \mathbb{R} \cup \{\infty\}$ to trigger. If $Q(s, s') > 0$ then $\tau_{s,s'}$ is exponentially distributed with rate $Q(s, s')$ and otherwise if $Q(s, s') = 0$ then we set $\tau_{s,s'} := \infty$. For a fixed $s \in S$ all these times $\tau_{s,s'}$ are independent and concurrently enabled. Therefore, they define a race among each other and only that $\tau_{s,s_0'}$ which triggers first, within a finite amount of time (i.e. $\tau_{s,s_0'} \le \tau_{s,s'}$ for all $s' \in S$), wins the race. In this case the system performs a transition from s to $s_0' \ne s$ and collects the impulse reward $i(s, s_0') \in \mathbb{R}$. The time τ_s that the system resides in state s (up to transition) is called the **sojourn time** and fulfills $\tau_s = \min \{\tau_{s,s'} \mid s' \ne s\}$. While the system is in state s the rate reward $r(s)$ is accumulated proportionally to the sojourn time τ_s. Thus, the accumulated reward in s for the sojourn time including the transition to state s_0' is given by $R(s) := i(s, s_0') + r(s)\tau_s$. The quantity $E(s) := \sum_{s' \ne s} Q(s, s')$ is called the **exit rate** in state s and by definition of the generator function Q it holds that $E(s) = -Q(s, s) \ge 0$. If $E(s) > 0$ then there is some state s' with $Q(s, s') > 0$ and due to the race condition it holds that τ_s is exponentially distributed with rate $E(s)$. The **transition probability** $P(s, s')$ is the probability that $\tau_{s,s'}$ wins the race and is given by $P(s, s') = P(\tau_{s,s'} = \tau_s) = \frac{Q(s,s')}{E(s)}$. Otherwise, if $E(s) = 0$ (or all $Q(s, s') = 0$) then $\tau_s = \infty$ and state s is absorbing. In this case we set $P(s, s') := \delta_{s,s'}$, i.e. $P(s, s') = 1$ if $s' = s$ and $P(s, s') = 0$ if $s' \ne s$. The function $P : S \to \mathcal{D}(S)$ with $(P(s))(s') := P(s, s')$ is called the **embedded transition probability** function. The model (S, P) can be considered as a discrete-time Markov chain, which models the transitions of the underlying CTMC and abstracts from the continuous time information. Similarly to DT-MRMs, $i(s) := \sum_{s' \ne s} P(s, s')i(s, s')$ will denote the state-based version of the transition-based impulse reward $i(s, s')$, i.e. $i(s)$ is the expected impulse reward from state s to some other state $s' \ne s$.

Example 4.1 (WSN node model). A wireless sensor network (WSN) consists of nodes that have to observe their environment by sensing activities and transmit information towards a destination. Each node consists of a battery unit with some initial capacity, a sensor and a transmitter. Furthermore, environmental events occur randomly. For the purposes of this section and in order to show how CTMRMs can be modelled, we assume a very simple WSN node model (see Fig. 4.1), which consists only of

- one sensor node, which randomly switches between "idle" and "listen" states after an exponentially distributed time and does not transmit any information and
- the environment, in which activities occur and endure in a memoryless way.

For simplicity we further assume that the node has infinite energy supply and does not consume any energy in idle mode. In case an environmental activity takes place and the node is listening, it must observe the activity at least until it stops. When the sensor switches from idle to listen it consumes instantaneously 5 energy units. While if the sensor is listening it consumes energy with rate 10. We want to measure the energy consumption in this model. Suitable measures could be the average energy consumption (per time unit) or some discounted energy consumption. □

environment   sensor   composed WSN node model

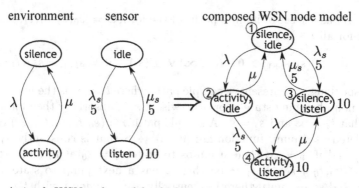

Fig. 4.1. A simple WSN node model, which consists of a single node which can listen to activities in the environment. The transition rates are $\lambda = 2$, $\mu = 4$, $\lambda_s = 4$, $\mu_s = 30$. If the sensor is listening, it uses 10 energy units per time. For every activation to "listen" or deactivation to "idle" an impulse energy of 5 units is employed.

Remark 4.1. Puterman [33] allows impulse rewards which are state-based and are gained in a state s if s is the initial state or when s is reached after some transition from s' to s ("arrival" point of view). In contrast, we have defined transition-based impulse rewards $i(s, s')$ that are gained when state s is left, i.e. a transition from s to s' is performed ("departure" point of view). Therefore, the impulse reward can be considered state-based as the expectation $i(s) = \sum_{s' \neq s} i(s, s') P(s, s')$ over transition probabilities. When considering the infinite-horizon total reward measure or the average reward measure, then both points of view lead to the same value functions and thus their distinction doesn't matter in this case. However, this difference is important when we are dealing with the finite-horizon total reward measure and the discounted reward measure.

Before being able to define and evaluate reward measures for the continuous-time case, we have to provide more theoretical background. The next section is devoted to this.

4.2   Probability Space for CTMCs

In the following, we want to formalize the transition behavior of a CTMC $\mathcal{M} = (S, Q)$ that we have informally introduced in Sect. 4.1. For this reason, we first define a suitable sample space Ω together with a Borel σ-algebra $\mathcal{B}(\Omega)$ consisting of those measurable events for which we will assign probabilities. Subsequently, we will define several stochastic processes upon Ω that are all induced by the CTMC $\mathcal{M}$. These processes will allow us to define a time-dependent transition probability matrix, which in turn will play an important role for the definition of reward measures for a CTMRM.

4.2.1   Sample Space
Since continuous time plays a role for these measures, we put this information along with the state space S into the sample space. Define the sample space $\Omega \subseteq$

$(S \times (0, \infty])^{\mathbb{N}}$ as the set of infinite paths of the form $\omega = (s_0, t_0, s_1, t_1, s_2, t_2, \dots)$ such that for all $i \in \mathbb{N}$:

$$(E(s_i) > 0 \Rightarrow Q(s_i, s_{i+1}) > 0 \wedge t_i < \infty) \vee (E(s_i) = 0 \Rightarrow s_{i+1} = s_i \wedge t_i = \infty).$$

Roughly speaking, ω represents a sample path, where $t_i < \infty$ is the finite sojourn time in a non-absorbing state s_i or otherwise if s_i is absorbing then for all $j \geq i$ it holds that $t_j = \infty$ and $s_j = s_i$. A sample path $\omega = (s_0, t_0, s_1, t_1, \dots)$ can also be considered as a jump function $\omega : [0, \infty) \to S$ that is constantly s_0 for all $t \in [0, t_0)$ and if $t_0 \neq \infty$ then ω jumps to state $s_1 \neq s_0$ at t_0 and $\omega(t) = s_1$ for all $t \in [t_0, t_0 + t_1)$. If $t_1 \neq \infty$ then ω has a next jump to state $s_2 \neq s_1$ at $t_0 + t_1$ and so on, until there is eventually a first index i with $t_i = \infty$ and therefore $\omega(t) = s_i$ for all $t \geq \sum_{k=0}^{i-1} t_k$. In order to define a probability space over Ω we transform Ω to the set $Path$ of finite absorbing and infinite paths as defined in [3]. Let $\psi : \Omega \to Path$ be the transformation that drops the artificial repetitions of absorbing states, i.e.

$$\psi(s_0, t_0, s_1, t_1, \dots) := \begin{cases} (s_0, t_0, s_1, t_1, \dots), & \text{if } \forall k \in \mathbb{N} : t_k < \infty \\ (s_0, t_0, s_1, t_1, \dots, s_l), & l := \min\{k \mid t_k = \infty\} < \infty \end{cases}$$

where $\min \emptyset := \infty$. Note that in the definition of ψ the two cases are disjoint. Since ψ is bijective the probability space $(Path, \mathcal{F}(Path), Pr_\alpha)$ as defined in [3] (where $\alpha \in \mathcal{D}(S)$ is a distribution over initial states) induces for each $s \in S$ a probability space $(\Omega, \mathcal{B}(\Omega), P_s)$ in a canonical way:

$$\mathcal{B}(\Omega) := \{A \subseteq \Omega \mid \psi(A) \in \mathcal{F}(Path)\} \quad \text{and} \quad P_s := Pr_{\delta_s} \circ \psi,$$

where we choose $\alpha := \delta_s$ with $\delta_s(s') := \delta_{s,s'}$ (i.e. s is the initial state). Before moving on, we want to mention that both sample spaces Ω and $Path$ are equivalent, since ψ is bijective (and measurable by definition of $\mathcal{B}(\Omega)$). The sample space $Path$ allows for an intuitive interpretation of sample paths ω regarded as jump functions $\omega : [0, \infty) \to S$ as described above. Every jump function that is constant on intervals of positive length has at most a finite or countably infinite number of jumps – this distinction is encoded in the sample paths of $Path$. However, this differentiation of cases would directly be transferred to a corresponding differentiation in the definition of stochastic processes that we will introduce in the sequel. For this reason, we have chosen Ω as the sample space which embeds these cases already in its definition and thus does not lead to an overload of notation in the definition of these processes.

4.2.2   Induced Stochastic Processes

The CTMC $\mathcal{M} = (S, Q)$ induces a number of stochastic processes over Ω. For $\omega = (s_0, t_0, s_1, t_1, \dots) \in \Omega$ define the

(i) discrete-time state process $(X_n)_{n \in \mathbb{N}}$ by

$$X_n(\omega) := s_n$$

(ii) sojourn time $(\tau_n)_{n \in \mathbb{N}}$, where

$$\tau_n(\omega) := t_n \leq \infty$$

(iii) total elapsed time $(T_n)_{n \in \mathbb{N}}$ for the first n transitions as

$$T_n(\omega) := \sum_{i=0}^{n-1} \tau_i(\omega)$$

(iv) number of transitions $(N_t)_{0 \leq t < \infty}$ up to time t as

$$N_t(\omega) := \max \{ n \mid T_n(\omega) \leq t \} \in \mathbb{N}$$

(note that with probability 1 the maximum is taken over a finite set and thus N_t is almost surely finite, i.e. $P(N_t < \infty) = 1$)

(v) continuous-time state process $(Z_t)_{0 \leq t < \infty}$, where

$$Z_t(\omega) := X_{N_t(\omega)}(\omega),$$

i.e. Z_t is the state of the system at point in time $t \geq 0$.

Remark 4.2. For all $t \in [0, \infty)$ and $n \in \mathbb{N}$ the following equalities of events hold:

$$\{ N_t = n \} = \{ T_n \leq t < T_{n+1} \} \quad \text{and} \quad \{ N_t \geq n \} = \{ T_n \leq t \}.$$

The discrete-time state process X_n represents the n-th visited state (or an absorbing state) and it fulfills the discrete-time Markov property as in (2.4), i.e. for all $s, s_0, s_1, \ldots s_k \in S$ and $0 < n_1 < \cdots < n_k < n$

$$P_{s_0}(X_n = s \mid X_{n_1} = s_1, \ldots, X_{n_k} = s_k) = P_{s_0}(X_n = s \mid X_{n_k} = s_k).$$

From $Z_t(\omega) = X_{N_t(\omega)}(\omega)$ and $N_t(\omega)$ non-decreasing for all ω it follows that the continuous-time state process Z_t also fulfills the Markov property, which reads as a continuous time version:

$$P_{s_0}(Z_t = s \mid Z_{t_1} = s_1, \ldots, Z_{t_k} = s_k) = P_{s_0}(Z_t = s \mid Z_{t_k} = s_k)$$

for all $s, s_0, s_1, \ldots s_k \in S$ and $0 \leq t_1 < \cdots < t_k < t$. Thus given knowledge about the state $Z_{t_k} = s_k$ of the process for any arbitrary point in time $t_k < t$, then the process Z_t does not depend on its history comprising the visited states before time t_k. It further holds that Z_t is homogeneous in time, i.e. the following property holds:

$$P_{s_0}(Z_{t+t'} = s' \mid Z_t = s) = P_s(Z_{t'} = s').$$

As in Sect. 2 we fix a representation of the state space S through indices $\{1, 2, \ldots, n\}, n := |S|$ such that functions $S \to \mathbb{R}$ can be represented by vectors in $\mathbb{R}^n$ and functions $S \times S \to \mathbb{R}$ as matrices in $\mathbb{R}^{n \times n}$. Define the **transient probability matrix** $P(t)$ as

$$P(t)(s, s') := P_s(Z_t = s'). \tag{4.1}$$

The matrix $P(t)$ is stochastic for all $t \geq 0$ and fulfills the property

$$P(t + t') = P(t)P(t') \quad \forall t, t' \geq 0,$$

which reads componentwise as $P(t+t')(s, s') = \sum_u P(t)(s, u) \cdot P(t')(u, s')$. This means that from state s the probability to be in state s' after $t + t'$ time units is the probability to be in some arbitrary state $u \in S$ after t time units and traverse from there within further t' time units to state s'. It can be shown that all entries of $P(t)$ are differentiable for all $t \geq 0$ and $P(t)$ is related to the generator matrix Q of the CTMC by the **Kolmogorov differential equations**

$$\frac{d}{dt}P(t) = QP(t) \quad \text{and} \quad \frac{d}{dt}P(t) = P(t)Q, \tag{4.2}$$

which read in componentwise notation as

$$\frac{d}{dt}(P(t)(s, s')) = \sum_u Q(s, u) \cdot P(t)(u, s') = \sum_v P(t)(s, v) \cdot Q(v, s').$$

All solutions to these equations are of the form $P(t) = e^{Qt}$ since $P(0) = I$ is the identity matrix, where for a matrix A the quantity e^A denotes the matrix exponential that is given by $e^A = \sum_{k=0}^{\infty} \frac{1}{k!}A^k$.

4.2.3 State Classification

As in Sect. 2.1.2 there is also a classification of states in case of continuous time Markov chains. Since this taxonomy is almost the same as in the discrete-time case, we only present it very briefly. The most important difference is that in the continuous-time setup there is no notion for periodicity of states and it can be shown that the matrix $P(t)$ converges as $t \to \infty$ (for finite state spaces). We denote the limit by $P^* := \lim_{t \to \infty} P(t)$. Note that in Definition 2.9 we denoted the corresponding discrete-time limiting matrix as P^∞ and its time-averaged version as P^* and mentioned in Proposition 2.2 that they both coincide if P^∞ exists. Since the existence of this limit in the continuous-time case is always guaranteed, we call this limit directly P^* instead of P^∞ in order to use similar notation. One can show that P^* is stochastic and fulfills the invariance conditions

$$P^*P(t) = P(t)P^* = P^*P^* = P^*.$$

Therefore, the probability distribution $P^*(s, \cdot) \in \mathcal{D}(S)$ in each row of P^* is a **stationary distribution** and since $P(t)(s, \cdot) \to P^*(s, \cdot)$ as $t \to \infty$ it is also the **limiting distribution** from state s. Furthermore, it holds that

$$P^*Q = QP^* = 0,$$

which can be derived from (4.2) and $\frac{d}{dt}P(t) \to 0$ as $t \to \infty$.

Let the random variable $M_s \in (0, \infty]$ denote the point in time when the state process Z_t returns to s for the first time (given $Z_0 = s$). A state s is **transient** if $P_s(M_s = \infty) > 0$ or equivalently $P^*(s, s) = 0$. In the other case, if $P_s(M_s < \infty) = 1$ then s is called **recurrent** and it holds equivalently that $P^*(s,s) > 0$. It can be shown that there is always at least one recurrent state if the state space is finite. A state s' is **reachable** from s if $P(t)(s,s') > 0$ for some $t \geq 0$. The states s and s' are communicating if s' is reachable from s and s is reachable from s'. This communication relation is an equivalence relation and partitions the set of recurrent states into **closed recurrent classes**. Therefore, the state space partitions into $S = \bigcup_{i=1}^{k} S_i^r \cup S^t$, where S^t denotes the set of transient states and S_i^r is a closed recurrent class for all $i = 1, \ldots, k$. For $s, s' \in S_i^r$ in the same recurrent class it holds that $P^*(s,s') > 0$. As in the discrete-time case P^* can be represented by

$$
P^* = \begin{pmatrix}
P_1^* & 0 & 0 & \ldots & 0 & 0 \\
0 & P_2^* & 0 & \ldots & 0 & 0 \\
\vdots & \vdots & \vdots & \ddots & \vdots & 0 \\
0 & 0 & 0 & \ldots & P_k^* & 0 \\
\widetilde{P}_1^* & \widetilde{P}_2^* & \widetilde{P}_3^* & \ldots & \widetilde{P}_k^* & 0
\end{pmatrix}
\tag{4.3}
$$

where P_i^* has identical rows for the stationary distribution in class S_i^r and $\widetilde{P}_i^*$ contains the trapping probabilities from transient states S^t into S_i^r. If a closed recurrent class consists of only one state s, then s is called **absorbing**. A CTMC is **unichain** if $k = 1$ and **multichain** if $k \geq 2$. A unichain CTMC is called **irreducible** or **ergodic** if $S^t = \emptyset$.

4.3   Model Transformations

In this section we present a set of model transformations, which will allow us to

- unify the different types of rewards (impulse reward and rate reward) in the reward accumulation process ("Continuization") and
- relate some continuous-time concepts to discrete-time Markov Reward Models from Sect. 2 ("Embedding" and "Uniformization").

These transformations simplify the evaluation process of all the reward measures and map the computation of the value functions for continuous-time models to the discrete-time case.

4.3.1   Embedding

As mentioned in Sect. 4.1, a CTMC (S, Q) defines for all states $s, s' \in S$ the embedded transition probabilities $P(s, s')$. The structure (S, P) can be considered as a discrete-time Markov chain and it induces on the sample space $\Omega' := \{(s_0, s_1, s_2, \ldots) \in S^{\mathbb{N}} \mid P(s_{i-1}, s_i) > 0 \text{ for all } i \geq 1\}$ as in (2.3) the state process X'_n (by Definition 2.3) given by $X'_n(s_0, s_1, \ldots) = s_n$. This stochastic process is related to the discrete-time state process $X_n : \Omega \to S$ by abstracting away from the time information, i.e. for all $n \in \mathbb{N}$

$$
X_n(s_0, t_0, s_1, t_1, \ldots) = X'_n(s_0, s_1, \ldots).
$$

This equation establishes the connection to DTMCs and thus X_n can be considered as the state process of the DTMC (S, P). Therefore, (S, P) is also called the **embedded discrete-time Markov chain** and X_n is the **embedded state process** of the CTMC (S, Q).

Now consider a CTMRM (S, Q, i, r) and define a function $R : S \times S \to \mathbb{R}$ where $R(s, s')$ denotes the expected accumulated rate reward $r(s)$ in state s over time including the impulse reward $i(s, s')$ gained for transition from s to some other state $s' \neq s$ (as in Sect. 4.1). If s is non-absorbing, then the sojourn time τ_s in s is exponentially distributed with rate $E(s) > 0$ and $R(s, s')$ is given by

$$R(s, s') := i(s, s') + \frac{r(s)}{E(s)}. \tag{4.4}$$

Otherwise, if s is absorbing, the embedding is only possible if $r(s) = 0$ and in this case we define $R(s, s') := 0$ for all s'.

It is very important to note that if we consider a reward measure on the CTMRM with value function V and a corresponding reward measure on the transformed DTMRM with value function V', then it is of course desirable that $V = V'$, i.e. the transformation should be **value-preserving**. This allows to compute the value V by applying the theory and algorithms for the discrete-time models as presented in Sect. 2. However, as we will see, such a model transformation needs in general the reward measure itself as input in order to be value-preserving. As an example, the integration of the rate reward $r(s)$ into the reward $R(s, s')$ is performed by total expectation over an infinite time-horizon, which gives the term $\frac{r(s)}{E(s)}$. If one considers a finite horizon for the continuous-time model, then $R(s, s')$ as defined is obviously not the appropriate reward gained in state s in the embedded discrete-time model.

4.3.2   Uniformization

We have seen in Sect. 4.1 that the quantities $Q(s, s')$ for $s' \neq s$ can be regarded as rates of exponentially distributed transition times $\tau_{s,s'}$. All these transition events define a race and only the fastest event involves a transition to another state $s' \neq s$. We can manipulate this race, by adding to the set of events $\{\tau_{s,s'} \mid s' \neq s\}$ of a state s an auxiliary exponentially distributed event $\tau_{s,s}$ with an arbitrary positive rate $L(s) > 0$ that introduces a self-loop (i.e. a transition from s to s), if it wins the race. The time up to transition is $\tau_s := \min \{\tau_{s,s'} \mid s' \in S\}$ and it is exponentially distributed with increased exit rate $\widetilde{E}(s) := E(s) + L(s)$. The probability that $\tau_{s,s}$ wins the race can be computed to $P(\tau_{s,s} \leq \tau_{s,s'} \; \forall s' \in S) = \frac{L(s)}{\widetilde{E}(s)} = 1 + \frac{Q(s,s)}{\widetilde{E}(s)}$ and for all $s'_0 \neq s$ it holds that $P(\tau_{s,s'_0} \leq \tau_{s,s'} \; \forall s' \in S) = \frac{Q(s,s')}{\widetilde{E}(s)}$. We can add such events $\tau_{s,s}$ to a set of states s and thus increase the exit rates for all these states simultaneously. Moreover, we can choose an arbitrary $\mu > 0$ with $\max \{E(s) \mid s \in S\} \leq \mu < \infty$ (called **uniformization rate**) such that $\widetilde{E}(s) \equiv \mu$ is constant for all $s \in S$. The uniformization rate μ allows to define a transformation to the μ-**uniformized**

DTMRM $\mathcal{M}^\mu := (S, P^\mu, R^\mu)$ where a transition from s to s' in $\mathcal{M}^\mu$ captures the event that $\tau_{s,s'}$ wins the race and thus

$$P^\mu(s, s') := \delta_{s,s'} + \frac{Q(s, s')}{\mu}. \tag{4.5}$$

Note that the probability to eventually leave state s to a state $s' \neq s$ is exactly the embedded transition probability $P(s, s') = \sum_{i=0}^\infty P^\mu(s, s)^i P^\mu(s, s')$. The reward $R^\mu(s, s')$ combines the accumulated rate reward in state s and the impulse reward up to transition to some state s'. In the CTMRM the rate reward $r(s)$ is accumulated for the complete sojourn time in s. Since self-loops are possible in the uniformized DTMRM the accumulation process stops when an arbitrary transition occurs. The expected value of the accumulated rate reward up to transition is given by $r(s) \cdot \frac{1}{\mu}$. Furthermore, the impulse reward $i(s, s')$ is only gained if a transition to another state $s' \neq s$ takes place. But since $i(s, s) = 0$ for all $s \in S$ by Definition 4.1 it follows that for all $s, s' \in S$ the total uniformized reward $R^\mu(s, s')$ is given by

$$R^\mu(s, s') := i(s, s') + \frac{r(s)}{\mu}. \tag{4.6}$$

This equation is similar to (4.4) with the difference that the exit rate $E(s)$ is replaced by the uniformization rate $\mu \geq E(s)$. A further difference comes into the picture when considering the accumulation of these rewards. Both rewards $R(s, s)$ and $R^\mu(s, s)$ for self-loops are possibly non-zero. In case of the embedded DTMRM the probability $P(s, s)$ for self-loops is 0 in non-absorbing states s and thus $R(s, s)$ is not accumulated, in contrast to the uniformized model where $P^\mu(s, s) > 0$ is possible.

So far we have defined the two transformations "Embedding" and "Uniformization" both discretizing the continuous time of a CTMRM and the accumulation of the rate reward over time. In contrast, the upcoming third transformation does not modify the time property itself, but rather merges the impulse rewards into the rate reward. In this way, the CTMRM model has no discrete contributions in the reward accumulation process, which allows to simplify the evaluations of the reward measures (as we will see in the upcoming sections).

4.3.3 Continuization

Let $\mathcal{M} = (S, Q, i, r)$ be a CTMRM and for a non-absorbing state s denote $R(s) := \sum_{s' \neq s} P(s, s') R(s, s')$, where $R(s, s')$ is as in (4.4) and $P(s, s')$ is the embedded transition probability. Thus

$$R(s) = \sum_{s' \neq s} P(s, s') i(s, s') + \frac{r(s)}{E(s)}$$

is the expected accumulated rate reward $r(s)$ in state s including the expected impulse reward $\sum_{s' \neq s} P(s, s') i(s, s') = i(s)$ gained for transition from s to some other state $s' \neq s$. Consider for the moment that $i(s, s') = 0$ for all s, s', i.e. there are no impulse rewards defined. Then $r(s) = R(s)E(s)$, which means that the rate reward $r(s)$ is the same as the expected reward $R(s)$ accumulated in s weighted by the exit rate $E(s)$. More generally, if the impulse rewards $i(s, s')$ were defined then from $P(s, s') = \frac{Q(s,s')}{E(s)}$ it follows that $R(s)E(s) = \sum_{s' \neq s} i(s, s')Q(s, s') + r(s)$. This means that we can transform the original CTMRM $\mathcal{M}$ with impulse rewards into a CTMRM $\overline{\mathcal{M}} = (S, Q, \overline{r})$ without impulse rewards by integrating the original impulse rewards into a new rate reward

$$\overline{r}(s) := \sum_{s' \neq s} i(s, s')Q(s, s') + r(s).$$

We call $\overline{r}$ the **_continuized_** rate reward since in the continuized model $\overline{\mathcal{M}}$ there is no discrete contribution to the reward accumulation process. As we will see in Theorem 4.1 this (rather heuristically deduced) transformation preserves the finite-horizon total reward measure and thus all the reward measures that are derived from the finite-horizon case.

Figure 4.2 shows a diagram with all the presented transformations and also some relations between them. It is interesting to note that this diagram commutes.

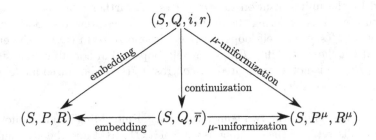

Fig. 4.2. Commuting model transformations

This means that instead of computing the embedded or uniformized DTMRM from the CTMRM (S, Q, i, r) it is possible to continuize the model before performing such a transformation and the resulting DTMRM is the same. We show the commutation of the transformation only for the μ-uniformization, since analogous arguments can be employed for the embedding. When performing the μ-uniformization on (S, Q, i, r) then $R^{\mu}(s, s') = i(s, s') + \frac{r(s)}{\mu}$ by (4.6). Also denote $\overline{R}^{\mu}(s)$ as the μ-uniformization of the continuized rate reward $\overline{r}(s)$. Due to the absence of impulse rewards in the continuized model it follows for all $s \in S$ that

$$\overline{R}^{\mu}(s) = \frac{\overline{r}(s)}{\mu} = \frac{1}{\mu} \left(\sum_{s' \neq s} i(s, s')Q(s, s') + r(s) \right) = \sum_{s' \neq s} i(s, s')P^{\mu}(s, s') + \frac{1}{\mu}r(s)$$

by definition of P^μ as in (4.5). Furthermore, since $i(s,s) = 0$ it follows

$$\overline{R}^\mu(s) = \sum_{s' \in S} i(s,s) P^\mu(s,s') + \frac{1}{\mu} r(s) = \sum_{s' \in S} R^\mu(s,s') P^\mu(s,s') = R^\mu(s).$$

Thus, the μ-uniformization of the continuized rate reward $\overline{R}^\mu(s)$ is exactly the state-based view on the μ-uniformized reward $R^\mu(s,s')$. Also note that the definition of recurrency and reachability in the discrete-time and continuous-time cases are similar. For this reason the classification of states into closed recurrent classes S_i^r and transient states S^t is invariant under the model transformations, since the directed graph structure of the model does not change.

In the following we are going to provide natural definitions for the value functions of the reward measures that we have also considered in the discrete-time case in Sect. 2. The most important question that we will consider is whether the transformations we have presented in this section are value-preserving. More clearly, let $\mathcal{R}$ be a reward measure with value function V on the CTMRM $\mathcal{M}$. We can also evaluate $\mathcal{R}$ on one of the transformed models, e.g. on $\mathcal{M}^\mu$ which gives a value function V^μ. Under what circumstances is $V = V^\mu$? This question will be answered in the forthcoming sections.

4.4   Total Reward Measure

With all the definitions and tools introduced in the preceding sections we are now set to define the total reward measure. We write $\mathbb{E}_s$ for the expectation operator if $X_0 = s$ (or $Z_0 = s$) is the initial state. For a random variable Y we also write $\mathbb{E}[Y]$ for the function $s \mapsto \mathbb{E}_s[Y] \in \mathbb{R}$, respectively for the vector in $\mathbb{R}^{|S|}$ consisting of the expected values $\mathbb{E}_s[Y]$.

Definition 4.2. *Let $T \in \mathbb{R}$, $T \geq 0$ be some finite real time horizon and N_T the random number of transitions up to time T. The **finite-horizon total value function** is defined as*

$$V_T(s) := \mathbb{E}_s \left[\sum_{k=1}^{N_T} i(X_{k-1}, X_k) \right] + \mathbb{E}_s \left[\int_0^T r(Z_t) \, dt \right], \qquad (4.7)$$

*if both expectations exist. If furthermore the expectations $\mathbb{E}_s\left[\sum_{k=1}^{N_T} |i(X_{k-1}, X_k)|\right]$ and $\mathbb{E}_s\left[\int_0^T |r(Z_t)| \, dt\right]$ converge as $T \to \infty$, then we also define the (**infinite-horizon**) total value function as*

$$V_\infty(s) := \lim_{T \to \infty} V_T(s).$$

In (4.7) the rate reward $r(Z_t)$ in state Z_t is continuously accumulated over the time interval $[0, T]$ by integration, whereas the impulse rewards $i(X_{k-1}, X_k)$ for the N_T transitions from states X_{k-1} to X_k for $k = 1, \ldots, N_T$ are discretely accumulated via summation. Note that the upper bound $N_T = \max\{n \mid T_n \leq T\}$

in the summation is random. If $\omega = (s_0, t_0, s_1, t_1, \dots) \in \Omega$ then $N_T(\omega) \in \mathbb{N}$ and the random variable $\sum_{k=1}^{N_T} i(X_{k-1}, X_k)$ takes the value $\sum_{k=1}^{N_T(\omega)} i(s_{k-1}, s_k) \in \mathbb{R}$. Furthermore, N_T has finite expectation (see Lemma A.3 in the Appendix). Since the state space is finite there exists $C \geq 0$ such that $|r(s)| \leq C$ and $|i(s, s')| \leq C$ for all $s, s' \in S$. Therefore

$$\mathbb{E}_s \left[\left| \int_0^T r(Z_t)\, dt \right| \right] \leq C \cdot T < \infty \quad \text{and} \quad \mathbb{E}_s \left[\left| \sum_{k=1}^{N_T} i(X_{k-1}, X_k) \right| \right] \leq C \cdot \mathbb{E}[N_T] < \infty$$

such that $V_T(s)$ is defined for all $T \geq 0$. In the prerequisites for the definition of the total value function V_∞ we require a more restrictive absolute convergence. However, this property is quite natural since it is equivalent to the (joint) integrability of the function $r(Z_t) : [0, \infty) \times \Omega \to \mathbb{R}$ with respect to the probability measure P_s on Ω for the expectation $\mathbb{E}_s$ and the Lebesgue measure for the integral over $[0, \infty)$.

Note that $\mathbb{E}_s[r(Z_t)] = \sum_{s'} P_s(Z_t = s')r(s')$ is the s-th row of the vector $P(t)r$. If we assume that $i(s, s') = 0$ for all $s, s' \in S$ then the finite-horizon total value function $V_T \in \mathbb{R}^S$ regarded as a vector in $\mathbb{R}^{|S|}$ can be computed by

$$V_T = \mathbb{E} \left[\int_0^T r(Z_t)\, dt \right] = \int_0^T P(t)r\, dt. \tag{4.8}$$

The following theorem generalizes this computation for the case with impulse rewards $i(s, s')$. Furthermore, it explains why the continuization transformation as defined in Sect. 4.3.3 preserves the finite-horizon total reward measure. Therefore, this can be considered as the main theorem in the section on CTMRMs.

Theorem 4.1 (Value Preservation of Continuization).
For a CTMRM $\mathcal{M} = (S, Q, i, r)$ let $\overline{\mathcal{M}} = (S, Q, \overline{r})$ be its continuization with

$$\overline{r}(s) = \sum_{s' \neq s} i(s, s')Q(s, s') + r(s).$$

For the finite-horizon total value function it holds that

$$V_T(s) = \mathbb{E}_s \left[\int_0^T \overline{r}(Z_t)\, dt \right].$$

V_T can be computed by

$$V_T = \int_0^T P(t)\overline{r}\, dt,$$

which reads in componentwise notation as

$$V_T(s) = \sum_{s' \in S} \overline{r}(s') \int_0^T P(t)(s, s')\, dt.$$

Proof. In (4.8) we have already shown the statement for the integral term in the definition of V_T in (4.7). It remains to show the statement for the summation term. We have already mentioned that N_T has finite expectation. By Lemma A.1 in the Appendix and the law of total expectation it follows for an arbitrary initial state $s_0 \in S$ that

$$\mathbb{E}_{s_0}\left[\sum_{k=1}^{N_T} i(X_{k-1}, X_k)\right] = \sum_{k=1}^{\infty} \mathbb{E}_{s_0}\left[i(X_{k-1}, X_k)\right] P_{s_0}(N_T \geq k) =$$

$$\sum_{k=1}^{\infty}\sum_{s,s'} i(s, s') P_{s_0}(X_{k-1} = s, X_k = s') P_{s_0}(N_T \geq k) = \sum_{s,s'} i(s, s') n_T(s, s'),$$

where

$$n_T(s, s') := \mathbb{E}_{s_0}\left[\sum_{k=1}^{N_T} \mathbb{1}_{\{X_{k-1}=s, X_k=s'\}}\right] = \sum_{k=1}^{\infty} P_{s_0}(X_{k-1} = s, X_k = s') P_{s_0}(N_T \geq k)$$

is the expected number of transitions from s to s' up to time T from initial state s_0. If we can show that

$$n_T(s, s') = Q(s, s') \cdot \int_0^T P_{s_0}(Z_t = s)\, dt$$

then we are done. The proof for this equation is outsourced to the Appendix. There, in Lemma A.2 we present a proof which uses the uniformization method and in Remark A.2 we sketch a more direct proof without the detour with uniformization which relies on facts from queueing theory.   □

Example 4.2. We come back to our WSN node model introduced in Example 4.1 and assume $\lambda = 2$ activities per hour and an average duration of 15 minutes, i.e. $\mu = 4$ and for the sensor $\lambda_s = 4$ and $\mu_s = 30$. Figure 4.3 shows the transient probabilities $P(t)(s_{\text{init}}, s)$ for the initial state $s_{\text{init}} = (\text{silence}, \text{idle})$ and the finite-horizon total value function

$$V_T(s_{\text{init}}) = (1, 0, 0, 0) \int_0^T e^{Qt}\bar{r}\, dt = \tag{4.9}$$

$$\frac{220}{7}T + \frac{10}{49} + \frac{5}{833}e^{-22T}\left(\left(13\sqrt{51} - 17\right)e^{-2\sqrt{51}T} - \left(13\sqrt{51} + 17\right)e^{2\sqrt{51}T}\right)$$

indicating the total energy consumption up to time T. The continuized rate reward is given by

$$\bar{r} = (5\lambda_s, 5\lambda_s, 10 + 5\mu_s, 10)^T = (20, 20, 160, 10)^T.$$ □

In the following we provide methods for the evaluation of the infinite-horizon total reward measure V_∞. We also show that the model transformations embedding, uniformization and continuization are value-preserving with respect to this

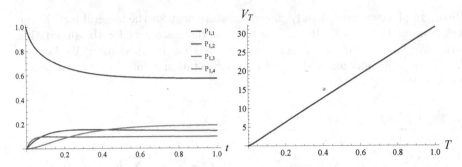

Fig. 4.3. Left: Transient probability functions for initial state $s_{\text{init}} = (\text{silence}, \text{idle})$ converging to the limiting distribution. Right: Total energy consumption during the first hour given by the finite-horizon total value function $V_T(s_{\text{init}})$ as a function of T.

measure. This enables us to provide several methods for the evaluation of V_∞. Before presenting Theorem 4.2, we start with an important proposition about the relation between the existence of the infinite-horizon total value function and the model data Q, i and r of a CTMRM. A proof can be found in the Appendix on page 237.

Proposition 4.1. For a CTMRM (S, Q, i, r) let $S = \bigcup_{i=1}^{k} S_i^r \cup S^t$ be the partitioning of S into the k closed recurrent classes S_i^r and transient states S^t. The infinite-horizon total value function V_∞ exists if and only if for all $i = 1, \ldots, k$ and for all $s, s' \in S_i^r$ it holds that

$$r(s) = 0 \quad \text{and} \quad i(s, s') = 0.$$

Theorem 4.2 (Total Reward Measure – Direct Evaluation and Embedding). If for a CTMRM (S, Q, i, r) the total value function V_∞ exists, then it is the unique solution to the system of linear equations

$$V_\infty(s) = R(s) + \sum_{s' \neq s} P(s, s') V_\infty(s') \quad \text{for } s \in S^t$$

$$V_\infty(s) = 0 \qquad \text{for } s \in S \setminus S^t. \tag{4.10}$$

Here, $P(s, s')$ are the embedded transition probabilities and $R(s)$ is the state-based embedded reward, i.e. $R(s) = \sum_{s' \in S} R(s, s') P(s, s')$ (see (4.4)). In vector notation this system of equation reads as

$$(I - P)V_\infty = R \quad \text{with} \quad V_\infty(s) = 0 \,\, \forall s \in S \setminus S^t.$$

If the impulse reward function i is represented as a matrix with entries $i(s, s')$ then this system of equations can be written in vector notation as

$$-QV_\infty = \text{diag}(iQ^T) + r, \tag{4.11}$$

where Q^T is the transpose of the matrix Q and $\text{diag}(iQ^T)$ is the diagonal of the matrix iQ^T. This equation represents the direct evaluation of the total reward measure (without performing the embedding).

Proof. In Sect. 4.5 on the discounted reward measure we establish similar equations to (4.10) which involve a discount rate parameter α. By setting α to 0 and noting that all occuring expectations exist (4.10) can be derived analogously. By multiplying (4.10) with $E(s)$ and rearranging terms one can directly deduce (4.11). The uniqueness holds since the $(|S^t| \times |S^t|)$-submatrix of Q with entries $Q(s, s')$ for transient states $s, s' \in S^t$ has full rank. $\qquad\square$

Corollary 4.1 (Total Reward Measure – Continuization).
Let $\mathcal{M} = (S, Q, i, r)$ be a CTMRM and $\overline{\mathcal{M}} = (S, Q, \overline{r})$ its continuization. If the total value function V_∞ for $\mathcal{M}$ exists then it also exists for $\overline{\mathcal{M}}$ and in this case they are equal, i.e. V_∞ is the unique solution to the system of linear equations

$$QV_\infty = -\overline{r} \tag{4.12}$$

with $V_\infty(s) = 0$ for all recurrent states s.

Proof. Let $S_i^r \subseteq S$ be a closed recurrent class of S and consider a recurrent state $s \in S_i^r$. If V_∞ exists for $\mathcal{M}$, then by Proposition 4.1 it holds that $r(s) = 0$ and $i(s, s') = 0$ for all $s' \in S_i^r$ in the same recurrent class. Furthermore, if $s' \in S \setminus S_i^r$ then $Q(s, s') = 0$ and therefore $\overline{r}(s) = \sum_{s' \neq s} i(s, s')Q(s, s') + r(s) = 0$. Thus, the total value function for $\overline{\mathcal{M}}$ denoted by $\overline{V}_\infty$ is also defined and it solves (4.11) which reads as $-Q\overline{V}_\infty = \overline{r}$. In order to show that $V_\infty = \overline{V}_\infty$ note that the s-th diagonal entry of the matrix iQ^T is $\sum_{s' \in S} i(s, s')Q(s, s') = \sum_{s' \neq s} i(s, s')Q(s, s')$ since $i(s, s) = 0$. Therefore, $\mathrm{diag}(iQ^T) + r = \overline{r}$ is the continuized rate reward and the conclusion follows since both V_∞ and $\overline{V}_\infty$ solve $-QX = \overline{r}$ and the solution is unique (with the property that both are 0 on recurrent states). $\qquad\square$

Corollary 4.2 (Total Reward Measure – Uniformization).
Let $\mathcal{M} = (S, Q, i, r)$ be a CTMRM and $\mathcal{M}^\mu = (S, P^\mu, R^\mu)$ the μ-uniformized DTMRM. If the total value function V_∞ for $\mathcal{M}$ exists then it also exists for $\mathcal{M}^\mu$ and in this case they are equal, i.e. V_∞ is the unique solution to

$$(I - P^\mu)V_\infty = R^\mu$$

with $V_\infty(s) = 0$ for all recurrent states s.

Proof. From (4.5) and (4.6) it holds that $R^\mu(s, s') = i(s, s') + \frac{r(s)}{\mu}$ and $P^\mu = I + \frac{1}{\mu}Q$. If s and s' are communicating recurrent states (i.e. in the same closed recurrent class) then $r(s) = 0$ and $i(s, s') = 0$ by Proposition 4.1 and therefore $R^\mu(s, s') = 0$. If V_∞^μ denotes the total value function for the μ-uniformized model $\mathcal{M}^\mu$ then V_∞^μ exists by Proposition 2.1 since $R^\mu(s, s') = 0$ for all states s and s' in the same closed recurrent class and by Theorem 2.1 V_∞^μ is also a solution of $(I - P^\mu)V_\infty^\mu = R^\mu$. It follows from (4.12) that

$$(I - P^\mu)V_\infty = -\frac{1}{\mu}QV_\infty = \frac{1}{\mu}\overline{r} = R^\mu$$

and since V_∞ and V_∞^μ are 0 on recurrent states, it follows that $V_\infty = V_\infty^\mu$. $\qquad\square$

4.5 Horizon-Expected and Discounted Reward Measure

In analogy to Sect. 2.3 we want to introduce the discounted reward measure in the continuous-time case. This reward measure can be formally deduced from the horizon-expected reward measure, which we are going to define first.

Definition 4.3. *Let $\mathcal{M} = (S, Q, i, r)$ be a CTMRM and consider a random horizon length T for $\mathcal{M}$, i.e. T is a non-negative continuous random variable that is independent of the state process Z_t of $\mathcal{M}$. Let $V_{(T)}$ denote the random finite-horizon total value function that takes values in $\{V_t \in \mathbb{R}^S \mid t \in [0, \infty)\}$. Define the **horizon-expected value function** as*

$$V(s) := \mathbb{E}\left[V_{(T)}(s)\right],$$

if the expectation exists for all $s \in S$, i.e. $|V_{(T)}(s)|$ has finite expectation.

The random variable $V_{(T)}(s)$ can be regarded as the conditional expectation

$$V_{(T)}(s) = \mathbb{E}_s\left[\sum_{k=1}^{N_T} i(X_{k-1}, X_k) + \int_0^T r(Z_t)\, dt \mid T\right] = \mathbb{E}_s\left[\int_0^T \bar{r}(Z_t)\, dt \mid T\right]$$

that takes the value $V_t(s)$ if $T = t$. Let P_T denote the probability measure of T. Due to the law of total expectation the horizon-expected value function $V(s)$ is the joint expectation with respect to the probability measures P_T of T and P_s of all the Z_t, i.e.

$$V(s) = \mathbb{E}\left[\mathbb{E}_s\left[\int_0^T \bar{r}(Z_t)\, dt \mid T\right]\right] = \mathbb{E}_s\left[\int_0^T \bar{r}(Z_t)\, dt\right], \qquad (4.13)$$

where $\mathbb{E}_s$ on the right hand side denotes the joint expectation.

Lemma 4.1. *Let T be a random horizon length with $\mathbb{E}[T] < \infty$ and probability measure P_T. Then the horizon-expected value function $V(s)$ exists and is given by*

$$V(s) = \mathbb{E}_s\left[\int_0^\infty \bar{r}(Z_t) P_T(T \geq t)\, dt\right] = \mathbb{E}_s\left[\sum_{n=0}^\infty \bar{r}(X_n) \int_{T_n}^{T_{n+1}} P_T(T \geq t)\, dt\right].$$

The proof can be found in the Appendix on page 237. Note that $V(s)$ can also be represented directly in terms of the impulse reward i and rate reward r (instead of the continuized rate reward $\bar{r}$) as

$$V(s) = \mathbb{E}_s\left[\sum_{n=0}^\infty \left(i(X_n, X_{n+1}) \cdot P_T(T \geq T_{n+1}) + r(X_n) \int_{T_n}^{T_{n+1}} P_T(T \geq t)\, dt\right)\right].$$

In this equation, one can also see that an impulse reward $i(X_n, X_{n+1})$ for the $(n + 1)$-st transition is only accumulated if the time horizon T is not exceeded by the total elapsed time T_{n+1} up to this transition.

Definition 4.4. *Let the horizon length T be exponentially distributed with rate $\alpha > 0$. In this case the horizon-expected reward measure is called **discounted reward measure** with **discount rate** α (or just α-discounted reward measure) and its value function will be denoted by $V^\alpha(s)$.*

The discounted value function V^α represented as a vector in $\mathbb{R}^{|S|}$ is given by

$$V^\alpha = \int_0^\infty e^{-\alpha t} P(t)\bar{r}\, dt. \tag{4.14}$$

This follows directly from Lemma 4.1 together with $P_T(T \geq t) = e^{-\alpha t}$ and $\mathbb{E}\left[\bar{r}(Z_t)\right] = P(t)\bar{r}$. As in Sect. 2.3 in the discrete-time setting we can also derive a system of linear equations which allows to compute the discounted value function V^α. The proof can be found in the Appendix on page 238.

Theorem 4.3 (Discounted Reward Measure – Continuization).
The discounted value function with discount rate $\alpha > 0$ is the unique solution to the system of linear equations

$$V^\alpha(s) = \frac{\bar{r}(s)}{\alpha + E(s)} + \sum_{s' \neq s} \frac{Q(s,s')}{\alpha + E(s)} V^\alpha(s'). \tag{4.15}$$

In vector notation this system of equations reads as

$$(Q - \alpha I)V^\alpha = -\bar{r}. \tag{4.16}$$

Note that in case the total value function V_∞ exists (and is thus finite) it is the limit of the α-discounted value function as α decreases to 0, i.e. for all $s \in S$ it holds that

$$V_\infty(s) = \lim_{\alpha \searrow 0} V^\alpha(s). \tag{4.17}$$

Example 4.3. Figure 4.4 shows the α-discounted value function $V^\alpha(s_{\text{init}})$ for the initial state $s_{\text{init}} = (\text{silence}, \text{idle})$ of the WSN node model from Example 4.1 dependent on the discount rate α. By solving (4.16) we get

$$V^\alpha(s_{\text{init}}) = \frac{20\left(\alpha^2 + 72\alpha + 440\right)}{\alpha\left(\alpha^2 + 44\alpha + 280\right)}.$$

Clearly, for increasing α the expected horizon length $\mathbb{E}\left[T\right] = \frac{1}{\alpha}$ decreases and thus the discounted value representing the expected energy consumption up to time T also decreases. On the other hand, if α decreases towards 0, then the discounted value increases and in our case it diverges to ∞. Note that the total value function V_∞ does not exist for this model.   □

Remember that one of our main goals in this section is to check, whether all the model transformations in Fig. 4.2 are value-preserving. For a CTMRM $(S, Q, \bar{r})$ with α-discounted value function V^α consider its μ-uniformization (S, P^μ, R^μ)

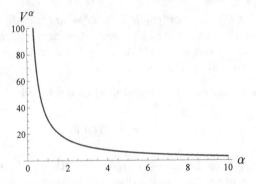

Fig. 4.4. The discounted value $V^\alpha(s_{\text{init}})$ for the initial state $s_{\text{init}} = (\text{silence}, \text{idle})$ as a function of the discount rate α

with γ-discounted value function V^γ. We show that there is no choice of $\gamma \in (0,1)$ such that $V^\alpha = V^\gamma$. Thus the μ-uniformization is not value-preserving with respect to the discounted reward measure. (As a special case it also follows, that the embedding is not value-preserving as well.)

Assume that there exists $\gamma \in (0,1)$ such that $V^\alpha = V^\gamma$. On the one hand V^α is the unique solution to $(Q - \alpha I)V^\alpha = -\bar{r}$ and thus $QV^\gamma = \alpha V^\gamma - \bar{r}$. On the other hand, V^γ is by Theorem 2.2 the unique solution to $(I - \gamma P^\mu)V^\gamma = R^\mu$ where $P^\mu = I + \frac{1}{\mu}Q$ and $R^\mu = \frac{1}{\mu}\bar{r}$. Thus

$$R^\mu = (I - \gamma P^\mu)\,V^\gamma = \left(I - \gamma\left(I + \frac{1}{\mu}Q\right)\right)V^\gamma = (1 - \gamma)\,V^\gamma - \gamma\frac{1}{\mu}\left(\alpha V^\gamma - \bar{r}\right).$$

By rearranging terms it follows that

$$\left(1 - \frac{\gamma\alpha}{(1 - \gamma)\mu}\right)V^\gamma = R^\mu,$$

which means that V^γ is a multiple of R^μ and is thus independent of the transition probabilities P^μ! However, we can save the value-preserving property by observing the following link between the discounted and the total reward measure in analogy to the discrete-time case as shown in Remark 2.1.

Remark 4.3. If $\mathcal{M} = (S, Q, i, r)$ is a CTMRM then extend $\mathcal{M}$ to a CTMRM $\mathcal{M}' = (S', Q', i', r')$ with an artificial absorbing reward-free state *abs* that is reachable with rate $\alpha > 0$ from every other state in S, i.e.

$$S' := S \cup \{abs\}, \quad Q' := \begin{pmatrix} Q - \alpha I & \alpha\mathbf{1} \\ 0 & 0 \end{pmatrix}, \quad i' := \begin{pmatrix} i & 0 \\ 0 & 0 \end{pmatrix} \quad \text{and} \quad r' := \begin{pmatrix} r \\ 0 \end{pmatrix}.$$

Since *abs* is the single recurrent state in $\mathcal{M}'$ it follows that the total value function V'_∞ for $\mathcal{M}'$ with $V'_\infty(abs) = 0$ is a solution to (4.12), i.e.

$$Q'V'_\infty = -\bar{r}',$$

where $\overline{r'}$ is the continuized rate reward of $\mathcal{M'}$. By definition of $\mathcal{M'}$ it holds for all $s \in S' \setminus \{abs\} = S$ that $i'(s, abs) = 0$ and $r'(s) = r(s)$ and it follows that

$$\overline{r'}(s) = \sum_{\substack{s' \in S' \\ s' \neq s}} i'(s, s')Q'(s, s') + r'(s) = \sum_{\substack{s' \in S \\ s' \neq s}} i(s, s')Q(s, s') + r(s) = \overline{r}(s).$$

Since V^α is the α-discounted value function for $\mathcal{M}$ it is the unique solution to $(Q - \alpha I)V^\alpha = -\overline{r}$ and thus

$$Q' \begin{pmatrix} V^\alpha \\ 0 \end{pmatrix} = \begin{pmatrix} Q - \alpha I & \alpha 1 \\ 0 & 0 \end{pmatrix} \begin{pmatrix} V^\alpha \\ 0 \end{pmatrix} = - \begin{pmatrix} \overline{r} \\ 0 \end{pmatrix} = -\overline{r'}.$$

Since V'_∞ is also a solution to $Q'V'_\infty = -\overline{r'}$ and also unique with the property $V'_\infty(abs) = 0$ it follows that $V^\alpha(s) = V'_\infty(s)$ for all $s \in S$.

This remark allows to provide a further method for the evaluation of the discounted value function by means of uniformization. Note that if $\mu \geq E(s)$ for all $s \in S$ is a uniformization rate for the original model $\mathcal{M}$ then $\mu + \alpha$ is a uniformization rate for the extended model $\mathcal{M'} = (S', Q', i', r')$. The following theorem states that the rates and the rewards have to be uniformized differently in order to be able to establish a connection between the α-discounted value function and a γ-discounted value function for some suitable DTMRM. For this reason, we refer to the transformation to that DTMRM as **separate uniformization**.

Theorem 4.4 (Discounted Reward Measure – separate Uniformization). Let $\mathcal{M} = (S, Q, i, r)$ be a CTMRM, $\mu > 0$ a uniformization rate for $\mathcal{M}$ and $\alpha > 0$ a discount rate. Then V^α is the unique solution to the system of linear equations

$$(I - \gamma P^\mu) V^\alpha = R^{\mu+\alpha},$$

where $P^\mu = I + \frac{1}{\mu}Q$ is the μ-uniformized transition probability matrix, $R^{\mu+\alpha} = \frac{1}{\mu+\alpha}\overline{r}$ is the $(\mu+\alpha)$-uniformized reward vector and $\gamma = \frac{\mu}{\mu+\alpha} \in (0, 1)$ is a discount factor. In other words the α-discounted value function V^α for the CTMRM $\mathcal{M}$ is precisely the γ-discounted value function for the DTMRM $\widetilde{\mathcal{M}} := (S, P^\mu, R^{\mu+\alpha})$ denoted by $\widetilde{V}^\gamma$, i.e.

$$V^\alpha = \widetilde{V}^\gamma.$$

The proof is straightforward and integrated in the following discussion on several relationships between models and value functions that can occur by the model transformations. Figure 4.5 shows transformations between Markov chains without rewards. A CTMC (S, Q) is uniformized into a DTMC (S, P^μ) and afterwards the model is extended with an auxiliary absorbing state abs as described in Remark 2.1 which leads to a DTMC $(S', (P^\mu)')$ with $S' = S \cup \{abs\}$ (γ-extension). On the other hand, (S, Q) can be directly extended with abs as described in Remark 4.3 to the model (S', Q') and then uniformized with rate $\mu + \alpha$ (α-extension). This diagram commutes, since

$$(P^\mu)' = \begin{pmatrix} \gamma P^\mu & (1-\gamma)\mathbf{1} \\ 0 & 1 \end{pmatrix} = \begin{pmatrix} \frac{\mu}{\mu+\alpha}(I + \frac{1}{\mu}Q) & \frac{\alpha}{\mu+\alpha}\mathbf{1} \\ 0 & 1 \end{pmatrix}$$

$$= \begin{pmatrix} I & 0 \\ 0 & 1 \end{pmatrix} + \frac{1}{\mu+\alpha} \begin{pmatrix} Q - \alpha I & \alpha\mathbf{1} \\ 0 & 0 \end{pmatrix} = (P')^{\mu+\alpha}.$$

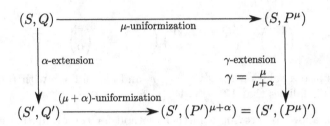

Fig. 4.5. Commuting model transformations on discrete-time and continuous-time Markov chains

In contrast, Fig. 4.6 shows the same transformations applied to (continuized) Markov reward models. This diagram does not commute, since in general

$$(R^\mu)' = \frac{1}{\mu}\begin{pmatrix} \overline{r} \\ 0 \end{pmatrix} \neq \frac{1}{\mu+\alpha}\begin{pmatrix} \overline{r} \\ 0 \end{pmatrix} = (R')^{\mu+\alpha}.$$

However, due to $(P')^{\mu+\alpha} = (P^\mu)'$ it is possible to compute the infinite-horizon total value function V_∞ on the DTMRM $(S', (P')^{\mu+\alpha}, (R')^{\mu+\alpha})$. Let us call its restriction on S as $\widetilde{V}^\gamma$. Since the uniformization is value-preserving with respect to the infinite-horizon total reward measure (see Corollary 4.2) and due to Remark 4.3 it follows that $V^\alpha = \widetilde{V}^\gamma$, which concludes the proof of Theorem 4.4.

4.6   Average Reward Measure

In Sect. 2.4 we defined the discrete-time average reward by considering a sequence of finite-horizon value functions V_N which were averaged over the horizon length N and the limit as $N \to \infty$ was considered. In complete analogy we define the average reward in the continuous-time case.

Definition 4.5. *Let* $\mathcal{M} = (S, Q, i, r)$ *be a CTMRM with finite-horizon total value function* V_T. *The **average reward** value function is defined as*

$$g(s) = \lim_{T\to\infty} \frac{1}{T} V_T(s),$$

if the limit exists for all $s \in S$.

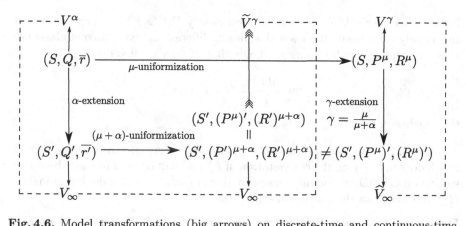

Fig. 4.6. Model transformations (big arrows) on discrete-time and continuous-time reward models that do not commute. A small arrow indicates an evaluation of a reward measure on a model. The dashed lines connect value functions that are related by equality. The value function $\widetilde{V}^\gamma$ is not directly evaluated on $(S', (P^\mu)', (R')^{\mu+\alpha})$ but is induced by V_∞ (feathered arrow) as a restriction from S' to S and it holds $V^\alpha = \widetilde{V}^\gamma$.

Example 4.4. In the WSN node model from Example 4.1 we saw in (4.9) that $V_T(s_{\text{init}}) = \frac{220}{7}T + f(T)$ with some function $f(T)$ such that $\frac{f(T)}{T} \to 0$ as $T \to \infty$. This result means, that on average over infinite time the energy consumption is $g(s) = \frac{220}{7}$ per hour (compare this with the slope of $V_T(s_{\text{init}})$ in Fig. 4.3).   □

In the following we want to provide methods for the computation of the average reward that do not rely on an explicit representation of V_T which is computed by integration over the transient probability matrix $P(t) = e^{Qt}$ as in Theorem 4.1. In Sect. 4.2.3 we mentioned that $P(t)$ converges to the limiting matrix P^*. Remind that P^* fulfills the properties

$$P(t)P^* = P^*P(t) = P^*P^* = P^* \quad \text{and} \quad P^*Q = QP^* = 0.$$

Proposition 4.2. *Let $\bar{r}$ be the continuized rate reward of a CTMRM (S, Q, i, r). Then the average reward can be computed by*

$$g = P^*\bar{r}.$$

Proof. By Theorem 4.1 it holds that

$$g = \lim_{T \to \infty} \frac{1}{T}V_T = \lim_{T \to \infty} \frac{1}{T}\int_0^T P(t)\bar{r}\,dt.$$

Fix two states s and s' and consider the monotonically increasing function $h(T) := \int_0^T P(t)(s, s')\,dt \geq 0$. If $h(T)$ is unbounded it follows by the rule of l'Hospital that

$$\lim_{T \to \infty} \frac{h(T)}{T} = \lim_{T \to \infty} P(T)(s, s') = P^*(s, s').$$

In the other case if $h(T)$ is bounded then clearly $P(t)(s, s')$ converges to 0. But this is only the case if either s and s' are in different closed recurrent classes or s' is transient and in both cases it holds that $P^*(s, s') = 0$. Thus from

$$\lim_{T \to \infty} \frac{h(T)}{T} = 0 = P^*(s, s')$$

the conclusion follows.   □

As in Sect. 2.4 we also show another possibility to evaluate the average reward which does not rely on the computation of P^* and will be used in the subsequent section on CTMDPs. For this reason we define the notion of a deviation matrix H and a bias h in the continuous-time case.

Definition 4.6. *For a CTMRM $\mathcal{M} = (S, Q, i, r)$ define the* **deviation matrix** *H as*

$$H := \int_0^\infty (P(t) - P^*) \, dt,$$

where integration is performed componentwise. Further define

$$h := H\bar{r} = \int_0^\infty (P(t)\bar{r} - g) \, dt$$

as the **bias** *of $\mathcal{M}$.*

Note that Q, H and P^* satisfy the following equations:

$$QH = HQ, \quad P^* = I + QH \quad \text{and} \quad HP^* = P^*H = 0. \tag{4.18}$$

That can be easily derived by the Kolmogorov equations (4.2).

In the following, we connect the discounted and the average reward measures. Consider for a fixed $s \in S$ the discounted value $V^\alpha(s)$ as a function of $\alpha \geq 0$. Then $V^\alpha(s)$ might have a pole at $\alpha = 0$ and can be extended as a Laurent series in α. For more information on the Laurent series expansion in continuous time we refer to Theorem A.1 in the Appendix. This theorem directly induces the Laurent series decomposition of the α-discounted value function as is stated in the following corollary.

Corollary 4.3 (Laurent Series of the Discounted Value Function). *The Laurent series expansion of V^α is given by*

$$V^\alpha = \alpha^{-1}g + \sum_{n=0}^\infty \alpha^n H^{n+1}\bar{r}.$$

Recall (4.17): In case the infinite-horizon total value V_∞ exists it follows for the average reward g and the bias h from the Laurent expansion for $\alpha \to 0$ that $g = 0$ and $h = V_\infty$. Thus on average no reward is gained over the infinite horizon which can also be seen by Proposition 4.1 since there are no rewards in recurrent states. By Definition 4.6 the bias h measures the total long-term deviation of the accumulated rewards from the average reward, i.e. $h = \lim_{T\to\infty} (V_T - g \cdot T)$. As in the discrete-time setting, the bias can also be seen as the excess of rewards $\bar{r}$ until the system reaches its steady-state. Moreover, h is also the infinite-horizon total value function for the CTMRM with average-corrected rewards $\bar{r} - g$ (if we also allow for non-absolute convergence in Definition 4.2). Thus, if $g = 0$ it follows that $h = V_\infty$.

Example 4.5. If we decompose the rational function for $V^\alpha(s_{\mathrm{init}})$ in Example 4.3 into a Laurent series at $\alpha = 0$ then

$$V^\alpha(s_{\mathrm{init}}) = \frac{20\left(\alpha^2 + 72\alpha + 440\right)}{\alpha\left(\alpha^2 + 44\alpha + 280\right)} = \frac{220}{7\alpha} + \frac{10}{49} - \frac{25\alpha}{343} + \frac{103\alpha^2}{9604} + O\left(\alpha^3\right).$$

We see that the average reward g for s_{init} is $\frac{220}{7}$ and the bias h is $\frac{10}{49}$. Compare these values also with (4.9). □

In the following we show two possibilities to compute the average reward by a system of linear equations. The first is a direct evaluation which uses the CTMRM model data Q, r and i and the second system of linear equations relies on the uniformized DTMRM.

Theorem 4.5 (Average Reward Measure – Direct Evaluation). *The average reward g and the bias h fulfill the following system of linear equations:*

$$\begin{pmatrix} -Q & 0 \\ I & -Q \end{pmatrix} \begin{pmatrix} g \\ h \end{pmatrix} = \begin{pmatrix} 0 \\ \bar{r} \end{pmatrix}. \tag{4.19}$$

Furthermore, a solution (u, v) to this equation implies that $u = P^*\bar{r} = g$ is the average reward and there exists $w \in \ker(I - P^*)$ such that $v - w = h$ is the bias.

Proof. Let $g = P^*\bar{r}$ be the average reward and $h = H\bar{r}$ the bias. From $QP^* = 0$ it follows that $Qg = 0$ and by using the Kolmogorov equations (4.2) it holds that

$$Qh = Q\int_0^\infty (P(t) - P^*)\bar{r}\, dt = \int_0^\infty P'(t)\bar{r}\, dt = (P^* - I)\bar{r} = g - \bar{r}$$

and thus (4.19) follows. Now let (u, v) be a solution to (4.19). Then clearly $0 = Qu = P(t)Qu = P'(t)u$ and by integrating $\int_0^t P'(\tau)u\, d\tau = 0$ and using $P(0) = I$ it follows that $P(t)u = u$ for all $t \geq 0$. Therefore, if $t \to \infty$ together with $u = \bar{r} + Qv$ and $P^*Q = 0$ it follows $u = P^*u = P^*(\bar{r} + Qv) = P^*\bar{r} = g$. Now

$$(I - P^*)v = -\int_0^\infty P'(t)v\, dt \qquad\qquad P(0) = I$$

$$= -\int_0^\infty P(t)Qv\, dt = -\int_0^\infty P(t)(g - \bar{r})\, dt \qquad (4.2), u - Qv = \bar{r}, u = g$$

$$= \int_0^\infty (P(t)\bar{r} - g)\, dt = \int_0^\infty (P(t) - P^*)\bar{r}\, dt \qquad P(t)g = g\ \forall t \geq 0$$

$$= H\bar{r} = (H - P^*H)\bar{r} = (I - P^*)h. \qquad (4.18), h = H\bar{r}$$

Therefore, $v = h + w$ for some $w \in \ker(I - P^*)$. □

In the special case, if $\mathcal{M}$ is unichain then $\ker(Q) = \mathbb{1}\mathbb{R}$ is one-dimensional and therefore $g = g_0\mathbb{1} \in \ker(Q)$ is constant with $g_0 \in \mathbb{R}$. This value can be computed by finding a solution to $g_0\mathbb{1} - Qh = \bar{r}$. Alternatively, in a unichain CTMRM the unique stationary distribution ρ fulfills $\rho Q = 0$ and $\rho\mathbb{1} = 1$ and thus $g_0 = \rho\bar{r}$.

Theorem 4.6 (Average Reward Measure – Uniformization). *Consider a CTMRM $\mathcal{M} = (S, Q, i, r)$ with average reward g and let μ be a uniformization rate. Then g is the unique solution to the system of equations*

$$\begin{pmatrix} I - P^\mu & 0 \\ \mu I & I - P^\mu \end{pmatrix}\begin{pmatrix} g \\ h \end{pmatrix} = \begin{pmatrix} 0 \\ R^\mu \end{pmatrix}.$$

If $\mathcal{M}^\mu = (S, P^\mu, R^\mu)$ is the μ-uniformized model and g^μ the average reward of the DTMRM $\mathcal{M}^\mu$ then

$$g = \mu g^\mu.$$

The statement of this theorem can be interpreted as follows: In the continuous-time model $g(s)$ is the average reward per time from initial state s, while in the corresponding μ-uniformized discrete-time model $g^\mu(s)$ is the average reward per transition. In the uniformized model the expected number of transitions per time unit is exactly the rate μ (which corresponds to Little's law) and thus $g(s) = \mu g^\mu(s)$. Note also that one can assume without loss of generality that all exit rates $E(s)$ satisfy $E(s) \le 1$ by changing the time scale. In this case, one can choose $\mu := 1$ and it follows that $g = g^\mu$. For this reason, the uniformization transformation (with $\mu = 1$ expected number of transitions per time unit) preserves the average reward measure.

Proof. We first show that $g = \mu g^\mu$. Theorem 4.4 allows to link for each discount rate $\alpha > 0$ the α-discounted continuous-time value V^α to the γ-discounted discrete-time value $\tilde{V}^\gamma$ of the separate uniformized DTMRM $(S, P^\mu, R^{\mu+\alpha})$ with discount factor $\gamma = \frac{\mu}{\mu+\alpha}$. From the continuous-time Laurent series in Corollary 4.3 it follows that $g = \lim_{\alpha \to 0} \alpha V^\alpha$. On the other hand, since $\lim_{\alpha \to 0} R^{\mu+\alpha} = R^\mu$ it follows from the discrete-time Laurent series in Theorem 2.3 that $g^\mu = \lim_{\rho \to 0} \frac{\rho}{1+\rho}\tilde{V}^\gamma$, where $\rho = \frac{1-\gamma}{\gamma} = \frac{\alpha}{\mu}$. Combining both gives

$$g^\mu = \lim_{\rho \to 0} \frac{\rho}{1+\rho}\tilde{V}^\gamma = \lim_{\alpha \to 0} \frac{\alpha}{\mu+\alpha}V^\alpha = \frac{1}{\mu}g$$

and the conclusion follows. The system of the linear equations can be directly established from Theorem 2.4 with $P^\mu = I + \frac{1}{\mu}Q$ and $R^\mu = \frac{1}{\mu}\bar{r}$. $\square$

4.7 Big Picture – Model Transformations

We summarize all the transformations and evaluation methods presented in this section in Fig. 4.7. Theorem 4.1 allows to continuize a CTMRM (S, Q, i, r) into a CTMRM $(S, Q, \bar{r})$ and hereby preserving all considered value functions. For

total reward measure

$(S, P, R) \xleftarrow{\quad \text{embedding} \quad} (S, Q, \bar{r}) \xrightarrow{\quad \text{uniformization} \quad} (S, P^\mu, R^\mu)$

$V_\infty = \sum_{n=0}^{\infty} P^n R \qquad V_\infty = \int_0^\infty e^{Qt} \bar{r} \, dt \qquad V_\infty = \sum_{n=0}^{\infty} (P^\mu)^n R^\mu$

$(I - P)V_\infty = R \qquad QV_\infty = -\bar{r} \qquad (I - P^\mu)V_\infty = R^\mu$

α-discounted reward measure

$(S, P, R) \xleftarrow{\quad \text{embedding} \quad} (S, Q, \bar{r}) \xrightarrow[\quad \text{uniformization} \quad]{\text{separate}} (S, P^\mu, R^{\mu+\alpha})$

via α-extension to $\qquad V^\alpha = \int_0^\infty e^{(Q-\alpha I)t} \bar{r} \, dt \qquad V^\alpha = \sum_{n=0}^{\infty} \left(\dfrac{\mu}{\mu+\alpha} P^\mu \right)^n R^{\mu+\alpha}$

$S' = S \cup \{abs\}$ and

value-preservation for V_∞

$V^\alpha(s) = V_\infty(s) \, \forall s \in S \qquad (Q - \alpha I)V^\alpha = -\bar{r} \qquad \left(I - \dfrac{\mu}{\mu+\alpha} P^\mu \right) V^\alpha = R^{\mu+\alpha}$

average reward measure

$(S, Q, \bar{r}) \xrightarrow{\quad \text{uniformization} \quad} (S, P^\mu, R^\mu)$

multichain: $\qquad\qquad\qquad$ multichain:

$g = P^* \bar{r} \qquad\qquad\qquad g = \mu(P^\mu)^* R^\mu$

$\begin{pmatrix} -Q & 0 \\ I & -Q \end{pmatrix} \begin{pmatrix} g \\ h \end{pmatrix} = \begin{pmatrix} 0 \\ \bar{r} \end{pmatrix} \qquad \begin{pmatrix} I - P^\mu & 0 \\ \mu I & I - P^\mu \end{pmatrix} \begin{pmatrix} g \\ h \end{pmatrix} = \begin{pmatrix} 0 \\ R^\mu \end{pmatrix}$

unichain: $\qquad\qquad\qquad$ unichain:

$g_0 = \rho \bar{r}, \quad \rho Q = 0, \, \rho \mathbf{1} = 1 \qquad g_0 = \mu \rho R^\mu, \quad \rho P = \rho, \, \rho \mathbf{1} = 1$

$g_0 \mathbf{1} - Qh = \bar{r} \qquad\qquad\qquad \dfrac{1}{\mu} g_0 \mathbf{1} + (I - P^\mu)h = R^\mu$

Fig. 4.7. Big Picture: Value-preserving transformations from the continuization $(S, Q, \bar{r})$ of a CTMRM (S, Q, i, r)

this reason, we omit the model (S, Q, i, r) in the figure. The embedded DTMRM (S, P, R) is defined by

$$P = I + E^{-1}Q \quad \text{and} \quad R = \text{diag}(iP^T) + E^{-1}r \in \mathbb{R}^{|S|},$$

where E^{-1} is defined as a diagonal matrix with entries $\frac{1}{E(s)}$ if $E(s) \neq 0$ and 0 otherwise. The vector $\text{diag}(iP^T)$ is the state-based view on the impulse rewards $i(s, s')$ collected in a matrix i. The μ-uniformized DTMRM (S, P^μ, R^μ) is defined by

$$P^\mu = I + \frac{1}{\mu}Q \quad \text{and} \quad R^\mu = \text{diag}\left(i\,(P^\mu)^T\right) + \frac{1}{\mu}r \in \mathbb{R}^{|S|}.$$

The total reward measure is value-preserving for both transformations embedding and uniformization. Therefore, all presented methods for computation of

V_∞ in continuous and discrete time can be used. In order to transform the discounted reward measure with discount rate α we need to consider an extended model (see Remark 4.3). The evaluation of the total reward measure on the extended model is equivalent to the evaluation of the discounted reward measure on the original model. For the average reward model, there is in general no simple direct method to compute the average reward g via embedding, since continuous time and the transition-counting time are not compatible when building averages over time.

We want to conclude this section with a remark on more general reward structures. Beyond impulse rewards or rate rewards as we defined, the authors of [22], [23] and [35] also analyze rewards that can vary over time. This variation can be homogeneous (depending on the length of a time interval) or non-homogeneous (depending on two points in time). These reward structures are mostly accompanied by the more general model class of Semi-Markov Reward Processes. Furthermore, [30] defines path-based rewards which can be analyzed by augmenting the model with special reward variables, such that the state space does not need to be extended for path information.

5   Continuous Time Markov Decision Processes

In this section we merge both model types MDP and CTMRM together into a CTMDP model. This section is rather short, because all of the necessary work has been already done in the preceding sections. For this reason, we establish connections to the previous results. Moreover, we also present an additional method for the computation of the average reward which directly works on CTMDPs.

5.1   Preliminaries and Retrospection

Definition 5.1. *A **continuous-time Markov Decision Process (CTMDP)** is a structure $\mathcal{M} = (S, Act, e, Q, i, r)$, where S is a finite state space, Act a finite set of actions, $e\colon S \to 2^{Act} \setminus \emptyset$ the action-enabling function, $Q\colon S \times Act \times S \to \mathbb{R}$ an action-dependent generator function, $i\colon S \times Act \times S \to \mathbb{R}$ the action-dependent impulse reward function with $i(s, a, s) = 0$ for all $a \in e(s)$ and $r\colon S \times Act \to \mathbb{R}$ the action-dependent rate reward function.*

Completely analogous to Sect. 3 we define the set of policies

$$\Pi := \{\pi\colon S \to Act \mid \pi(s) \in e(s)\}.$$

Applying π to a CTMDP $\mathcal{M}$ induces a CTMRM $\mathcal{M}^\pi = (S, Q^\pi, i^\pi, r^\pi)$, where

$$Q^\pi(s, s') := Q(s, \pi(s), s'), \quad i^\pi(s, s') := i(s, \pi(s), s') \quad \text{and} \quad r^\pi(s) := r(s, \pi(s)).$$

A reward measure $\mathcal{R}$ for the CTMDP $\mathcal{M}$ induces for each policy π a value V^π for $\mathcal{M}^\pi$.

Definition 5.2. *Let $\mathcal{M}$ be a CTMDP with reward measure $\mathcal{R}$ and for each $\pi \in \Pi$ let V^π be the value of π with respect to $\mathcal{R}$. The value V^* of $\mathcal{M}$ is defined as*

$$V^*(s) := \sup_{\pi \in \Pi} V^\pi(s).$$

A policy $\pi^* \in \Pi$ is called optimal if

$$\forall s \in S \; \forall \pi \in \Pi : \; V^{\pi^*}(s) \geq V^\pi(s).$$

In order to optimize the CTMDP we can transform $\mathcal{M}$ by embedding or uniformization into an MDP and by continuization into another CTMDP. The transformations follow the Big Picture as presented in Sect. 4.7 (Fig. 4.7) with the difference that all action-dependent quantities (i.e. Q, i and r) are transformed in an action-wise manner. The following theorem states that these transformations preserve both the optimal value and the optimal policies.

Theorem 5.1. *Let $\mathcal{M}$ be a CTMDP with policy space Π, optimal value V^* and a set of optimal policies $\Pi^* \subseteq \Pi$. Further let $\widehat{\mathcal{M}}$ be a transformed model (MDP or CTMDP) as in Fig. 4.7 with policy space $\widehat{\Pi}$, value $\widehat{V}^*$ and optimal policies $\widehat{\Pi}^* \subseteq \widehat{\Pi}$. Then*

$$V^* = \widehat{V}^* \quad \text{and} \quad \Pi^* = \widehat{\Pi}^*.$$

Proof. Note that V^* and $\widehat{V}^*$ are defined over policies, i.e.

$$V^* = \sup_{\pi \in \Pi} V^\pi \quad \text{and} \quad \widehat{V}^* = \sup_{\pi \in \widehat{\Pi}} \widehat{V}^\pi.$$

All the transformations in Fig. 4.7 do not transform S, Act and e, thus $\Pi = \widehat{\Pi}$. Furthermore, for each $\pi \in \Pi$ the transformations preserve the value V^π, i.e. $V^\pi = \widehat{V}^\pi$ and thus $V^* = \sup_{\pi \in \Pi} V^\pi = \sup_{\pi \in \widehat{\Pi}} \widehat{V}^\pi = \widehat{V}^*$. In order to show that $\Pi^* = \widehat{\Pi}^*$ let $\pi^* \in \Pi^*$. Then for all s and for all $\pi \in \Pi$ by definition of π^* it holds that

$$V^{\pi^*}(s) \geq V^\pi(s) = \widehat{V}^\pi(s) \quad \text{and} \quad V^{\pi^*}(s) = \widehat{V}^{\pi^*}(s).$$

and therefore π^* is optimal for $\widehat{\mathcal{M}}$, i.e. $\pi^* \in \widehat{\Pi}^*$. In complete analogy it follows that $\widehat{\Pi}^* \subseteq \Pi^*$ and the equality for the sets of optimal policies follows. $\quad\square$

5.2   Average Reward Measure

All the necessary work has already been done for analyzing CTMDPs by transformation to MDPs. It remains to provide optimality equations for the average reward and algorithms which can be used directly on CTMDPs. Consider a CTMDP (S, Act, e, Q, i, r) with average reward measure and let

$$\bar{r}(s, a) = \sum_{s' \neq s} i(s, a, s') Q(s, a, s') + r(s, a)$$

denote the continuized rate reward. Define the Bellman operators $\mathcal{B}_{\mathrm{av}} : \mathbb{R}^S \to \mathbb{R}^S$ and $\mathcal{B}^g_{\mathrm{bias}} : \mathbb{R}^S \to \mathbb{R}^S$ (parametrized by $g \in \mathbb{R}^S$) as follows:

$$(\mathcal{B}_{\mathrm{av}} g)(s) := \max_{a \in e(s)} \left\{ \sum_{s' \in S} Q(s, a, s') g(s') \right\}$$

$$(\mathcal{B}^g_{\mathrm{bias}} h)(s) := \max_{a \in e^g(s)} \left\{ \overline{r}(s, a) + \sum_{s' \in S} Q(s, a, s') h(s') \right\} - g(s)$$

$$\text{where } e^g(s) := \left\{ a \in e(s) \mid \sum_{s' \in S} Q(s, a, s') g(s') = 0 \right\}$$

These operators look similar to the Bellman operators (3.10) and (3.11) in the discrete-time case. The difference is that instead of searching for fixed-points we need to search for zeros of $\mathcal{B}_{\mathrm{av}}$ and $\mathcal{B}^g_{\mathrm{bias}}$ (see (4.19)). This gives the first and the second Bellman optimality equations

$$\max_{a \in e(s)} \left\{ \sum_{s' \in S} Q(s, a, s') g(s') \right\} = 0 \tag{5.1}$$

$$\max_{a \in e^g(s)} \left\{ \overline{r}(s, a) + \sum_{s' \in S} Q(s, a, s') h(s') \right\} - g(s) = 0. \tag{5.2}$$

The following existence theorem is the analogue version of Theorem 5.2 for discrete-time MDPs.

Theorem 5.2 (Existence Theorem).

(i) *The average optimal value function g^* is a solution to (5.1), i.e $\mathcal{B}_{\mathrm{av}} g^* = 0$. For $g = g^*$ there exists a solution h to (5.2), i.e. $\mathcal{B}^{g^*}_{\mathrm{bias}} h = 0$. If g and h are solutions to (5.1) and (5.2) then $g = g^*$.*

(ii) *There exists an optimal policy π^* and it holds that $g^{\pi^*} = g^*$.*

(iii) *For any solution h to (5.2) with $g = g^*$ an optimal policy π^* can be derived from*

$$\pi^*(s) \in \operatorname*{argmax}_{a \in e^{g^*}(s)} \left\{ \overline{r}(s, a) + \sum_{s' \in S} Q(s, a, s') h(s') \right\}.$$

For a direct proof we refer to [17]. We propose here another proof sketch based on uniformization and its value-preserving property.

Proof. Without loss of generality we assume that $E(s, a) \leq 1$ and set the uniformization rate $\mu := 1$ such that the uniformization is value-preserving. The μ-uniformized MDP is given by $\mathcal{M}^\mu = (S, Act, e, P^\mu, R^\mu)$ where

$$P^\mu(s, a, s') = \delta_{s, s'} + Q(s, a, s') \quad \text{and} \quad R^\mu(s, a) = \overline{r}(s, a).$$

If $(g^\mu)^*$ denotes the optimal average reward for $\mathcal{M}^\mu$ then by Theorem 5.1 it holds that $g^* = (g^\mu)^*$. Since finding a fixed point of some operator $\mathcal{T}$ is equivalent to finding a zero of the operator $\mathcal{B} = \mathcal{T} - id$, where id is the identity operator, part (i) follows. Furthermore, Theorem 3.7 guarantees the existence of an optimal policy for $\mathcal{M}^\mu$ and by Theorem 5.1 also for $\mathcal{M}$ such that parts (ii) and (iii) follow. □

We restate the policy iteration algorithm from [17] since our CTMDP model as introduced in Definition 5.1 allows also impulse rewards.

Theorem 5.3 (Policy Iteration). *Let* $\mathcal{M} = (S, Act, e, Q, i, r)$ *be a CTMDP and* $\bar{r}(s, a)$ *the continuized rate reward. For an initial policy* $\pi_0 \in \Pi$ *define the following iteration scheme:*

1. ***Policy evaluation:*** *Compute a solution* $(g^{\pi_n}, h^{\pi_n}, w)^T$ *to*

$$\begin{pmatrix} -Q^{\pi_n} & 0 & 0 \\ I & -Q^{\pi_n} & 0 \\ 0 & I & -Q^{\pi_n} \end{pmatrix} \begin{pmatrix} g^{\pi_n} \\ h^{\pi_n} \\ w \end{pmatrix} = \begin{pmatrix} 0 \\ \bar{r}^{\pi_n} \\ 0 \end{pmatrix}$$

2. ***Policy improvement:*** *Define for each state* s *the set of improving actions*

$$B_{n+1}(s) := \left\{ a \in e(s) \;\middle|\; \begin{array}{l} \sum_{s'} Q(s, a, s') g^{\pi_n}(s') > 0 \;\vee \\ (\sum_{s'} Q(s, a, s') g^{\pi_n}(s') = 0 \\ \Rightarrow \bar{r}(s, a) + \sum_{s'} Q(s, a, s') h^{\pi_n}(s') > g^{\pi_n}(s)) \end{array} \right\}$$

and choose an improving policy π_{n+1} *such that*

$$\pi_{n+1}(s) \in B_{n+1}(s) \text{ if } B_{n+1}(s) \neq \emptyset \quad \text{or} \quad \pi_{n+1}(s) := \pi_n(s) \text{ if } B_{n+1}(s) = \emptyset.$$

Termination: If $\pi_{n+1} = \pi_n$ *then* π_n *is an optimal policy. Otherwise go to the policy evaluation phase with* π_{n+1}.

The values g^{π_n} *are non-decreasing and policy iteration terminates in a finite number of iterations with an optimal policy* π_n *and optimal average reward* g^{π_n}.

The policy evaluation phase in this algorithm can be derived from the evaluation phase of the policy iteration algorithm in Theorem 3.8 for the uniformized model. However, the main difference between these algorithms is the policy improvement phase. Here $B_{n+1}(s)$ provides all actions which lead to at least some improvement in the policy π_n whereas in Theorem 3.8 a greedy maximal improving policy is chosen: $G_{n+1}(s)$ respectively $H_{n+1}(s)$. Note that $G_{n+1}(s) \cup H_{n+1}(s) \subseteq B_{n+1}(s)$. Of course, the choice of π_{n+1} in Theorem 5.3 can also be established by the greedy improving policy.

Example 5.1 (Bridge circuit). Consider a brige circuit as outlined in the reliability block diagram in Fig. 5.1.

The system is up, if there is at least one path of working components from s to t and it is down if on every path there is at least one failed component. Each working component $C \in \{L_1, L_2, B, R_1, R_2\}$ can fail after an exponentially distributed time with rate λ_C and there is a single repair unit, which can fix a failed component C after an exponentially distributed time with rate μ_C.

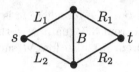

Fig. 5.1. The reliability block diagram of the bridge circuit system. An edge represents a component, which can be working or failed.

We assume that the components L_1 and L_2 (respectively R_1 and R_2) are identical and the parameter values for all components are

$$\lambda_{L_i} = 1.0 \quad \lambda_B = 0.1 \quad \lambda_{R_i} = 2.0$$
$$\mu_{L_i} = 10.0 \quad \mu_B = 100.0 \quad \mu_{R_i} = 10.0.$$

The action model allows the repair unit to be assigned to a failed component or to decide not to repair. We further assume that repair is preemptive, i.e. if during repair of a failed component another component fails, then the repair unit can decide again which component to repair. Note that due to the memory-less property of the exponential repair distribution, the remaining repair time in order to complete the repair does not depend on the elapsed time for repair. We want to find optimal repair policies, in order to pursue the following two goals:

(G1): maximize the MTTF (mean time to failure)
(G2): maximize the availability (i.e. the fraction of uptime in the total time).

Figure 5.2 shows an excerpt of the state space (with 32 states), which we apply to both goals (G1) and (G2). Note that for both measures (MTTF and availability) we define the reward structure which consists only of the rate reward r, which is 1 on up states and 0 on down states. The difference between both goals affects the state space as follows: For (G1) the 16 down states are absorbing (for every policy), while for (G2) a repair of failed components is also allowed in down system states.

We optimize (G1) by transforming the CTMDP by embedding into a discrete-time SSP (S, P, R) (cf. Definition 3.4 and Fig. 4.7) and hereby aggregate all absorbing down states to the goal state for the SSP. By embedding transformation, the reward $R(s, a)$ is the expected sojourn time in state s under action a in the CTMDP model, i.e. for all $a \in e(s)$

$$R(s, a) = \begin{cases} \frac{1}{E(s,a)}, & \text{for } s \neq goal \\ 0, & \text{for } s = goal \end{cases},$$

where $E(s, a)$ is the exit rate. Table 5.1 shows the resulting optimal policy and its corresponding maximal MTTF value function.

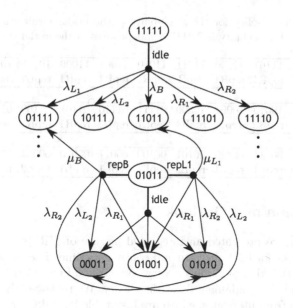

Fig. 5.2. State space of the bridge circuit CTMDP model. State encoding 01011 represents (from left to right) that L_1 has failed, L_2 is working, B has failed, R_1 is working and R_2 is working. From every state where at least some component has failed, there are repair actions and from each state there is also the idle action indicating that the repair unit can choose not to repair. Shadowed states represent down system states.

Table 5.1. Optimal policy and maximal MTTF value function of the bridge ciruit system (Example 5.1)

11111	11110	11101	11011	11010	11001	10111	10110	10101
idle	repR2	repR1	repB	repB	repB	repL2	repR2	repR1
1.449	1.262	1.262	1.448	1.234	1.234	1.324	1.095	1.087

10011	10010	01111	01110	01101	01011	01001	goal
repB	repB	repL1	repR2	repR1	repB	repB	idle
1.291	1.073	1.324	1.087	1.095	1.291	1.073	0.000

Problem (G2) is optimized by applying the CTMDP policy iteration algorithm for the average reward as outlined in Theorem 5.3. Beginning with the initial policy constantly *idle*, policy iteration converged in 6 iteration steps to the optimal policy given in Table 5.2.

Note that for (G2) the model can be shown to be unichain. Thus, the average reward g^π is constant for all policies π such that $Q^\pi g^\pi = 0$. For this reason, the policy evaluation and improvement phases in the policy iteration algorithm can be simplified.   □

Table 5.2. Optimal policy for the availability of the bridge ciruit system (Example 5.1). The maximal availability is 0.917757 independent of the initial state.

11111	11110	11101	11100	11011	11010	11001	11000	10111	10110	10101
idle	repR2	repR1	repR1	repB	repR2	repR1	repR1	repL2	repR2	repR1

10100	10011	10010	10001	10000	01111	01110	01101	01100	01011	01010
repR1	repB	repR2	repB	repR1	repL1	repR2	repR1	repR2	repB	repB

01001	01000	00111	00110	00101	00100	00011	00010	00001	00000
repR1	repR2	repL1	repL1	repL2	repL1	repL1	repL1	repL2	repL2

6   Conclusion and Outlook

In this tutorial, we presented an integrated picture of MRMs and MDPs over finite state spaces for both discrete and continuous time. The theory and application area for this kind of models is very popular and broad. For this reason, we just focussed on the fundamentals of the theory. We reviewed the most important basic facts from literature, which are inevitable for a deeper understanding of Markovian models. Furthermore, we set up the theory step by step from discrete-time MRMs up to continuous-time MDPs and pointed out important links between these theories. We also connected these models by a number of model transformations and highlighted their properties. In order to show the applicability of these models, we introduced small prototypical examples, coming from the domain of performance and dependability evaluation and optimization. Of course, many current applications in the optimization of dependable systems suffer from the curse of dimensionality. However, there are established techniques which can be used in order to overcome this curse and make evaluation and optimization for large practical models accessible, e.g. approximate solutions (e.g. approximate dynamic programming), simulative approaches (reinforcement learning) or the use of structured models. There are also several important extensions to the Markov model types we could not address in this introductory tutorial, such as partially observable MDPs (used especially in the area of AI), denumerable state and action spaces (e.g. for queueing systems) or even continuous state and action spaces (leading directly to control theory).

A   Appendix

A.1   Lemmata and Remarks

Lemma A.1.

(i) Let N be a non-negative discrete random variable with expected value and $(a_n)_{n \in \mathbb{N}}$ a bounded sequence. Then

$$\mathbb{E}\left[\sum_{n=0}^{N} a_n\right] = \sum_{n=0}^{\infty} a_n P(N \geq n).$$

(ii) Let T be a non-negative continuous random variable with expected value and $a : [0, \infty) \to \mathbb{R}$ an integrable and bounded function. Then

$$\mathbb{E}\left[\int_0^T a(t)\, dt\right] = \int_0^\infty a(t) P(T \geq t)\, dt.$$

Proof. We show only (i) – the proof for (ii) is analogous when summation is replaced by integration. Since a_n is bounded and $\mathbb{E}[N] = \sum_{n=0}^\infty P(N \geq n) < \infty$ it follows that $\sum_{n=0}^\infty a_n P(N \geq n)$ converges absolutely. From

$$\sum_{n=0}^\infty a_n P(N \geq n) = \sum_{n=0}^\infty \sum_{k=n}^\infty a_n P(N = k) = \sum_{n=0}^\infty \sum_{k=0}^\infty a_n P(N = k) \mathbb{1}_{\{k \geq n\}}(n, k)$$

we can interchange both infinite summations by the Fubini theorem. It follows

$$\sum_{n=0}^\infty a_n P(N \geq n) = \sum_{k=0}^\infty \sum_{n=0}^\infty a_n P(N = k) \mathbb{1}_{\{n \leq k\}}(n, k)$$

$$= \sum_{k=0}^\infty \sum_{n=0}^k a_n P(N = k) = \mathbb{E}\left[\sum_{n=0}^N a_n\right]. \qquad \square$$

Remark A.1. As presented in Sect. 4.2.2 a CTMC $\mathcal{M} = (S, Q)$ induces the stochastic processes X_n, τ_n, T_n, N_t and Z_t. If we fix a uniformization rate $\mu > 0$ then $\mathcal{M}$ also induces the **uniformized stochastic processes** $\widetilde{X}_n$, $\widetilde{\tau}_n$, $\widetilde{T}_n$, $\widetilde{N}_t$ and $\widetilde{Z}_t$ over $\Omega = (S \times (0, \infty])^{\mathbb{N}}$. Here $\widetilde{X}_n$ is the n-th visited state in the uniformized DTMC and $\widetilde{\tau}_n$ is the time up to transition in state $\widetilde{X}_n$, i.e. all $\widetilde{\tau}_n$ are independent and exponentially distributed with rate μ. Moreover, the total elapsed time $\widetilde{T}_n := \sum_{k=0}^{n-1} \widetilde{\tau}_k$ for the first n transitions is Erlang distributed with n phases and rate μ and the number $\widetilde{N}_t := \max\{n \mid \widetilde{T}_n(\omega) \leq t\}$ of (uniformized) transitions up to time $t \geq 0$ is Poisson distributed with parameter μt. Note also that $\widetilde{Z}_t := \widetilde{X}_{\widetilde{N}_t} = X_{N_t} = Z_t$ for all $t \geq 0$. Thus, when uniformization is considered as adding exponentially distributed self-loop transitions to states of the CTMC $\mathcal{M}$, then the continuous-time state process Z_t is not modified at all. Therefore, the probability measures P_s of the CTMC for all $s \in S$ are left invariant under uniformization and especially the transient probability matrix $P(t)$ as defined in (4.1). $\qquad \square$

Lemma A.2. *Consider a CTMC $\mathcal{M} = (S, Q)$ with discrete-time and continuous-time state processes X_n and Z_t. Let*

$$n_T(s, s') := \mathbb{E}_{s_0}\left[\sum_{k=1}^{N_T} \mathbb{1}_{\{X_{k-1}=s, X_k=s'\}}\right]$$

be the expected number of transitions from state s to state $s' \neq s$ within the time interval $[0, T]$ from a fixed initial state $X_0 = s_0$. Then

$$n_T(s, s') = Q(s, s') \int_0^T P_{s_0}(Z_t = s)\, dt. \tag{A.1}$$

Proof. If s is absorbing then clearly (A.1) holds and we can assume in the following that $E(s) > 0$. We abbreviate in the following the notation by $\mathbb{E} := \mathbb{E}_{s_0}$ and $P := P_{s_0}$. Define

$$n_T(s) := \sum_{s' \neq s} n_T(s, s') = \mathbb{E}\left[\sum_{k=1}^{N_T} \mathbb{1}_{\{X_{k-1}=s\}}\right] = \sum_{k=1}^{\infty} P(X_{k-1} = s)P(N_T \geq k)$$

as the number of complete visits to state s, such that s has been left before time T. From $P(X_{k-1} = s, X_k = s') = P(X_k = s'|X_{k-1} = s)P(X_{k-1} = s)$ it follows that

$$n_T(s, s') = P(s, s') \sum_{k=1}^{\infty} P(X_{k-1} = s)P(N_T \geq k) = P(s, s')n_T(s),$$

where $P(s, s') = \delta_{s,s'} + \frac{Q(s,s')}{E(s)}$ is the embedded transition probability. We use uniformization as means for proof with a uniformization rate $\mu \geq \max_{s \in S} E(s)$. Let $\tilde{X}_k$, $\tilde{\tau}_k$, $\tilde{T}_k$ and $\tilde{N}_t$ be the uniformized stochastic processes as defined in Remark A.1. Then $\tilde{T}_k$ is Erlang distributed with density $f_{\tilde{T}_k}(t) = e^{-\mu t}\frac{\mu^k t^{k-1}}{(k-1)!}$ for $t \geq 0$ and $\tilde{N}_t$ has the Poisson probabilities $P(\tilde{N}_t = k) = e^{-\mu t}\frac{(\mu t)^k}{k!}$. The total accumulated time for complete visits in s up to time T fulfills

$$\frac{1}{E(s)}\mathbb{E}\left[\sum_{k=1}^{N_T} \mathbb{1}_{\{X_{k-1}=s\}}\right] = \frac{1}{\mu}\mathbb{E}\left[\sum_{k=1}^{\tilde{N}_T} \mathbb{1}_{\{\tilde{X}_{k-1}=s\}}\right]$$

and therefore

$$n_T(s) = \frac{E(s)}{\mu}\mathbb{E}\left[\sum_{k=1}^{\tilde{N}_T} \mathbb{1}_{\{\tilde{X}_{k-1}=s\}}\right]$$

is a fraction of the number of uniformized transitions up to time T from state s to some arbitrary state s'. It follows that

$$n_T(s) = \frac{E(s)}{\mu}\mathbb{E}\left[\sum_{k=1}^{\tilde{N}_T} \mathbb{1}_{\{\tilde{X}_{k-1}=s\}}\right] = \frac{E(s)}{\mu}\sum_{k=1}^{\infty} P(\tilde{X}_{k-1} = s)P(\tilde{N}_T \geq k)$$

$$= \frac{E(s)}{\mu}\sum_{k=0}^{\infty} P(\tilde{X}_k = s)P(\tilde{T}_{k+1} \leq T)$$

$$= \frac{E(s)}{\mu}\sum_{k=0}^{\infty} P(\tilde{X}_k = s)\int_0^T e^{-\mu t}\frac{\mu^{k+1}t^k}{k!}\,dt$$

$$= E(s)\int_0^T e^{-\mu t}\sum_{k=0}^{\infty} P(\tilde{X}_k = s)\frac{(\mu t)^k}{k!}\,dt = E(s)\int_0^T P(Z_t = s)\,dt$$

since

$$P(Z_t = s) = \sum_{k=0}^{\infty} P(Z_t = s \mid \tilde{N}_t = k)P(\tilde{N}_t = k) = \sum_{k=0}^{\infty} P(\tilde{X}_k = s)e^{-\mu t}\frac{(\mu t)^k}{k!}.$$

Thus, (A.1) follows from $P(s, s') = \frac{Q(s,s')}{E(s)}$ for $s' \neq s$ and $n_T(s, s') = P(s, s')n_T(s)$.

□

Remark A.2. In the proof of Lemma A.2, we have applied uniformization as a detour in order to show that

$$n_T(s, s') = Q(s, s') \int_0^T P_{s_0}(Z_t = s)\, dt.$$

There is also a more direct way to show this equation by an argument that is used in the proof of the PASTA property ("Poisson Arrivals See Time Averages") [41]. The PASTA property is a tool that is frequently used in the theory of queueing systems. Consider a system that is represented by the Markov chain $\mathcal{M} = (S, Q)$ with state process Z_t for $t \geq 0$ and fix two states s and s' with $Q(s, s') > 0$. Let $\tau_{s,s'}$ be the exponentially distributed time with rate $Q(s, s')$ that governs the transition from s to s' as shown in Sect. 4.1. Further define an independent sequence of such random variables $\tau_{s,s'}^{(n)}$, $n \in \mathbb{N}$ with same distribution as $\tau_{s,s'}$. Then the process $A_t := \max\left\{n \mid \sum_{k=0}^{n-1} \tau^{(k)} \leq t\right\}$ is a Poisson process with rate $Q(s, s')$ and is regarded as a stream of arriving jobs to the system. Since $\tau_{s,s'}$ is memoryless it holds that when the system is in state s and an arrival occurs then the system performs a transition to s'. Therefore, the counting process $Y_t := \sum_{k=1}^{N_t} \mathbb{1}_{\{X_{k-1}=s, X_k=s'\}}$ is precisely the number of arrivals of A_t to the system (up to time t) that find the system in state s. Further let $U_t := \mathbb{1}_{\{Z_t=s\}}$ be the process that indicates whether the system is in state s at time t. It holds that Y_t can be represented as a stochastic Riemann-Stiltjes integral of the process U_t with respect to the arrival process A_t, i.e. for all $T \geq 0$ it holds that $Y_T = \int_0^T U_t\, dA_t$ with probability 1. Note that for each $t \geq 0$ the set of future increments $\{A_{t+s} - A_s \mid s \geq 0\}$ and the history of the indicator process $\{U_s \mid 0 \leq s \leq t\}$ are independent. Thus the "lack of anticipation assumption" as needed for [41] is satisfied and it follows that

$$n_T(s, s') = \mathbb{E}_{s_0}[Y_T] = Q(s, s') \cdot \mathbb{E}_{s_0}\left[\int_0^T U_t\, dt\right] = Q(s, s') \int_0^T P_{s_0}(Z_t = s)\, dt.$$

□

Lemma A.3. *Let T be a non-negative continuous random horizon length for a CTMRM $\mathcal{M} = (S, Q, i, r)$ and independent of the state process Z_t of $\mathcal{M}$. Further let $N_T = \max\{n \mid T_n \leq T\}$ be the random number of transitions up to time T. If the k-th moment of T exists, then it also exists for N_T.*

In order to prove this theorem we need the following definition.

Definition A.1. *For two random variables X and Y with distributions F_X and F_Y we say that X is **stochastically smaller** then Y (denoted by $X \preceq Y$) if $F_X(x) \geq F_Y(x)$ for all $x \in \mathbb{R}$.*

It follows that if $X \preceq Y$ then $\mathbb{E}[g(X)] \leq \mathbb{E}[g(Y)]$ for a monotonically increasing function g.

Proof. Let X_n, τ_n, T_n and N_t be the stochastic processes as defined in Sect. 4.2.2. Further choose $\mu := \max\{E(s) \mid s \in S\}$ as a uniformization rate and $\widetilde{X}_n, \widetilde{\tau}_n, \widetilde{T}_n$ and $\widetilde{N}_t$ the uniformized processes as in Remark A.1. First we show that $N_T \preceq \widetilde{N}_T$: From $\mu \geq E(s)$ for all s it follows that $\widetilde{\tau}_n \preceq \tau_n$ for all n and thus $\widetilde{T}_n \preceq T_n$. Therefore

$$N_T = \max\{n \mid T_n \leq T\} \preceq \max\{n \mid \widetilde{T}_n \leq T\} = \widetilde{N}_T$$

and thus $\mathbb{E}[N_T^k] \leq \mathbb{E}[\widetilde{N}_T^k]$. In order to show that $\mathbb{E}[N_T^k]$ is finite we show $\mathbb{E}[\widetilde{N}_T^k] < \infty$. It holds that $P(\widetilde{N}_T = n) = P(\widetilde{T}_n \leq T < \widetilde{T}_{n+1})$ and therefore

$$\mathbb{E}[\widetilde{N}_T^k] = \sum_{n=0}^{\infty} n^k P(\widetilde{T}_n \leq T < \widetilde{T}_{n+1}).$$

We show that the sequence $P(\widetilde{T}_n \leq T < \widetilde{T}_{n+1})$ is decreasing fast enough.

$$P(\widetilde{T}_n \leq T < \widetilde{T}_{n+1}) = \int_{z=0}^{\infty} f_T(z) \int_{v=0}^{z} f_{\widetilde{T}_n}(v) \int_{u=z-v}^{\infty} f_{\widetilde{\tau}_n}(u)\, du\, dv\, dz =$$

$$\int_{z=0}^{\infty} f_T(z) \int_{v=0}^{z} f_{\widetilde{T}_n}(v) e^{-\mu(z-v)}\, dv\, dz = \int_{z=0}^{\infty} f_T(z) e^{-\mu z} \int_{v=0}^{z} \frac{\mu^n v^{n-1}}{(n-1)!}\, dv\, dz =$$

$$\int_{z=0}^{\infty} f_T(z) \frac{e^{-\mu z}(\mu z)^n}{n!}\, dz.$$

Therefore

$$\mathbb{E}[\widetilde{N}_T^k] = \sum_{n=0}^{\infty} n^k \int_{z=0}^{\infty} f_T(z) \frac{e^{-\mu z}(\mu z)^n}{n!}\, dz = \int_{z=0}^{\infty} f_T(z) \mathbb{E}[\widetilde{N}_z^k]\, dz,$$

where $\widetilde{N}_z$ is the number of uniformized transitions up to time z which is Poisson distributed with parameter μz. Now the k-th moment $g(z) := \mathbb{E}[\widetilde{N}_z^k]$ is a polynomial in z of degree k and therefore

$$\mathbb{E}[\widetilde{N}_T^k] = \int_{z=0}^{\infty} g(z) f_T(z)\, dz < \infty,$$

as a polynomial of degree k in the moments of T. $\square$

Remark A.3. In the case $k = 1$ it holds that $\mathbb{E}[\widetilde{N}_T] = \mu \mathbb{E}[T]$ represents exactly Little's law: If jobs enter a queue at rate μ and if their mean residence time in the queue is $\mathbb{E}[T]$, then there are on average $\mu \mathbb{E}[T]$ jobs in the queue. For $k \geq 2$ the theorem generalizes Little's law and allows to compute $\mathbb{E}[\widetilde{N}_T]$ analytically since the coefficients of g can be computed analytically.

A.2   Laurent Series Expansion for Continuous Time Models

Proposition A.1. *Let $Q \in \mathbb{R}^{n \times n}$ be the generator matrix of a CTMC over a finite state space and $P(t) = e^{Qt}$ the transient probability matrix. Then the*

CTMC is exponentially ergodic, i.e. there exists $\delta > 0$ and $L > 0$, such that

$$\|P(t) - P^*\| \leq Le^{-\delta t}$$

for all $t \geq 0$, where $\|A\| := \max_i \sum_j |A_{i,j}|$ is the matrix maximum norm.

In [18] an equivalent statement is described for finite state DTMCs. We transfer and modify the proof to the continuous-time case.

Proof. Since $P_{i,j}(t) \to P_{i,j}^*$ for all i, j, it follows that for an arbitrary fixed $\varepsilon > 0$ there exists $T > 0$, such that for all i

$$\sum_j |P_{i,j}(T) - P_{i,j}^*| \leq e^{-\varepsilon} < 1$$

and therefore $\|P(T) - P^*\| \leq e^{-\varepsilon}$. Now split $t = n_t T + s_t$ with $n_t \in \mathbb{N}$ and $s_t \in [0, T)$.

$$\begin{aligned}
\|P(t) - P^*\| &= \|P(T)^{n_t} P(s_t) - P^*\| & P(s+t) = P(s)P(t) \\
&= \|(P(T)^{n_t} - P^*)(P(s_t) - P^*)\| & P(t)P^* = P^*, \ P^*P^* = P^* \\
&\leq \|P(T)^{n_t} - P^*\| \cdot \|P(s_t) - P^*\| & \text{subadditivity of norm} \\
&= \|(P(T) - P^*)^{n_t}\| \cdot \|P(s_t) - P^*\| \\
&\leq \|P(T) - P^*\|^{n_t} \cdot \|P(s_t) - P^*\| \\
&\leq e^{-\varepsilon n_t} \|P(s_t) - P^*\| \\
&= e^{-\varepsilon t/T} e^{\varepsilon s_t/T} \|P(s_t) - P^*\|.
\end{aligned}$$

Defining

$$\delta := \frac{\varepsilon}{T} \quad \text{and} \quad L := \sup_{s \in [0,T)} \left(e^{\varepsilon s/T} \|P(s) - P^*\|\right) < \infty$$

gives $\|P(t) - P^*\| \leq Le^{-\delta t}$. $\square$

Define the transient deviation matrix $\Delta(t) := P(t) - P^*$ and the total deviation matrix $H := \int_0^\infty \Delta(t)\, dt$ (componentwise integration). From Proposition A.1 it follows that the integral defining H converges since $\|\Delta(t)\| \leq Le^{-\delta t}$.

Theorem A.1. *For $\alpha > 0$ let $W(\alpha) := \int_0^\infty e^{-\alpha t} P(t)\, dt$ be the Laplace transform of $P(t)$. Then there exists $\delta > 0$, such that for all $0 < \alpha < \delta$ the Laurent series of $W(\alpha)$ is given by*

$$W(\alpha) = \alpha^{-1} P^* + \sum_{n=0}^{\infty} (-\alpha)^n H^{n+1}.$$

Proof. Since $P(t)P^* = P^*P(t) = P^*P^* = P^*$ it follows that $\Delta(t+s) = \Delta(t)\Delta(s)$ for all $s, t \geq 0$. Now

$$W(\alpha) = \int_0^\infty e^{-\alpha t}(P(t) - P^* + P^*)\, dt$$

$$= \alpha^{-1}P^* + \int_0^\infty e^{-\alpha t}\Delta(t)\, dt$$

$$= \alpha^{-1}P^* + \int_0^\infty \left(\sum_{n=0}^\infty \frac{(-\alpha t)^n}{n!}\right)\Delta(t)\, dt$$

$$= \alpha^{-1}P^* + \sum_{n=0}^\infty (-\alpha)^n \int_0^\infty \frac{t^n}{n!}\Delta(t)\, dt,$$

where the last equality follows from Lebesgue's dominated convergence theorem, since for all i, j the sequence $\sum_{n=0}^N \frac{(-\alpha t)^n}{n!}\Delta_{i,j}(t)$ can be dominated by the integrable function $Ce^{(\alpha-\delta)t}$ (for $\delta > 0$ from Proposition A.1 and some $C > 0$) for all $0 < \alpha < \delta$. We show by induction that

$$\int_0^\infty \frac{t^n}{n!}\Delta(t)\, dt = H^{n+1}. \tag{A.2}$$

For $n = 0$ this is true by definition of H. Let (A.2) be true for an arbitrary $n \in \mathbb{N}$. Then

$$\int_0^\infty \frac{t^{n+1}}{(n+1)!}\Delta(t)\, dt = \int_0^\infty \left(\int_0^t \frac{s^n}{n!}\, ds\right)\Delta(t)\, dt = \int_0^\infty \frac{s^n}{n!}\int_s^\infty \Delta(t)\, dt\, ds$$

$$= \int_0^\infty \frac{s^n}{n!}\int_0^\infty \Delta(s+t)\, dt\, ds = \int_0^\infty \frac{s^n}{n!}\Delta(s)\int_0^\infty \Delta(t)\, dt\, ds$$

$$= \left(\int_0^\infty \frac{s^n}{n!}\Delta(s)\, ds\right)H = H^{n+1}$$

and the Laurent series follows. □

A.3   Collection of Proofs

Proof (of Theorem 2.1). (i) For an arbitrary $s_0 \in S$ it holds

$$V_N(s_0) = \mathbb{E}_{s_0}\left[\sum_{i=1}^N R(X_{i-1}, X_i)\right] = \sum_{s_1,\dots,s_N}\left(\left(\sum_{i=1}^N R(s_{i-1}, s_i)\right)\prod_{i=1}^N P(s_{i-1}, s_i)\right)$$

$$= \sum_{s_1} P(s_0, s_1)\left(R(s_0, s_1)\sum_{s_2,\dots,s_N}\prod_{i=2}^N P(s_{i-1}, s_i) +\right.$$

$$\left.\sum_{s_2,\dots,s_N}\sum_{i=2}^N R(s_{i-1}, s_i)\prod_{i=2}^N P(s_{i-1}, s_i)\right).$$

Now since $\sum_{s_1} R(s_0, s_1)P(s_0, s_1) = R(s_0)$ and for each $s_1 \in S$ it holds that

$$\sum_{s_2,\dots,s_N} \prod_{i=2}^{N} P(s_{i-1}, s_i) = 1 \quad \text{and}$$

$$\sum_{s_2,\dots,s_N} \left(\sum_{i=2}^{N} R(s_{i-1}, s_i) \right) \prod_{i=2}^{N} P(s_{i-1}, s_i) = \mathbb{E}_{s_1}\left[\sum_{i=2}^{N} R(X_{i-1}, X_i) \right] = V_{N-1}(s_1)$$

it follows that

$$V_N(s_0) = R(s_0) + \sum_{s_1} P(s_0, s_1)V_{N-1}(s_1).$$

In case V_∞ exists then $V_\infty(s) = \lim_{N \to \infty} V_N(s)$ for all $s \in S$ and statement (ii) follows from (i) by taking the limit on both sides. □

Proof (of Proposition 2.1). We are going to sketch a proof for this fact in case the Markov chain is aperiodic. Since V_∞ exists for the model (S, P, R) if and only if it exists for $(S, P, |R|)$ we can assume without loss of generality that $R(s, s') \geq 0$ for all $s, s' \in S$. Note that here $|R|$ has to be interpreted as the transition-based reward function with $|R|(s, s') := |R(s, s')|$. The reason is that the state-based view on the absolute reward values $\sum_{s' \in S} P(s, s')|R(s, s')|$ in general differs from $|\sum_{s' \in S} P(s, s')R(s, s')|$ which is the absolute value of the state-based view on the reward values!

"$\Rightarrow$": Assume that V_∞ exists and $R(\tilde{s}, \tilde{s}') > 0$ for some states $\tilde{s}, \tilde{s}' \in S_i^r$ and thus $R(\tilde{s}) = \sum_{s' \in S} P(\tilde{s}, s')R(\tilde{s}, s') > 0$. For all $k \in \mathbb{N}$ it holds that $\mathbb{E}_s[R(X_{k-1}, X_k)]$ is the reward gained for the k-th transition when starting in s. Therefore

$$\mathbb{E}_{\tilde{s}}[R(X_{k-1}, X_k)] = \sum_{s' \in S} P^{k-1}(\tilde{s}, s') \sum_{s'' \in S} P(s', s'')R(s', s'')$$

$$= \sum_{s' \in S} P^{k-1}(\tilde{s}, s')R(s') \geq P^{k-1}(\tilde{s}, \tilde{s})R(\tilde{s}).$$

Since P is aperiodic and $\tilde{s}$ is recurrent it follows that $P^{k-1}(\tilde{s}, \tilde{s})$ converges to $\rho_{\tilde{s}}(\tilde{s}) > 0$, where $\rho_{\tilde{s}}$ is the limiting distribution from $\tilde{s}$ (see Sect. 2.1.2). Therefore the sequence $\mathbb{E}_{\tilde{s}}\left[\sum_{k=1}^{N} |R(X_{k-1}, X_k)| \right] \geq \sum_{k=1}^{N} P^{k-1}(\tilde{s}, \tilde{s})R(\tilde{s})$ is unbounded, which is a contradiction to the existence of V_∞.

"$\Leftarrow$": Assume that $R(s, s') = 0$ for all $s, s' \in S_i^r$ and all $i = 1, \dots, k$. We anticipate a result from Proposition 2.2 in Sect. 2.4, which states that the limiting matrix $P^\infty := \lim_{n \to \infty} P^n$ exists since P is aperiodic. In [18] it is shown that P is geometric ergodic, i.e. there exists $n_0 \in \mathbb{N}$, $c > 0$ and $\beta < 1$ such that

$$\|P^n - P^\infty\| \leq c\beta^n$$

for all $n \geq n_0$, where $\|.\|$ is the maximum norm. (This result as stated holds for unichain models, but it can also be directly extended to the multichain case). First of all, we want to show that $P^\infty R = 0$, i.e. for all $s \in S$ it holds that

$$(P^\infty R)(s) = \sum_{s' \in S} P^\infty(s, s') \sum_{s'' \in S} P(s', s'')R(s', s'') = 0.$$

If $s \in S_i^r$ is recurrent then we only have to consider those terms in the summation for which s' and s'' are in the same closed recurrent class S_i^r. But since both $s', s'' \in S_i^r$ it follows that $R(s', s'') = 0$ and thus $(P^\infty R)(s) = 0$. On the other hand if $s \in S^t$ is transient then $P^\infty(s, s') = 0$ for all $s' \in S^t$ and otherwise if s' is recurrent then again $P(s', s'') = 0$ or $R(s', s'') = 0$ dependent on whether s' and s'' are in the same closed recurrent class. (Compare this also to the representation of $P^\infty = P^*$ in (2.14).) Combining together it follows for all $s \in S$ and $k \geq n_0 + 1$ that

$$\mathbb{E}_s\left[R(X_{k-1}, X_k)\right] = \sum_{s' \in S} P^{k-1}(s, s') \sum_{s'' \in S} P(s', s'') R(s', s'') = \sum_{s' \in S} P^{k-1}(s, s') R(s')$$

$$\leq \max_{s \in S}\left\{\sum_{s' \in S} P^{k-1}(s, s') R(s')\right\} = ||P^{k-1} R|| = ||P^{k-1} R - P^\infty R|| \leq c\beta^{k-1} ||R||.$$

Therefore

$$\mathbb{E}_s\left[\sum_{k=1}^N R(X_{k-1}, X_k)\right] \leq \sum_{k=1}^N c\beta^{k-1} ||R||$$

converges as $N \to \infty$ since $\beta < 1$. □

Proof (of Theorem 3.2). The convergence of V_n to $(V^\gamma)^*$ has been already remarked in Remark 3.2. It further holds

$$||V^{\pi_\varepsilon} - (V^\gamma)^*|| \leq ||V^{\pi_\varepsilon} - V_{n+1}|| + ||V_{n+1} - (V^\gamma)^*||.$$

From (2.11) it follows that for every policy π the linear operator T^π defined by $T^\pi V := R^\pi + \gamma P^\pi V$ is also a contraction with the same Lipschitz constant $q := \gamma < 1$ as for $\mathcal{T}$. Let $V^{\pi_\varepsilon} = T^{\pi_\varepsilon} V^{\pi_\varepsilon}$ be the fixed point of T^{π_ε}. By definition of π_ε in (3.7) (i.e. $\pi_\varepsilon(s)$ is a maximizing action) it follows that $T^{\pi_\varepsilon} V_{n+1} = \mathcal{T} V_{n+1}$. Thus, for the first term it holds

$$||V^{\pi_\varepsilon} - V_{n+1}|| \leq ||V^{\pi_\varepsilon} - \mathcal{T} V_{n+1}|| + ||\mathcal{T} V_{n+1} - V_{n+1}||$$
$$= ||T^{\pi_\varepsilon} V^{\pi_\varepsilon} - T^{\pi_\varepsilon} V_{n+1}|| + ||\mathcal{T} V_{n+1} - \mathcal{T} V_n||$$
$$\leq q||V^{\pi_\varepsilon} - V_{n+1}|| + q||V_{n+1} - V_n||.$$

Therefore

$$||V^{\pi_\varepsilon} - V_{n+1}|| \leq \frac{q}{1-q} ||V_{n+1} - V_n||.$$

In analogy it follows for the second term

$$||V_{n+1} - (V^\gamma)^*|| \leq q||V_n - (V^\gamma)^*|| \leq q\left(||V_n - V_{n+1}|| + ||V_{n+1} - (V^\gamma)^*||\right)$$

and thus

$$||V_{n+1} - (V^\gamma)^*|| \leq \frac{q}{1-q} ||V_{n+1} - V_n||.$$

By combining the inequalities together it follows that

$$||V^{\pi_\varepsilon} - (V^\gamma)^*|| \leq \frac{2q}{1-q} ||V_{n+1} - V_n||.$$

Hence the conclusion follows from $||V_{n+1} - V_n|| < \frac{1-\gamma}{2\gamma}\varepsilon$ for $q = \gamma$. □

Proof (of Proposition 4.1). The proof is analogous to the proof of Proposition 2.1 in the discrete-time setting. By Definition 4.2 the value function V_∞ is defined if and only if it holds for all $s \in S$ that both terms $\mathbb{E}_s \left[\sum_{k=1}^{N_T} |i(X_{k-1}, X_k)| \right]$ and $\mathbb{E}_s \left[\int_0^T |r(Z_t)| \, dt \right]$ converge as $T \to \infty$. For simplicity, we only sketch the proof for the rate reward. Without loss of generality we assume that $r(s) \geq 0$ for all $s \in S$.

"$\Rightarrow$": V_∞ is defined if and only if $\int_0^T \mathbb{E}_s [r(Z_t)] \, dt$ converges with $T \to \infty$ and thus $\mathbb{E}_s [r(Z_t)] \to 0$ as $t \to \infty$. But if s is recurrent then $\lim_{t\to\infty} P(t)(s,s) = P^*(s,s) > 0$ and from $\mathbb{E}_s [r(Z_t)] = \sum_{s' \in S} P(t)(s,s')r(s') \geq P(t)(s,s)r(s)$ it follows that $r(s) = 0$.

"$\Leftarrow$": Let $r(s) = 0$ for all recurrent states s. As in the discrete-time case, one can show that the transient probability matrix $P(t)$ of the finite-state CTMC (S, Q) is exponentially ergodic, i.e. there exists $L > 0$ and $\delta > 0$ such that $||P(t) - P^*|| \leq Le^{-\delta t}$ for all $t \geq 0$ where $||.||$ is the maximum norm (see Proposition A.1). We first show that

$$(P^*r)(s) = \sum_{s' \in S} P^*(s,s')r(s') = 0$$

for all $s \in S$ (see also the representation of P^* in (4.3)). If $s \in S_i^r$ is recurrent then $P^*(s,s') = 0$ if $s' \in S \setminus S_i^r$ and $r(s') = 0$ if $s' \in S_i^r$. Otherwise, if $s \in S^t$ is transient then $P^*(s,s') = 0$ for all transient states $s' \in S^t$ and $r(s') = 0$ for all recurrent states $s' \in S \setminus S^t$. It follows for all $s \in S$ that

$$\mathbb{E}_s \left[\int_0^T r(Z_t) \, dt \right] = \int_0^T \sum_{s' \in S} P(t)(s,s')r(s') \, dt \leq \int_0^T ||P(t)r|| \, dt$$
$$= \int_0^T ||P(t)r - P^*r|| \, dt \leq \int_0^T Le^{-\delta t} ||r|| \, dt$$

converges as $T \to \infty$.   □

Proof (of Lemma 4.1). We show the first equality by regarding the representation of $V(s)$ in (4.13) as a total expectation. We can interchange both expectations in the middle term by the Fubini theorem (or law of total expectation), i.e.

$$V(s) = \mathbb{E} \left[\mathbb{E}_s \left[\int_0^T \bar{r}(Z_t) \, dt \mid T \right] \right] = \mathbb{E}_s \left[\mathbb{E} \left[\int_0^T \bar{r}(Z_t) \, dt \mid Z_t \right] \right].$$

Here $\mathbb{E} \left[\int_0^T \bar{r}(Z_t) \, dt \mid Z_t \right]$ is a conditional expectation given knowledge of all the Z_t for $t \geq 0$, i.e. it is a random variable over Ω that takes the values $\mathbb{E} \left[\int_0^T \bar{r}(Z_t(\omega)) \, dt \right]$ for $\omega \in \Omega$. Since the state space is finite it holds that the map $t \mapsto \bar{r}(Z_t(\omega))$ is bounded for all $\omega \in \Omega$ and it follows by Lemma A.1 that

$$V(s) = \mathbb{E}_s \left[\int_0^\infty \bar{r}(Z_t)P_T(T \geq t) \, dt \right].$$

For the second equality of $V(s)$ in Lemma 4.1 note that $Z_t(\omega) = X_{N_t(\omega)}(\omega)$ and $N_t(\omega)$ piecewise constant in t for all $\omega \in \Omega$. Therefore $r(Z_t(\omega)) = r(X_n(\omega))$ for all $t \in [T_n(\omega), T_{n+1}(\omega))$ and it follows that

$$\mathbb{E}_s \left[\int_0^\infty \bar{r}(Z_t) P_T(T \geq t)\, dt \right] = \mathbb{E}_s \left[\sum_{n=0}^\infty \bar{r}(X_n) \int_{T_n}^{T_{n+1}} P_T(T \geq t)\, dt \right]. \qquad \square$$

Proof (of Theorem 4.3). Equation (4.16) can be established by multipliying (4.15) with $\alpha + E(s)$ and using $E(s) = -Q(s,s)$ when rearranging terms. Thus we only have to show (4.15). If s is absorbing then $Q(s,s') = 0$ for all s', $E(s) = 0$ and $P(t)(s,s') = \delta_{s,s'}$. The conclusion follows from (4.14) since $V^\alpha(s) = \int_0^\infty \bar{r}(s) e^{-\alpha t}\, dt = \frac{\bar{r}(s)}{\alpha}$. Assume in the following that s is non-absorbing and thus $E(s) > 0$. From Lemma 4.1 it holds that

$$V^\alpha(s) = \mathbb{E}_s \left[\sum_{n=0}^\infty \bar{r}(X_n) \int_{T_n}^{T_{n+1}} e^{-\alpha t}\, dt \right] = \mathbb{E}_s \left[\sum_{n=0}^\infty e^{-\alpha T_n} \bar{r}(X_n) \int_0^{T_n} e^{-\alpha t}\, dt \right],$$

since $T_{n+1} = T_n + \tau_n$. Define $R(X_n, \tau_n) := \bar{r}(X_n) \int_0^{\tau_n} e^{-\alpha t}\, dt$. Because τ_0 given $X_0 = s$ is exponentially distributed with rate $E(s) > 0$ it follows by Lemma A.1 that

$$\mathbb{E}_s \left[R(X_0, \tau_0) \right] = \frac{\bar{r}(s)}{\alpha + E(s)}$$

and thus

$$V^\alpha(s) = \mathbb{E}_s \left[\sum_{n=0}^\infty e^{-\alpha \sum_{k=0}^{n-1} \tau_k} R(X_n, \tau_n) \right]$$

$$= \mathbb{E}_s \left[R(X_0, \tau_0) \right] + \mathbb{E}_s \left[e^{-\alpha \tau_0} \sum_{n=1}^\infty e^{-\alpha \sum_{k=1}^{n-1} \tau_k} R(X_n, \tau_n) \right]$$

$$= \frac{\bar{r}(s)}{\alpha + E(s)} + \mathbb{E}_s \left[e^{-\alpha \tau_0} \sum_{n=0}^\infty e^{-\alpha \sum_{k=0}^{n-1} \tau_{k+1}} R(X_{n+1}, \tau_{n+1}) \right]$$

$$= \frac{\bar{r}(s)}{\alpha + E(s)} + \mathbb{E} \left[e^{-\alpha \tau_0} V^\alpha(X_1) \mid X_0 = s \right],$$

where $V^\alpha(X_1)$ is the random variable representing the discounted value when the process starts in X_1. Now since $V^\alpha(X_1)$ is independent of τ_0 (given $X_0 = s$) it follows that

$$\mathbb{E} \left[e^{-\alpha \tau_0} V^\alpha(X_1) \mid X_0 = s \right] = \mathbb{E} \left[e^{-\alpha \tau_0} \mid X_0 = s \right] \mathbb{E} \left[V^\alpha(X_1) \mid X_0 = s \right]$$

$$= \int_0^\infty e^{-\alpha t} \cdot E(s) e^{-E(s)t}\, dt \cdot \sum_{s' \neq s} V^\alpha(s') P(s,s') = \frac{E(s)}{\alpha + E(s)} \sum_{s' \neq s} V^\alpha(s') P(s,s')$$

$$= \sum_{s' \neq s} \frac{Q(s,s')}{\alpha + E(s)} V^\alpha(s'),$$

where the last equation follows from $Q(s,s') = P(s,s')E(s)$. $\qquad \square$

References

1. Altman, E.: Constrained Markov Decision Processes. Chapman & Hall (1999)
2. Altman, E.: Applications of Markov Decision Processes in Communication Networks. In: Feinberg, E.A., Shwartz, A. (eds.) Handbook of Markov Decision Processes. International Series in Operations Research & Management Science, vol. 40, pp. 489–536. Springer, US (2002)
3. Baier, C., Haverkort, B., Hermanns, H., Katoen, J.-P.: Model-Checking Algorithms for Continuous-Time Markov Chains. IEEE Transactions on Software Engineering 29(6), 524–541 (2003)
4. Bäuerle, N., Rieder, U.: Markov Decision Processes with Applications to Finance. Springer, Heidelberg (2011)
5. Bellman, R.: Dynamic Programming. Princeton University Press, Princeton (1957)
6. Benini, L., Bogliolo, A., Paleologo, G.A., De Micheli, G.: Policy Optimization for Dynamic Power Management. IEEE Transactions on Computer-Aided Design of Integrated Circuits and Systems 18, 813–833 (1998)
7. Bertsekas, D.: Dynamic Programming and Optimal Control, 3rd edn., vol. I. Athena Scientific (1995) (revised in 2005)
8. Bertsekas, D.: Dynamic Programming and Optimal Control, 4th edn., vol. II. Athena Scientific (1995) (revised in 2012)
9. Bertsekas, D., Tsitsiklis, J.: An analysis of stochastic shortest path problems. Mathematics of Operations Research 16(3), 580–595 (1991)
10. Bertsekas, D., Tsitsiklis, J.: Neuro-Dynamic Programming, 1st edn. Athena Scientific (1996)
11. Beynier, A., Mouaddib, A.I.: Decentralized Markov decision processes for handling temporal and resource constraints in a multiple robot system. In: Proceedings of the 7th International Symposium on Distributed Autonomous Robotic System, DARS (2004)
12. Bolch, G., Greiner, S., de Meer, H., Trivedi, K.S.: Queueing Networks and Markov Chains - Modelling and Performance Evaluation with Computer Science Applications, 2nd edn. Wiley (2006)
13. Cassandra, A.R.: A survey of POMDP applications. In: Working Notes of AAAI 1998 Fall Symposium on Planning with Partially Observable Markov Decision Processes, pp. 17–24 (1998)
14. Diz, F.J., Palacios, M.A., Arias, M.: MDPs in medicine: opportunities and challenges. In: Decision Making in Partially Observable, Uncertain Worlds: Exploring Insights from Multiple Communities, IJCAI Workshop (2011)
15. Fox, B.L., Landi, D.M.: An algorithm for identifying the ergodic subchains and transient states of a stochastic matrix. Communications of the ACM 11(9), 619–621 (1968)
16. Gouberman, A., Siegle, M.: On Lifetime Optimization of Boolean Parallel Systems with Erlang Repair Distributions. In: Operations Research Proceedings 2010 - Selected Papers of the Annual International Conference of the German Operations Research Society, pp. 187–192. Springer (January 2011)
17. Guo, X., Hernandez-Lerma, O.: Continuous-Time Markov Decision Processes - Theory and Applications. Springer (2009)
18. Heidergott, B., Hordijk, A., Van Uitert, M.: Series Expansions For Finite-State Markov Chains. Probability in the Engineering and Informational Sciences 21(3), 381–400 (2007)

19. Hou, Z., Filar, J.A., Chen, A. (eds.): Markov Processes and Controlled Markov Chains. Springer (2002)
20. Howard, R.A.: Dynamic Programming and Markov Processes. John Wiley & Sons, New York (1960)
21. Hu, Q., Yue, W.: Markov Decision Processes with their Applications. Springer (2008)
22. Janssen, J., Manca, R.: Markov and Semi-Markov Reward Processes. In: Applied Semi-Markov Processes, pp. 247–293. Springer, US (2006)
23. Janssen, J., Manca, R.: Semi-Markov Risk Models for Finance, Insurance and Reliability. Springer (2007)
24. Jensen, A.: Markoff chains as an aid in the study of Markoff processes. Skandinavisk Aktuarietidskrift 36, 87–91 (1953)
25. Stidham Jr., S., Weber, R.: A survey of Markov decision models for control of networks of queues. Queueing Systems 13(1-3), 291–314 (1993)
26. Mahadevan, S.: Learning Representation and Control in Markov Decision Processes: New Frontiers. Foundations and Trends in Machine Learning 1(4), 403–565 (2009)
27. Mahadevan, S., Maggioni, M.: Proto-value Functions: A Laplacian Framework for Learning Representation and Control in Markov Decision Processes. Journal of Machine Learning Research 8, 2169–2231 (2007)
28. Mausam, Kolobov, A.: Planning with Markov Decision Processes: An AI Perspective. Synthesis Lectures on Artificial Intelligence and Machine Learning. Morgan & Claypool Publishers (2012)
29. Momtazi, S., Kafi, S., Beigy, H.: Solving Stochastic Path Problem: Particle Swarm Optimization Approach. In: Nguyen, N.T., Borzemski, L., Grzech, A., Ali, M. (eds.) IEA/AIE 2008. LNCS (LNAI), vol. 5027, pp. 590–600. Springer, Heidelberg (2008)
30. Obal, W.D., Sanders, W.H.: State-space support for path-based reward variables. In: Proceedings of the Third IEEE International Performance and Dependability Symposium on International Performance and Dependability Symposium, IPDS 1998, pp. 233–251. Elsevier Science Publishers B. V. (1999)
31. Ott, J.T.: A Markov Decision Model for a Surveillance Application and Risk-Sensitive Markov Decision Processes. PhD thesis, Karlsruhe Institute of Technology (2010)
32. Powell, W.B.: Approximate Dynamic Programming - Solving the Curses of Dimensionality. Wiley (2007)
33. Puterman, M.L.: Markov Decision Processes - Discrete Stochastic Dynamic Programming. John Wiley & Sons INC. (1994)
34. Qiu, Q., Pedram, M.: Dynamic power management based on continuous-time Markov decision processes. In: Proceedings of the 36th Annual ACM/IEEE Design Automation Conference, DAC 1999, pp. 555–561. ACM (1999)
35. Sanders, W.H., Meyer, J.F.: A Unified Approach for Specifying Measures of Performance, Dependability, and Performability. Dependable Computing for Critical Applications 4, 215–238 (1991)
36. Schaefer, A.J., Bailey, M.D., Shechter, S.M., Roberts, M.S.: Modeling medical treatment using Markov decision processes. In: Brandeau, M.L., Sainfort, F., Pierskalla, W.P. (eds.) Operations Research and Health Care. International Series in Operations Research & Management Science, vol. 70, pp. 593–612. Kluwer Academic Publishers (2005)
37. Sutton, R.S., Barto, A.G.: Reinforcement Learning: An Introduction. A Bradford Book. MIT Press (March 1998)

38. Trivedi, K.S., Malhotra, M.: Reliability and Performability Techniques and Tools: A Survey. In: Messung, Modellierung und Bewertung von Rechen- und Kommunikationssystemen. Informatik aktuell, pp. 27–48. Springer, Heidelberg (1993)
39. Tsitsiklis, J.N.: NP-Hardness of checking the unichain condition in average cost MDPs. Operations Research Letters 35(3), 319–323 (2007)
40. White, D.J.: A Survey of Applications of Markov Decision Processes. The Journal of the Operational Research Society 44(11), 1073–1096 (1993)
41. Wolff, R.W.: Poisson Arrivals See Time Averages. Operations Research 30(2), 223–231 (1982)

Applying Mean-Field Approximation
to Continuous Time Markov Chains

Anna Kolesnichenko[1], Valerio Senni[3],
Alireza Pourranjabar[2], and Anne Remke[1]

[1] DACS, University of Twente, The Netherlands
{a.v.kolesnichenko,a.k.i.remke}@utwente.nl
[2] LFCS, University of Edinburgh, UK
a.pourranjbar@sms.ed.ac.uk
[3] IMT Institute for Advanced Studies, Lucca, Italy
valerio.senni@imtlucca.it

Abstract. The mean-field analysis technique is used to perform analysis of a system with a large number of components to determine the emergent deterministic behaviour and how this behaviour modifies when its parameters are perturbed. The computer science performance modelling and analysis community has found the mean-field method useful for modelling large-scale computer and communication networks. Applying mean-field analysis from the computer science perspective requires the following major steps: (1) describing how the agent populations evolve by means of a system of differential equations, (2) finding the emergent deterministic behaviour of the system by solving such differential equations, and (3) analysing properties of this behaviour. Depending on the system under analysis, performing these steps may become challenging. Often, modifications of the general idea are needed. In this tutorial we consider illustrating examples to discuss how the mean-field method is used in different application areas. Starting from the application of the classical technique, moving to cases where additional steps have to be used, such as systems with local communication. Finally, we illustrate the application of existing model checking analysis techniques.

1 Introduction

Mean Field Approximation originated in statistical physics [1] and is a technique developed within the field of probability theory. This technique is useful to study the behaviour of stochastic processes with a very large state space (e.g. in the study of systems with a large number of particles), where Monte Carlo simulations are impractical. In those systems, a first approximation of the behaviour is obtained by replacing the effect of the other particles over a given particle by a single averaged effect and studying this two-body problem [23,31]. Beyond physics, this approximation technique is applied in studies of epidemics models [24], queueing theory [6,1], and network performance [30,11].

In this tutorial, the stochastic systems we are interested in typically consist of a relatively small number of particle types. The particles of each type often have

A. Remke and M. Stoelinga (Eds.): ROCKS Autumn School 2012, LNCS 8453, pp. 242–280, 2014.

a simple behaviour and are replicated many times to form large populations. Their interaction may give rise to a complex behaviour and patterns that can not be found considering the single particle, but emerge by their interaction. Mean-field approximation is used to model and analyse efficiently the so-called *emergent behaviour* of such large-scale systems. Classical applications of this technique generally require two abstractions. The first is that when studying the system, one abstracts away from the particles' identities, and instead of capturing the behaviour of each instance, the system's behaviour is observed at the level of populations [22]. The second abstraction suggests that the spatial distribution of the agents across the system locations is ignored, and the particles are assumed to be uniformly spread across the system space (in chemistry this idea is embodied in the notion of well-stirred chemical reaction [17,37]). In this tutorial we illustrate both a classical application (Section 3) and a more sophisticated modelling where space inhomogeneity has a significant impact on the system's emergent behaviour (Section 4).

The core idea of the mean-field method is to approximate the dynamics of a Markov population process through a system of differential equations [27]. The result is a reliable approximation when the population size is sufficiently large, since under specific conditions the behaviour of the system tends to the deterministic dynamics captured by the differential equations. In this case, one additional important property is the *decoupling assumption*; that is the joint probability distribution associated with the system can be expressed as the product of the marginals. This property allows to study the behaviour of individual particles within the whole system in an efficient way.

A closely related approximation technique is known as *moment closure* [16]. This technique allows to estimate the first few moments of a stochastic process by a closed system of equations. Mean-field approximation can be seen as a form of moment closure where the second moment (variance) and the higher moments have been set to zero. The first-order approximation is often very coarse and can potentially lead to misleading results [33]. In practice, however, it can be used to gain some insights about the average or the global behaviour of the system at a relatively low cost.

When first-order or mean-field approximation is applied, the resulting model can be described in terms of a deterministic system, as mentioned previously. In the literature this is often referred to as *deterministic approximation* [4,9].

Another related technique is called *linear noise approximation*, which is frequently used to find approximate solutions of the Chemical Master Equation by giving an estimate of the second moment of this equation [37].

Continuous Time Markov Chains are often used to provide a stochastic semantics to process algebra used in performance modelling of computer systems [20]. However, stochastic process algebra models of realistic size can easily result in very large and intractable state spaces. In that context a technique called *fluid-flow approximation* [21] has been used to construct a continuous state-space representation of the underlying discrete state-space, and ordinary differential equations are used to describe their dynamics. This technique is justified by results on mean-field approximation of Continuous Time Markov Chains [36,22,19]. Indeed, the notion of fluid approximation has been used in various contexts such as Petri Nets, and relies on the idea that a discrete variable can be approximated using a continuous variable [34].

In our tutorial we focus on CTMC models and their continuous-time approximation using ordinary differential equations. The goal of this paper is to provide an example-guided tutorial to the application of fluid approximation, including fluid model checking [8]. The interested reader can find very complete and detailed tutorials in [9], treating both Continuous Time Markov Chains and Discrete Time Markov Chains. A more technical survey of the topic and related mathematical results can be found in [13].

2   Preliminaries

In this paper we consider systems consisting of large populations of interacting objects. Such systems are common in biology and chemistry, as well as in telecommunications and queueing theory [3,12,22,35]. Due to the problem of state space explosion, the models of such systems are often unmanageable for the purpose of analysis and are not suitable for direct application of classic analysis techniques such as simulation and model checking. In this tutorial we address the modelling and analysis of such models using *mean-field method.*

The main idea of the mean-field analysis is to describe the evolution of a population that is composed of many similar objects via a deterministic behaviour. It states that under certain assumptions on the dynamics of the system and when the size of the population grows, the ratio of the system's *variance* to the size of the state space tends to zero. Therefore, when the population is large, the stochastic behaviour of the system can be studied through the *unique solution* of a system of Ordinary Differential Equations (ODE) defined by using the limit dynamics of the whole system.

Since the purpose of this tutorial is to provide the guided examples of the application of the mean-field method, we will not be discussing the detailed theoretical background of the mean-field method (see, e.g. [9]). Instead, we present the modelling procedure from the practical point of view. We build the model of the whole population based on the behaviour of the random individual object.

2.1   Model Definition

Let us start with a random individual object in the large population. We assume that the size of the population N is constant and do not distinguish between the classes of the individual objects for the simplicity of the notation. However, this assumptions can be relaxed, see, e.g., Section 4 of the current tutorial.

The behaviour of such an object can be described by defining the states or "modes" this object experiences during its lifetime, and the transitions between these states. Formally, the individual or local model (the model of the random object in the population) is defined as follows:

Definition 1 (Local model). *A local model $\mathcal{X}$ describing the behaviour of one object is constructed as a tuple $(S, \mathbf{Q}, L)$ that consists of a finite set of K local states $S = \{s_1, s_2, ..., s_K\}$; the infinitesimal generator matrix $\mathbf{Q}$ which may depend on the overall system state; and the labelling function $L : S \rightarrow 2^{LAP}$ that assigns local atomic propositions from a fixed finite set of Local Atomic Properties (LAP) to each state.* □

Self-loops are assumed to be eliminated. The generator matrix $\mathbf{Q}$ is a matrix $S \times S$, whose entries describe the rate at which an individual object changes states. The $\mathbf{Q}$ may potentially depend on the system's overall state. We discuss the transitions rates of the individual objects later in this section.

Given the large number N of objects, we build the overall model of the whole population. Instead of modelling each object individually, which would lead to the state-space explosion problem, we (i) lump the state space; (ii) normalize the population, and (iii) check whether the convergence of the behaviour to the deterministic limit holds and build *the overall mean-field model X*, using the local model $\mathcal{X}$. Let us first provide the explanations on the way this model is built, which will be followed by the definition of the overall (or global) model.

If the identity of each object is preserved, the state space of the model of the whole population $\mathcal{X}^{(N)}$ will potentially consists of K^N states, where K is the number of states of the local model. However, due to the identical and unsynchronized behaviour of the individual objects the *counting abstraction* is applied to find the stochastic process X, whose states capture the distribution of the individual objects across the states of the local model $\mathcal{X}$. In general, the transition rates may depend on the state of the overall model, $\overline{X}(t)$. Therefore, using the counting abstraction the generator matrix $\mathbf{Q}(\overline{X}(t))$ is constructed as in [6]:

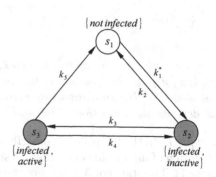

Fig. 1. The model describing computer virus spread

$$\mathbf{Q}_{i,j}(\overline{X}(t)) = \begin{cases} \lim_{\Delta \to 0} \frac{1}{\Delta} \text{Prob}(\mathcal{X}(t+\Delta)) = j | \mathcal{X}(t) = i, \overline{X}(t)), & \text{if } X_i(t) > 0, \\ 0, & \text{if } X_i(t) = 0, \\ -\sum_{h \in S, j \neq i} \mathbf{Q}_{i,h}(\overline{X}(t)), & \text{for } i = j, \end{cases}$$

where $\mathcal{X}(t)$ is a state of the local model at time t.

The first step for the construction of the mean field model is to normalize the state vector. The normalized state space is as follows: $\overline{x}(t) = \overline{X}(t)/N$, where $0 \leq \overline{x}_i(t) \leq 1$; and the related transition rates are $Q_{i,j}^{(N)}(\overline{x}(t)) = \mathbf{Q}_{i,j}(N \cdot \overline{x}(t))$.

In this tutorial we only consider models which satisfy a condition known as *density dependence*. This condition requires that there exists a matrix of rate functions that is constant for all the normalised models in a sequence of models with increasing sizes. This means that transition rates scale together with the model population, so that in the normalized models they are independent of the population. Formally, in the limit of $N \to \infty$, the matrix of rate functions (generator matrix $Q_{i,j}(\overline{x}(t))$) satisfies $Q_{i,j}(\overline{x}(t)) = Q_{i,j}^{(N)}(\overline{x}(t))$ for all $N > 1$.

The existence and properties of $Q_{i,j}(\overline{x}(t))$ play a crucial role in the applicability of the mean-field theory to the given sequence of local models and building the overall model. In the context of the models which satisfy density dependence, the rate functions are required to be Lipschitz-continuous. Secondly, the model should satisfy *convergence of the initial occupancy vector*. The limit theorem which relies on these assumptions will be covered later. First, let us state the construction of the mean-field model.

Definition 2 (Overall mean-field model). *An overall mean-field model X describes the limit behaviour of $N \to \infty$ identical objects, each modelled by $\mathcal{X}$, and is defined as a tuple (X, Q), that consists of an infinite set of states*

$$X = \{\overline{x} = (x_1, x_2, \ldots, x_K) | (\forall j \in \{1, \ldots, K\}, x_j \in [0,1] \wedge \sum_{i=1}^{K} x_i = 1)\},$$

where $\overline{x}$ is called occupancy vector, and $\overline{x}(t)$ is the value of the occupancy vector at time t; x_j denotes the fraction of the individual objects that are in state s_j of the local model $\mathcal{X}$. The transition rate matrix $Q(\overline{x}(t))$ consists of entries $Q_{s,s'}(\overline{x}(t))$ that describe the transition of the system from state s to state s'. □

Example 1. In the following we describe a simple model of the virus spread in the population of interacting computers of size N. We start with the local model (see Figure 1). The states of $\mathcal{X}$ represent the modes of an individual computer, which can be *not-infected, infected and active* or *infected and inactive*. An infected computer is *active* when it is spreading the virus and *inactive* when it is not. This results in the finite local state space $S = \{s_1, s_2, s_3\}$ with $|S| = K = 3$ states. They are labelled as *infected, not infected, active* and *inactive*, as indicated in Figure 1.

Given a system of N such computers, we can model the limiting behaviour of the whole system through the overall mean-field model, which has the same underlying structure as the individual model (see Figure 1), however, with state space $\overline{x} = \{x_1, x_2, x_3\}$, where x_1 denotes the fraction of not-infected computers, and x_2 and x_3 denote the fraction of active and inactive infected computers, respectively. For example, a system without infected computers is in state $\overline{x} = (1, 0, 0)$; a system with 50% not infected computers and 40% and 10% of inactive and active infected computers, respectively, is in state $\overline{x} = (0.5, 0.4, 0.1)$.

The transition rates k_1^*, k_2, k_3, k_4, k_5 represent the following: the infection rate k_1^*, the recovery rate for an inactive infected computer k_2, the recovery rate for an active infected computer k_5, and the rates with which computers become active k_3 and return to the inactive state k_4. Rates $k_2, k_3, k_4,$ and k_5 are specified by the individual computer and computer virus properties and do not depend on the overall system state. The infection rate k_1^* does depend on the fraction of computers that is infected and active and the fraction of not-infected computers. We discuss the generator matrix in the next example.

2.2   Mean-Field Analysis

We stated $\mathcal{X}$ represents the behaviour of each object and X represents the limiting behaviour of N identical objects. The model respects the density dependence condition. Here we express a reformulation of the Kurtz's theorem which relates the behaviour of the sequence of models with increasing sizes to the limit behaviour. Assuming that functions in $Q_{i,j}(\overline{x}(t))$ are Lipschitz-continuous and for increasing values of the system size, the initial occupancy vectors converge to $\overline{x}(0)$, then when $N \to \infty$, the sequence of local models *converges almost surely* [5] to the occupancy vector $\overline{x}$.

Theorem 1 (Mean-field convergence theorem). *The normalized occupancy vector $\overline{x}(t)$ at time $t < \infty$ tends to be deterministic in distribution and satisfies the following differential equations when N tends to infinity:*

$$\frac{d\overline{x}(t)}{dt} = \overline{x}(t) \cdot Q(\overline{x}(t)), \ \ given \ \overline{x}(0). \tag{1}$$

$\square$

The ODE (1) is called *limit ODE*. It provides the results for $N \to \infty$, which is not the case for a real-life models. When the number of objects in the population is finite, but sufficiently large the limit ODE provides an accurate approximation of the mean of the occupancy vector $\overline{x}(t)$ over time.

The transient analysis of the overall system behaviour can be performed using the above system of differential equations (1), i.e., the fraction of the objects in each state of $\mathcal{X}$ at every time t is calculated, starting from some given initial occupancy vector $\overline{x}(0)$.

For models considered in practice, however, the assumption of density dependence may be too restrictive [13]. Furthermore, also the assumption of (global) Lipschitz continuity of transition rates can be unrealistic [7]. Therefore, this assumptions can be relaxed and a more general version of the mean-field approximation theorem, having less strict requirements and applied to *prefixes* of trajectories rather than to full model trajectories, can be obtained. We will not be focusing on the reformulation of the convergence theorem here, instead we refer to [9], and provide the following example.

Example 2. In the following we provide an example of applying the mean-field method to the virus spread model, as in Example 1. We explain how to obtain the ODEs, describing the behaviour of the system and produce performance evaluation measures.

As was discussed in the previous example, all transition rates of a single computer model are constant, but k_1^*. This rate depends on how often a not infected computer gets attacked. In this example we assume that the virus is "smart enough" to attack not infected computers only. The infection rate then might be seen as the number of attacks performed by all active infected computers, which is distributed over all not-infected computes in a chosen group:

$$k_1^*(\overline{x}(t)) = k_1 \cdot \frac{x_3(t)}{x_1(t)},$$

where $\overline{x}(t) = (x_1(t), x_2(t), x_3(t))$ represents the fraction of computers in each state at time t, and k_1 is the attack rate of a single active infected computer.

The transition rates are collected to the generator matrix:

$$Q(\overline{X}(t)) = \begin{pmatrix} -k_1^*(\overline{x}(t)) & k_1^*(\overline{x}(t)) & 0 \\ k_2 & -(k_2 + k_3) & k_3 \\ k_5 & k_4 & -(k_4 + k_5) \end{pmatrix} \tag{2}$$

Then Theorem 1 is used to derive the system of ODEs (1), that describes the mean-field model:

$$\begin{cases} \dot{x}_1(t) = -k_1 x_3(t) + k_2 x_2(t) + k_5 x_3(t), \\ \dot{x}_2(t) = (k_1 + k_4) x_3(t) - (k_2 + k_3) x_2(t), \\ \dot{x}_3(t) = k_3 x_2(t) - (k_4 + k_5) x_3(t). \end{cases} \tag{3}$$

To obtain the distribution of the objects between the states of the model over time the above ODEs have to be solved.

The convergence theorem does not explicitly cover the asymptotic behaviour (i.e. limit in time). However, when certain assumptions hold, the mean-field equations allow to perform various studies including steady state analysis of the population models as well as model checking [8]. We will not cover the details here and the interested reader is referred to [3]. We will use mean-field for steady state analysis in Section 4.

3   Mean-Field Analysis of a Botnet

In this section we discuss the applicability of the mean-field method to modelling peer-to-peer botnet, as in [26] . In Section 3.1 we discuss the characteristics of the botnet, which are important for modelling. Section 3.2 describes the mean-field model of the botnet spread. The performance evaluation results are presented in Section 3.3, together with an example of wider usability of the mean-field model.

3.1   Description of the System

Let us describe the steps each computer goes through during the botnet spread. These are similar to the examples in the previous section, however, the current

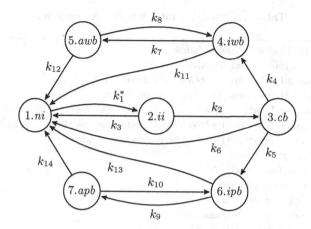

Fig. 2. Possible states of a computer in the network. The shorthand names are defined as follows: $ni=NotInfected$, $ii=InitialInfection$, $cb=ConnectedBot$, $iwb=InactiveWorkingBot$, $awb=ActiveWorkingBot$, $ipb=InactivePropagationBot$, and $apb=ActivePropagationBot$.

Botnet model is more detailed (see Figure 2) and comply the realistic botnet behaviour.

The computer which is in the *NotInfected* state (S_1) enters the *InitialInfection* (S_2) state with rate k_1^*. Then, it attempts to connect to the other bots in the botnet; if the connection is successful the computer goes tot he *ConnectedBot* state (S_3) with rate k_2. The initially infected computer recovers and returns to the state S_1 with rate k_3. After connecting to the botnet, computer downloads a malware and joins the botnet either as *InactiveWorkingBot* (S_4) or as *InactivePropagationBot* (S_6) with rates k_4 and k_5, respectively; otherwise, the computer recovers from the connected state with the rate k_6.

Once the bot becomes either an *InactiveWorkingBot* or an *InactivePropagationBot* it never switches between the *Working-* or *Propagation-* classes. In order not to be detected, the bot is inactive most of the time and it only becomes active for a very short period of time. Transitions from *InactivePropagationBot* to *ActivePropagationBot* (S_7) and back occur with rates k_9 and k_{10}, respectively. The transition rates for moving from *InactiveWorkingBot* to *ActiveWorkingBot* (S_5) and back are denoted k_7 and k_8, respectively.

The computer can recover from its infection, e.g., if an anti-malware software discovers the virus, or if the computer is physically disconnected from the network. In these cases, it leaves the *InactivePropagationBot* or the *ActivePropagationBot* state and moves to the *NotInfected* state with rates k_{13}, k_{14}, respectively. The same holds for the working bots: the recovery rates from *InactiveWorkingBot* and *ActiveWorkingBot* are k_{11}, k_{12}, respectively.

The model we construct considers several computers in a network, each of them being in one of the above mentioned states $S_1, .., S_7$, depicted also in Figure 2. The rates of transitions between states may depend on several factors, e.g., probability of a successful connection between initially infected computer and

Table 1. Transition rates for a single computer

k_1	RateOfAttack · ProbInstallInitialInfection
k_1^*	Rate depends on k_1 and the environment
k_2	RateConnectBotToPeers · ProbConnectToPeers
k_3	RateConnectBotToPeers · (1 − ProbConnectToPeers)
k_4	RateSecondaryInjection · ProbSecondaryInjectionSuccess · (1 − ProbPropagationBot)
k_5	RateSecondaryInjection · ProbSecondaryInjectionSuccess · ProbPropagationBot
k_6	RateSecondaryInjection · (1 − ProbSecondaryInjectionSuccess)
k_7	RateWorkingBotWakens
k_8	RateWorkingBotSleeps
k_9	RatePropagationBotWakens
k_{10}	RatePropagationBotSleeps
k_{11}	RateInactiveWorkingBotRemoved
k_{12}	RateActiveWorkingBotRemoved
k_{13}	RateInactivePropagationBotRemoved
k_{14}	RateActivePropagationBotRemoved

another infected computer, while moving from the state *InitialInfection* to the *ConnectedBot* state; or the probability of *ConnectedBot* to become *Working* or *Propagation* bot, respectively. Table 1 provides the description of the transition rates for one computer model, while numerical values are given in Table 2. Rates $k_2 \ldots k_{14}$ are constant for each computer, while rate k_1^* to move from the *Not-Infected* state (S_1) to the *InitialInfection* state (S_2) is not constant. This rate depends on k_1 and on the number of computers in the *ActivePropagationBot* state, which are responsible of spreading the malware.

3.2 Mean-Field Model

We study the spread of the botnet in a network of N computers by using the mean-field approximation method for finding the (average) deterministic dynamics of the system. The mean-field model captures the number of objects in a particular state, rather than considering the state of each single object. The mean-field state vector $\overline{X} = \langle X_1, X_2, \ldots X_7 \rangle$ counts how many computers are in states $S_1, \ldots, S_7$. The occupancy measure is found by normalizing $\overline{X}$ into $\overline{x}$.

We first construct the rate matrix, which collects the rates with which possible transitions take place. Transition rates may depend on time as well as on the state $\overline{x}(t)$ of the system. The rate matrix $R(\overline{x}(t))$ of the model is given as:

$$R(\overline{x}(t)) = \begin{pmatrix} 0 & k_1^* & 0 & 0 & 0 & 0 & 0 \\ k_3 & 0 & k_2 & 0 & 0 & 0 & 0 \\ k_6 & 0 & 0 & k_4 & 0 & k_5 & 0 \\ k_{11} & 0 & 0 & 0 & k_7 & 0 & 0 \\ k_{12} & 0 & 0 & k_8 & 0 & 0 & 0 \\ k_{13} & 0 & 0 & 0 & 0 & 0 & k_9 \\ k_{14} & 0 & 0 & 0 & 0 & k_{10} & 0 \end{pmatrix} \qquad (4)$$

The $|S| \times |S|$ infinitesimal generator matrix $Q(\overline{x}(t))$ is given as follows: Q_{s_1,s_2} is equal to the transition rate R_{s_1,s_2} to move from the state s_1 to the state s_2 and $Q_{s,s}$ is equal to the negative the sum of all the rates in row s. In a given example the only rate which depends on a state of the system is the infection rate $k_1^*(\overline{x}(t))$, which depends on the number of computers (bots) actively spreading infection. The total rate of infections produced by all bots that are in the active propagation state is $k_1 \cdot x_7(t)$. These infections are spread out randomly over all not-yet infected computers[1], whose number is denoted by $x_1(t)$. Hence, the infection rate k_1^* perceived by each individual computer is given by the ratio:

$$k_1^*(\overline{x}(t)) = \frac{k_1 \cdot x_7(t)}{x_1(t)}. \tag{5}$$

Once we have constructed the infinitesimal generator matrix $\mathbf{Q}$, we can use it to construct the set of Ordinary Differential Equations whose solution represents the average dynamics of the system. Therefore, the initial value problem we study is defined as follows:

$$\frac{d\overline{x}(t)}{dt} = \overline{x}(t)Q(\overline{x}(t)), \qquad \text{with initial condition } \overline{x}(0). \tag{6}$$

The system of equations we obtain is:

$$\begin{cases} \dot{x}_1(t) = & k_3 x_2(t) + k_6 x_3(t) + k_{11} x_4(t) \\ & + k_{12} x_5(t) + k_{13} x_6(t) + (k_{14} - k_1) x_7(t) \\ \dot{x}_2(t) = & -(k_2 + k_3) x_2(t) + k_1 x_7(t) \\ \dot{x}_3(t) = & k_2 x_2(t) - (k_4 + k_5 + k_6) x_3(t) \\ \dot{x}_4(t) = & k_4 x_3(t) - (k_7 + k_{11}) x_4(t) + k_8 x_5(t) \\ \dot{x}_5(t) = & k_7 x_4(t) - (k_8 + k_{12}) x_5(t) \\ \dot{x}_6(t) = & k_5 x_3(t) - (k_9 + k_{13}) x_6(t) + k_{10} x_7(t) \\ \dot{x}_7(t) = & k_9 x_6(t) - (k_{10} + k_{14}) x_7(t) \end{cases} \tag{7}$$

The equations can be solved analytically, however the closed forms are impractically large. We used Wolfram Mathematica [39] to obtain the analytical solution.

3.3   Results

In this section we discuss the mean-field results in detail and compare them to the simulation results, the chosen parameters for all these experiments are given in Table 2. We essentially experimented considering different infection rates, denoting possible user behaviours, and their impact on the system behaviour.

The simulation of the model was done using the Möbius tool [14] as in [38]. Each experiment covered one week of simulated time; it was replicated 1000

---

[1]   In the considered example the propagation bots are "smart" enough to spread infection via not infected computers only.

Table 2. Setup for the three experiments. Bold indicates differences w.r.t. baseline.

Parameter	Baseline	Exper 1	Exper 2
ProbInstallInitialInfection	0.1	**0.06**	**0.04**
ProbConnectToPeers	1	1	1
ProbSecondaryInjectionSuccess	1	1	1
ProbPropagationBot	0.1	0.1	0.1
RateOfAttack	10.0	10.0	10.0
RateConnectBotToPeers	12.0	12.0	12.0
RateSecondaryInjection	14.0	14.0	14.0
RateWorkingBotWakens	0.001	0.001	0.001
RateWorkingBotSleeps	0.1	0.1	0.1
RatePropagationBotWakens	0.001	0.001	0.001
RatePropagationBotSleeps	0.1	0.1	0.1
RateInactiveWorkingBotRemoved	0.0001	0.0001	0.0001
RateActiveWorkingBotRemoved	0.01	0.01	0.01
RateInactivePropagationBotRemoved	0.0001	0.0001	0.0001
RateActivePropagationBotRemoved	0.01	0.01	0.01

The header spans: Experiments (Baseline | Exper 1 | Exper 2).

times; the mean values and 95% confidence intervals of the measures of interest are obtained. The initial conditions for each experiment are as follows: 200 computers are located in the place *ActivePropagationBots*.

We use Wolfram Mathematica [39] to obtain solutions for the set of differential equations (7) coupled with the transition rates from Table 2. Given an overall population of $N = 10^7$, the fraction of computers in the state *NotInfected* is initialized as $x_1(0) = (N - 200)/N$, the fraction of computers in the state *ActivePropagationBot* is initialized as $x_7(0) = 200/N$, and the fractions of computers in all other states are initialized as zero.

We first consider Baseline experiment. Figure 3 shows the number of the propagation bots along time. The number of propagation bots (both active and inactive) has been taken as measure of interest since they actively infect "healthy" computers. A logarithmic scale has been chosen for the number of propagation bots, in order to better visualize the exponential growth. The figure depicts the mean-field results of the Baseline experiment together with the 95% confidence intervals of the Möbius simulation. As can be seen, the mean-field results are very accurate in this case, since they lie mostly within the confidence intervals, even though the confidence intervals are very narrow.

To investigate how a reduced infection spread would influence the growth of botnets, Experiments 1 and 2 were done in [38]. The "user factor" (*ProbInstalInfection*) is reduced to 60% and 40%, respectively, as compared to the Baseline experiment to represent a lower probability of, e.g., opening infected files. The results are, together with those from the Baseline experiment, presented in

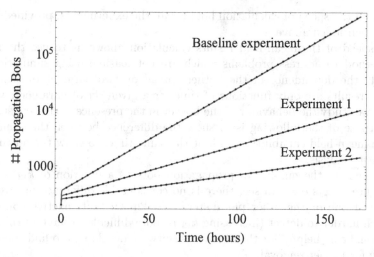

Fig. 3. Number of propagation bots over time in the Baseline experiment and experiments 1 ad 2 obtained from mean-field approximation together with the confidence intervals (black bars) obtained from the simulation

Table 3. Time spent on simulation and mean-field approximation

Experiment	Simulation	Mean-field
Baseline	5 d 3 h 25 min	1 sec
Exp. 1	9 h 51 min	1 sec
Exp. 2	5 h 37 min	1 sec

Figure 3. For both experiments, the results obtained with the mean-field model are very accurate and lie well within the confidence intervals most of the time.

One of the advantages of the mean-field method is that the time, needed for obtaining the means of the model is much smaller than the time, needed for the simulation, as shown in Table 3. The timings were obtained on a i7 processor with 3 GB RAM and 4 hyper-threading cores. The baseline experiment took 5 days 3 hours and 25 minutes, while the mean-field analysis was completed in one second. The difference between the simulation time for the different experiments is due to the dependency of the rates on a number of computers in *ActivePropagationBots* state. In the Baseline experiment the number of these computers is large, hence, the rate of infection becomes very large and more time is needed to simulate the resulting large number of events. The time spent on the simulation of the experiments with a lower number of computers involved is reasonably smaller; however the mean-field approximation is still much faster in all cases.

We do not provide all the experiments from [38] and [26] since they lie out of the scope of interest of this tutorial. Note, however, that the accuracy of the results and the speed of calculation hold for all the experiments, provided in the papers, mentioned above.

The speed of the mean-field results calculation allows us to use the mean-field method to address problems which are not feasible using simulation: (i) we study the dependence of the botnet spread on two parameters, while the previous results are only functions of time for a given set of parameter values, (ii) and we study the behaviour of the botnet in the presence of cost constraints. The purpose of the following is to show the difference between the simulation and the mean-field capabilities, and, at the same time, to show the advantages of the fast analysis.

We calculate the number of propagation bots as a function of k_{13} and k_{14} (see Figure 4). As one can see, there is no considerable difference in a relative increase of one or the other parameter. It is known that inactive computers are much harder to detect (increasing k_{13} is more difficult), therefore the above results might be helpful for the anti-virus software developers to find the better strategy for botnet removal.

Next, we introduce a cost concept to analyse the economical side of an infection. Two types of costs are considered: (i) the cost of a computer being infected, for example, due to the loss of information or productivity, and (ii) the cost of more frequent checking with anti-virus software. On one hand the number of infected computers, and hence their cost grows if computers are not frequently checked. On the other hand, if computers are checked too often the botnet is not growing, but running the anti-virus software becomes very expensive. We analyse this trade-off in more detail in the following. We calculate the cumulative cost between t_0 and t_1 as follows:

$$C(t_0, t_1, RR, D_1, D_2) \;=\; \int_{t_0}^{t_1} \left(D_1 \cdot \mathrm{IC}(t, RR) + D_2 \cdot RR \cdot \mathrm{AC} \right) dt \qquad (8)$$

where RR is the change in removal rates $k_{11}, ..., k_{14}$ with respect to the rates in the baseline experiment, i.e. $k_{11} = RR \cdot k_{11,baseline}$ (similarly for k_{12}, k_{13}, k_{14}); D_1 is the cost of infection; $\mathrm{IC}(t, RR)$ is the number of infected computers for a given RR, at time t, including active and inactive working and propagation bots; D_2 is the cost of one computer being checked, which probably is much lower than the cost of infection (D_1); AC is the number of the computers in the network. We calculate the cumulative cost of the system performance for three days. For RR from the interval $[0.001, 10]$ we calculate the cost as a function of time for given D_1 and D_2. Results are depicted in Figure 5. The cost grows exponentially with time and almost linearly with decreasing RR if the computers are not checked frequently (for the RR between 0 and 1). However, if anti-malware software is used too often (RR above 2), the cost grows linearly with RR.

We see that the mean-field method can be easily used for finding the removal rates which minimize the cost at a given moment of time. It can help network managers with careful decision-making, based on the situation at hand. Even though not all parameters might be known in reality, such analysis can help to obtain a better understanding of the characteristics of botnet spread.

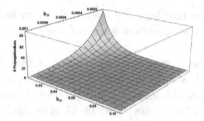
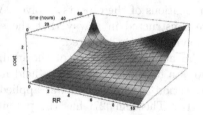

Fig. 4. Number of propagation bots for $(k_{13}, k_{14}) \in [8 \cdot 10^{-5}; 10^{-3}] \times [8 \cdot 10^{-3}; 10^{-1}]$ at time $T = 3days$, all other parameters are the same as for baseline experiment (see Table 2)

Fig. 5. Cost of the system performance for $D_1 = 0.01, D_2 = 4 \cdot 10^{-5}$

In this section the basic mean-field example was described together with the possible extensive use of the mean-field model. An example of using mean-field approximation for more sophisticated systems is given in the next sections.

4 Spatial Mean-Field Models

The mean-field analysis was firstly used in the fields of physics (when studying gas dynamics) and systems biology (studying how concentrations of reactants behave in a solution). In those domains, the assumption is made that the spatial distribution of particles/molecules across the system is homogeneous and the interacting entities are spread across the space uniformly. Such systems are often referred to as *spatially homogeneous*, in physics, and *well-stirred*, in chemistry. When analysing them, regardless of their spatial structure a single rate is assigned for each type of particle-to-particle interaction and these interactions respectively have the same probability to take place at different locations. Therefore, the effect the locations may potentially have on the overall dynamics is abstracted away.

In this section we focus on the appropriateness of the abstraction with respect to the spacial aspects in the context of modelling computer and communication networks. Indeed, depending on the system under study abstracting from the space might be a suitable simplifying step. For example, in the previous section the state vector only counted how many computers are in different local states, regardless of their locations across the geographical space (as a result, the transition rate functions did not depend on the computers' locations). Although this abstraction is reasonable in certain systems, but there exist those whose dynamics and emergent behaviours are significantly dependent on the locations of the constituent interacting objects. For those systems, the model should take into account the spatial aspects (the location of the entities, their distance, etc.) or else, the system's behaviour may not be captured effectively.

In this section, we consider an example of a large-scale peer-to-peer gossip network [11] where the emergent behaviour of the system significantly depends on locations of the objects involved. We describe how the mean-field equations are constructed in a way that the effect the locations have on the system's behaviour is also captured.

An additional feature of the example we review in this section is that it shows a case where the mean-field method is applied to a *uncountable* space. In Section 3, the method was applied to a finite-domain CTMC. Nevertheless, Kurtz's Theorem [27] has the potential to be applied also to Markov chains defined over uncountable domains [32]. As we will express, in the model we consider some of the state variables range over positive real numbers and this complicates the process of applying the method as the mean-field equations consists of partial differential equations. Here, we will review how the mean-field equations are practically constructed and avoid the proof of convergence. The more interested reader can refer to [11] for that purpose.

4.1 The Age of Gossip

We consider the example in [11], a model proposed for a peer-to-peer opportunistic communication network. Two types of entities are present in this network: some are mobile agents and can move through different locations, and some others are the stationary base stations. The base stations transmit fresh updates on a piece of data by the wireless medium and these updates are received by the mobile agents when they are close to one of the base stations. The data the base stations send is time-stamped. The age of a piece of data an agent holds is defined to be the time elapsed since it was transmitted by one of the base stations. Therefore, the age of data *just* received is zero. The age of an agent is defined to be the age of the data it holds. In addition to the data exchanges with the base stations, the mobile agents are capable of radio communication between themselves. If two such agents are close enough, the one who has the most recent version transmits its data to the other. This mechanism helps the agents receive updated data even if they have not directly visited a base station.

The system consists of a number of *locations* through which the mobile agents move. We assume that the base stations in each location can establish radio communication only with agents who are in the same location. The data exchange between two mobile agents can take place either when they both belong to the same location or when they are in two different locations. The latter captures the situation when agents are close to the borders of their location and can potentially exchange data with agents of the other locations.

Formal Model Description. Let $L = \{1, 2, \ldots, C\}$ be the set of locations and N denote the number of mobile agents. For the ith agent, we define $X_i \in \mathbb{R}^+$ to denote its age and $c_i \in L$ to represent its location. Hence, the state vector is $\boldsymbol{\xi} = \langle X_1, X_2, \ldots X_N, c_1, c_2 \ldots c_N \rangle$. Now we define the transitions which affect the system's state and the rate functions associated with these transitions.

1. **Mobility.** An agent moves from location c to c' with rate $\rho_{c,c'}, c \neq c'$. If there are N_c agents in c, the total rate at which agents from c move to c' is $N_c \times \rho_{c,c'}$.

2. **Contact with Base Station.** An agent i with age X_i in $c \in L$ may contact a base station in c and get fresh data. As the result, $X_i = 0$. For each location c a parameter μ_c describes the rate at which an agent in c receive data directly from base stations in c. If no base stations are in c, then $\mu_c = 0$.

3. **Opportunistic Contact within Locations.** An agent i in a location c opportunistically communicates with any of the other $N-1$ agents with rate $2\eta_c/(N-1)$. The total rate of communications observed between mobile agents in c is determined by two factors: the number of agents the location contains and its topological structure. The larger the number of agents is, the higher the frequency of the communication. However, when two locations have exactly the same number of agents, the respective rates of the meetings may not be the same, as the structural properties of one might encourage agent-to-agent interaction more than the other. Hence, for each $c \in L$, a parameter η_c is defined, which captures how effectively the location's structure encourages the such interactions. If there are N_c agents in location c, the total rate at which agents communicate between themselves is:

$$\binom{N_c}{2} \times \frac{2\eta_c}{(N-1)} = \frac{(N_c) \times (N_c - 1)}{N-1}\eta_c. \tag{9}$$

4. **Opportunistic Contact across Locations.** A mobile agent in a location c may communicate with a mobile agent from a different location c'. This interaction happens with rate $2\beta_{c,c'}/(N-1)$. For each c and c', $(c \neq c')$, $\beta_{c,c'}$ is a constant which affects the rate at which the agents in c communicate with the agents in c'.

The ages of the agents continuously grow unless they communicate with one of the base stations or receive fresher data from other mobile agents. At any point of time and for each location, one can derive the age distribution for the agents in that location. The aim is to construct the network in a way that an acceptable distribution of ages is maintained across all locations.

State Space Representation. The state vector used for capturing the state of a system depends on the system under study and the modelling goals. In the peer-to-peer network we consider, the age of the agents is one of their key properties. Therefore, let the configuration of the system at any time t be captured by a *continuous* distribution $\xi''(z,t)$, $z \in \mathbb{R}^+$, where $\xi''(j,t)$ denotes how many agents have age j at time t. Using this state representation, a *partial differential equations* over the dimensions z and t is formed to effectively study how the age distribution of the agents evolves. However, the modelling suffers from the fact that the mobility of the agents is abstracted away and the effect their locations potentially have on the system's emergent behaviour is not realised. The dynamics of the system is faithfully captured if the state vector takes into account both properties of the agents, i.e. their age and their locations.

Consider $c \in L$. For the ith agent with age X_i, we define the distribution δ_{X_i}, a Dirac mass at X_i. At a time t, the age distribution of agents in c across $\mathbb{R}^+$ is denoted by distribution $M_c^N(t) = \sum_{i=1}^N 1_{\{c_i=c\}} \delta_{X_i^N(t)}$, which is a continuous distribution denoting the number of agents who have any age z at location c at time

t. The *vector of such distributions* $\mathbf{M}^N(t) = \langle M_1^N(\cdot, t), M_2^N(\cdot, t), \ldots, M_C^N(\cdot, t) \rangle$ is capable of capturing both the locations and ages of the agents, and is used in the rest of this section for state state representation of the mean-field analysis.

4.2   Mean-Field Limit Behaviour

In order to derive the deterministic limit behaviour, first we focus on the mobility of the agents across locations and then we consider message propagation.

Mobility of Agents. Let $\mathbf{U}^N(t) = \langle U_1^N(t), U_2^N(t), \ldots, U_C^N(t) \rangle$ capture the number of agents in different locations at time t, assuming that there are N agents in the system. Thus, the *location occupancy measure* is defined as: $\bar{\mathbf{U}}^N(t) = \frac{\mathbf{U}^N(t)}{N} = \langle \bar{U}_0^N(t), \bar{U}_1^N(t), \ldots, \bar{U}_C^N(t) \rangle$ where each $U_c^N(t)_{c \in L}$ represents the fraction of the agents which are in location c at time t. Assume that, when $N \to \infty$, the sequence $\bar{U}_c^N(0)$ converges to a unique limit:

$$\lim_{N \to \infty} \bar{\mathbf{U}}^N(0) = \lim_{N \to \infty} \frac{U(0)}{N} = \left\langle \frac{U_1(0)}{N}, \frac{U_2(0)}{N}, \ldots, \frac{U_C(0)}{N} \right\rangle = \langle \bar{u}_1^0, \bar{u}_2^0, \ldots \bar{u}_C^0 \rangle = \bar{\mathbf{u}}^0$$

Since the convergence of initial occupancy measure holds and the system satisfies density dependence (rate functions in the normalised system is independent of N), we use Kurtz's Theorem [28] to prove that, *at any time point $t>0$, if $N \to \infty$, then* process $\bar{\mathbf{U}}^N(t)$ converges to a deterministic limit $\bar{\mathbf{u}}(t) = \langle \bar{u}_1(t), \bar{u}_2(t), \ldots \bar{u}_C(t) \rangle$, where $\bar{u}_c(t)$ is the solution of the following initial value problem:

$$\forall c \in L, \quad \frac{\partial \bar{u}_c(t)}{\partial t} = \left(\sum_{c' \neq c} \rho_{c',c} \bar{u}_{c'} \right) - \left(\sum_{c' \neq c} \rho_{c,c'} \right) \bar{u}_c, \quad \bar{u}_c(0) = \bar{u}_c^0 \quad (10)$$

The first term on the right hand side indicates the increase of $\bar{u}_c$ due to the agents coming from adjacent locations to c, and the second term indicates the decrease of $\bar{u}_c$ due to c agents leaving for the adjacent locations.

By the Cauchy-Lipschitz theorem, for any initial condition $\bar{\mathbf{u}}^0 = \langle \bar{u}_c^0 \rangle_{c \in L}$, Equation 10 admits a unique solution [11]. Let $\bar{u}_c(t \mid \bar{\mathbf{u}}^0)$ denotes the deterministic value of $\bar{u}_c$ at time t given the initial condition $\bar{\mathbf{u}}^0$. The stationary location occupancy measure can be derived using the fixed point method:

$$\forall c \in L, \frac{\partial \bar{u}_c(t)}{\partial t} = 0 \implies \forall c \in L, \tilde{u}_c \left(\sum_{c' \neq c} \rho_{c',c} u_{c'} \right) = \left(\sum_{c' \neq c} \rho_{c,c'} \right) \tilde{u}_c, \quad \sum_{c \in C} \tilde{u}_c = 1.$$

Evolution of Age Distributions. Consider $\mathbf{M}^N$, the state vector stated above. Assume that there are N agents in the network. The system's occupancy measure is defined as $\bar{\mathbf{M}}^N(t) = \frac{\mathbf{M}^N(t)}{N} = \langle \bar{M}_1(\cdot, t), \bar{M}_2(\cdot, t), \ldots, \bar{M}_C(\cdot, t) \rangle$, where

$\forall c \in L, \bar{M}_c^N(z,t)$ denotes the *density* of agents in c with age z at time t. We also define $F_c^N(z,t)$, the cumulative distribution function over $\bar{M}_c^N(t)$:

$$\forall c \in L, \ F_c^N(z,t) = M_c^N(t)[0:t] = \int_0^z \bar{M}_c^N(s,t) \, ds.$$

$\forall c \in L, \forall z, t \in \mathbb{R}^+$, $F_c^N(z,t)$ shows the proportion of N in c with age *less than or equal to* z. We assume that when $N \to \infty$, the initial occupancy measures $\bar{\mathbf{M}}^N(0)$ converge to a unique limit $\bar{\mathbf{m}}^0$: $\lim_{N\to\infty} \bar{\mathbf{M}}^N(0) = \bar{\mathbf{m}}^0$. This implies that $\forall c \in L$, $\lim_{N\to\infty} \bar{M}_c^N(0) = \bar{m}_c^0$.

The rate functions related to the data propagation satisfy the density dependence condition. Therefore, for any $t > 0$ and for all $c \in L$, when $N \to \infty$, $\bar{M}_c^N(t)$ converges to $\bar{m}_c(t)$, where $\bar{m}_c(t)$ is the solution of the following partial differential equation [11]. Here, $\bar{u}_c(t)$ is derived by solving Equation (10) for t.

$$\bar{m}_c(0,t) = \mu_c \times \bar{u}_c(t) \tag{11}$$

$$\frac{\partial \bar{m}_c(z,t)}{\partial t} = -\frac{\partial \bar{m}_c(z,t)}{\partial z} - \mu_c \bar{m}_c(z,t) + \sum_{c' \neq c} \rho_{c',c} \bar{m}_{c'}(z,t) - \left(\sum_{c' \neq c} \rho_{c,c'} \right) \bar{m}_c(z,t) \tag{12}$$

$$+ 2\eta_c \left[(+1) \times (u_c(t) - F_c(z,t)) \cdot \bar{m}_c(z,t) + (-1) \times \bar{m}_c(z,t) \cdot F_c(z,t) \right]$$

$$+ \sum_{c' \neq c} 2\beta_{c,c'} \left[(+1) \times (u_c(t) - F_c(z,t)) \cdot \bar{m}_{c'}(z,t) + (-1) \times \bar{m}_c(z,t) \cdot F_{c'(z,t)} \right]$$

We propose an intuitive explanation for forming Equation 12 by considering how much each $\bar{m}_c(z,t)_{c\in C}$ changes in an small time interval ∂t (the left hand side). Consider $c \in L$. During ∂t, agents with age z (accounted for by $\bar{m}_c(z,t)$) grow older and need to be removed from $\bar{m}_c(z,t)$. Additionally, agents with age $z - \triangle z$ become older and the density $\bar{m}_c(z - \triangle z, t)$ need to be added to $\bar{m}_c(z,t)$. Hence, the rate of change of $m_c(z,t)$ caused only by *aging* is (first term on the right hand side of Eq. 12):

$$\lim_{\triangle z \to 0} \frac{|\, \bar{m}_c(z - \triangle z, t) - \bar{m}_c(z,t) \,|}{\triangle z} = \frac{\partial \bar{m}_c(z,t)}{\partial z}.$$

The second term reflects the communication of agents, accounted by $\bar{m}_c(z,t)$, with one of the base stations. If, there are $\bar{m}_c(z,t)$ agents in c, given that the rate of communication with base stations in c is μ_c, then in ∂t, $\mu_c \times \bar{m}_c(z,t) \times \partial t$ communications take place and the agents involved leave $\bar{m}_c(z,t)$. Therefore, the rate of the change is $\mu_c \times \bar{m}_c(z,t)$.

The third expression shows the increase of $\bar{m}_c(z,t)$ as a result of agents with age z moving from other locations c' into c. The rate of the increase due to the flow from any $c' \neq c$ is $\rho_{c,c'} \bar{m}_{c'}(z,t)$. Conversely, the fourth term reflects the movement of agents contained in $\bar{m}_c(z,t)$ out of c into the adjacent locations. The decrease in $\bar{m}_c(z,t)$ due to this flow happens at rate $\sum_{c' \neq c} \rho_{c,c'}$.

The fifth term has two parts. The first shows the rate of the flow into $\bar{m}_c(z,t)$ due to agents with age z in c communicating with agents of higher age in c.

The total density of agents in c at time t is $\bar{u}_c(t)$ and of those with age less than z is $F_c(z,t)$. Therefore, $(u_c(t) - F_c(z,t))$ is the density of agents older than z. In the normalised system, by Equation 9, the rate of communication between the fraction with age z and those with higher ages is: $2\eta_c(u_c(t) - F_c(z,t))\bar{m}_c(z,t)$. The second part, $-2\eta_c(\bar{m}_c(z,t))F_c(z,t)$, reflects the drift out of $\bar{m}_c(z,t)$ as a result of agents with age z in c communicating with agents of lower age in c.

The sixth term is similar to fifth, with the difference that it shows the change of $\bar{m}_c(z,t)$ due to the agents from c communicating with agents from $c' \neq c$.

We simplify Equation 12 by integrating over z to obtain:

$$\forall c \in L: \frac{\partial F_c(z,t)}{\partial t} = -\frac{\partial F_c(z,t)}{\partial z} + \left(\sum_{c' \neq c} \rho_{c',c}\, F_{c'}(z,t)\right) - \left(\sum_{c' \neq c} \rho_{c,c'}\right) F_c(z,t) \qquad (13)$$

$$+ \left(u_c(t|d) - F_c(z,t)\right)\left(2\eta_c F_c(z,t) + \mu_c\right) + \left(u_c(t|d) - F_c(z,t)\right) \sum_{c' \neq c} 2\beta_{c,c'} F_{c'}(z,t)$$

$$\forall c \in L, \forall t \geq 0 : F_c(0,t) = 0 \quad , \quad \forall c \in L, \forall z \geq 0 : F_c(z,0) = F_c(z)$$

In this modelling, the set of ODEs (10) are constructed and solved independently, as the agents' mobility is not assumed to be dependent on the data propagation.

4.3   Solution of the Equations

Here we consider how Equation 13 is solved, for the case where there is only *one* *location* in the system and at $t = 0$, every agent has age zero.

The solution is obtained by introducing a change of variables. Let the space $\mathcal{A} = \{(x,y) \in \mathbb{R} \times \mathbb{R} \mid x \geq 0, x+y \geq 0\}$ and $G(x,y) : \mathcal{A} \to [0,1]$, $G(x,y) = F(x,x+y)$. In order to find $F(z,t)$ it is enough to derive $G(z, t-z)$. For function G we have:

$$\frac{\partial G(x,y)}{\partial x} = \frac{\partial F(z,t)}{\partial z}\bigg|_{(x,x+y)} + \frac{\partial F(z,t)}{\partial t}\bigg|_{(x,x+y)}.$$

Rearranging the terms in Equation (13), we obtain:

$$\frac{\partial G(x,y)}{\partial x} = (1 - G(x,y))(2\eta\, G(x,y) + \mu) \qquad G(0,y) = 0 \qquad (14)$$

The assumption that at time $t = 0$, no gossip exists, implies that $\forall t\ z < t$ and $y = t - z > 0$. For anu $y \in \mathbb{R}^+$, let us define $g_y : x \mapsto G(x,y)$. Therefore:

$$\frac{\partial g_y(x)}{\partial x} = (1 - g_y(x))(2\eta g_y(x) + \mu) \qquad g_y(0) = 0$$

By Cauchy-Lipschitz Theorem, this equation has a solution. The value obtained for $g_y(x)$ leads to the corresponding $F(z,t)$.

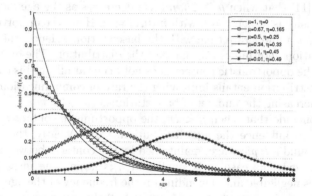

Fig. 6. The density at age z for different values of η and μ when $z \leq t$

Single Location - Analytical Solution. In this case, Equation 14 can be analytically solved to obtain the following solution:

$$F(z,t) = \begin{cases} 1 - \dfrac{2\eta + \mu}{2\eta + \mu e^{(\mu+2\eta)z}} & \text{if } z \leq t \\[4mm] 1 - \dfrac{2\eta + \mu}{2\eta + \frac{2\eta F(z-t,0)+\mu}{1-F(z-t,0)}e^{(\mu+2\eta)t}} & \text{if } z > t \end{cases} \tag{15}$$

We illustrated the reasoning behind the first case of the solution (when $z \leq t$). The second case ($z > t$), concerns the situation where in the initial configuration some agents have age greater than zero. Therefore, at any time t, it is possible to have agents with ages higher than t. The proportion of the agents who at time t have age $z > t$ depends on the proportion whose age was at least $(z - t)$ in the system's initial configuration. We skip the solution explanation for this case.

The solution allows us to study important aspects of the peer to peer network. In terms of performance, the network is well designed if with a high probability, the majority of agents remain within relatively low ranges of age. One way to satisfy this performance requirement is to deploy a relatively large number of base stations in each location; the agents frequently communicate with the base stations and receive fresh copies of the data. We introduce the term *infrastructure dominant* here. A location where the associated age distribution is mainly formed by the agent-to-base-station communication is said to be infrastructure dominant. In such a location, the agent-to-agent communication has less impact.

A location that does not enjoy strong infrastructure may still exhibit a satisfactory age distribution. In this case, the frequent and improved agent-to-agent communication is the main contributing factor in information dissemination. A location where the opportunistic contact determines the shape of the age distribution is referred to as *opportunistic dominant*.

Figure 6 shows the results of the analysis of the model when the system consists of only one location. Different values for the parameters μ, η capture different degrees of dominance of the infrastructure or of the opportunistic contacts.

We conclude [11] that when $\mu \geq 2\eta$, $m(z, t)$ decreases as the age increases. The maximum density is at age $z = 0$ with $m(0, t) = \mu$. Here, the opportunistic contacts happen at a lower rate than with the base stations. Hence, the latter type of communication determines the shape of the distribution. The extreme case is when $\eta = 0$; the opportunistic contact does not occur at all. In this case, improving the age distribution entails improving the rate of communications with base stations by increasing the number of base stations.

We also conclude that when $\mu < 2\eta$, the opportunistic contact rate becomes large enough to influence the age distribution. Consequently, there emerges a large mass around a *typical* age, maintained by the contacts between the mobile agents. In the extreme case, μ is small and η is large. The mass around age $z = 0$ becomes negligible and depending on the frequency of the agent meetings, the dominant age is centred at some age $z > 0$. In order to improve the age distribution in such a network without changing μ, one needs to improve η.

Multiple Locations. We explain the steps in the solution phase when the network contains multiple locations. Let us assume that the system has reached its equilibrium; $\forall c \in L, \frac{\partial F_c(z,t)}{\partial t} = 0$, $u_c(t) \to \tilde{u}_c$. Using Equation (13), we obtain:

$$\forall c \in L, \quad \frac{d\,F_c(z)}{dz} = +\tilde{u}_c\mu_c + \left(\tilde{u}_c 2\eta_c - \mu_c - \sum_{c' \neq c} \rho_{c,c'}\right) F_c(z) \tag{16}$$

$$+ \sum_{c' \neq c} (\rho_{c',c} + \tilde{u}_c 2\beta_{c,c'}) F_{c'(z)} - \sum_{c' \neq c} 2\beta_{c,c'} F_c(z).F_{c'(z)} - 2\eta_c (F_c(z))^2$$

with the initial condition $\forall c \in L$, $F_c(0) = 0$. In contrast with the previous case, this system of ODEs is multi-dimensional and non-linear, and has no simple analytical solution. Nevertheless, when $z \to 0$ or z is very large, it can be approximately solved. If $z \to 0$, then $F_c(z) \to 0$ and the factors $F_c(z)F_{c'}(z)$ and $(F_c(z))^2$ become negligible compared to the rest of the expression and can be ignored to find the following system shown in the matrix form:

$$F' = FA + B \tag{17}$$

$$A_{c,c} = \tilde{u}_c 2\eta_c - \mu_c - \sum_{c' \neq c} \rho_{c,c'} \quad, \quad A_{c,c'} = \rho_{c,c'} + \tilde{u}_{c'} 2\beta_{c,c'} \quad, \quad B = (\mu_0 \tilde{u}_0, \ldots, \mu_C \tilde{u}_C)$$

For $c \in L$ and $z \to 0$, $m_c(z) \approx \mu_c \tilde{u}_c$. The derivative of $m_c(z)$ is:

$$\frac{d\,\bar{m}_c(z)}{dz} = \mu_c \tilde{u}_c (\tilde{u}_c 2\eta_c - \mu_c - \sum_{c' \neq c} \rho_{c,c'}) + \sum_{c' \neq c} \mu_{c'} \tilde{u}_{c'} (\rho_{c',c} + \tilde{u}_c 2\beta_{c',c})$$

If $\forall\, c, c' \in L : \beta_{c,c'=0}$, then:

$$\frac{d\,\bar{m}_c(z)}{dz} = \mu_c \tilde{u}_c (\tilde{u}_c 2\eta_c - \mu_c) + \sum_{c' \neq c} (\mu_{c'} - \mu_c) \tilde{u}_{c'} \rho_{c',c} \tag{18}$$

Equation (18) is used to determine for each c, whether its is an infrastructure dominant or opportunistic contact dominant. If $\forall c, \mu_c = \mu$ (μ_c is the same in all locations), c has a dominant infrastructure (respectively, dominant opportunistic contact) if $2\eta_c < \mu_c$ (respectively, $2\eta_c > \mu_c$). For the case when the base stations are installed in non-neighbouring locations, then c with a base station has a dominant opportunistic contact if $2\eta_c \tilde{u}_c > \mu_c + \sum_{c' \neq c} \rho_{c,c'}$. In any location with no base stations, the age distribution will be dominated by the opportunistic contacts. The most general case happens when each location has its own specific μ_c and the base stations are distributed arbitrarily across the locations. In this case, the nature of each location can be decided only after plugging the parameters into Equation (18) and observing the sign of the derivative at $z = 0$.

For the case when the modeller is interested in high values of age ($z \to \infty$), a similar technique can be used to simplify the equations [11].

4.4   Model Validation

We reviewed how the model was developed and analysed [11]. Now we focus on *model validation*. This task has three steps. First, by using the data on the executions of the real system (eg. time series) the model's parameters are found. Then, a version of the model with concrete values for the parameters is constructed. Second, using a classical approach such as the stochastic simulation, the model is analysed and the observations are compared (qualitatively/quantitatively) against the real executions to check whether the model effectively captures the age distributions. Finally, the mean-field solution is obtained to check whether this particular method is suitable for the analysis of the model.

Validation Platform. CabSpotting [10] is a project where the San Francisco taxi company traces the location of its yellow cabs as they operate in the Bay Area (SFBA). Using GPS, each cab reports its location every minute and the data is stored in a database. By using the cabs' movement traces and introducing some realistic networking assumptions, one can construct a realistic opportunistic peer-to-peer network, similar to the model considered in Section 4.1, where the cabs and base stations are responsible for propagating data in the network. The realistic scenario, built in this manner, is used in the model validation.

Assume that SFBA is divided into 16 locations. There are a number of base stations which frequently transmit fresh copies of a piece of data. Each base station has a specific transmission range. The network consists also of a relatively larger number of taxi cabs. Each cab is equipped with a radio device to communicate with base stations or other cabs. Each cab scans its surrounding once per minute and when another entity is detected (another cab or one of the base stations), it tries to initiate a data exchange. The radio devices are assumed to have the range of 200m. A *meeting* or successful data exchange happens if the communicating entities remain in 200-meter proximity for at least 10 seconds (10 sec guarantees a data exchange). The goal of the meetings is to propagate updated copies of the data throughout the network. The age of a cab is equal to the time elapsed since the data it holds was sent by one of the base stations.

The CabSpotting database stores the cabs' movement traces. By using these traces and making the networking assumptions stated above, we can generate the *contact traces*. The latter not only captures the occurrence of the meetings, but also how the age of the cabs change as the result of such meetings. Therefore, the contact traces record how the cabs' ages change and can be used to observe how the age distributions evolve in different locations. In [11], contact traces were generated for dates between May, the 17th and June, the 15th, 2008 and for the time period between 8:00am till midnight, each day. They were then used for validation steps.

Extracting Model Parameters. The following quantities were measured using the contact traces generated. $N(t)$: total number of cabs in time slot t (time unit = one minute); $N_c(t)_{c\in\{1,2,3,...16\}}$: number of cabs in location c during time slot t; $N_{c,ub}(t)$: number of contacts between a mobile agent and a base station in c during time slot t; $N_{c,uu}(t)$: number of contacts between any two mobile agents in c during t; $N_{c,c',uu}(t)_{c\neq c'}$: number of contacts between an agent from c and another from c' during t.

Given the contact traces, one can calculate $\bar{\mu}_c(t) = \frac{N_{c,ub}(t)}{N_c(t)}$ as the rate at which an agent in c communicates with one of the base stations in that location during t. If, at t there are $N_c(t)$ agents in c, then on average $\bar{\mu}_c(t) \times N_c(t)$ meetings are expected in the following time unit. The average μ_c for an hour is calculated by averaging $\bar{\mu}_c(t)_{t\in[0,59]}$: $\mu_c = \frac{1}{60}\sum_{t=t_0}^{t_0+59} \bar{\mu}_c(t)$. This parameter is used in the model. Let us now focus on how other parameters are calculated.

In the model, for $c \in L$ the rate at which an agent in c meets another agent in c is $\frac{2\eta_c}{N-1}$. Consequently, the rate at which meetings occur in c is:

$$\binom{N_c}{2} \times \frac{2\eta_c}{(N-1)} = \frac{(N_c) \times (N_c - 1)}{N-1} \times \eta_c. \tag{19}$$

During the time unit t, the traces capture $N_{c,uu}(t)$ meetings which can be expressed using Equation 19. We assume that that at t, $\bar{\eta}_c(t)$ affects the rate of the meetings. Therefore, in time unit t we expect to observe $\frac{N_c(t)\times(N_c(t)-1)}{(N(t)-1)}\bar{\eta}_c(t)$ meetings. Thus:

$$N_{c,uu}(t) = \frac{(N_c(t))(N_c(t)-1)}{N(t)-1}\bar{\eta}_c(t) \Rightarrow \bar{\eta}_c(t) = \frac{N_{c,uu}(t)}{\frac{N_c(t)}{N(t)-1}(N_c(t)-1)} \approx \frac{N_{c,uu}(t)}{u_c(t)(N_c(t)-1)}.$$

The model's η_c is obtained by averaging $\bar{\eta}_c(t)$ for one hour; $\eta_c = \frac{1}{60}\sum_{t=t_0}^{t_0+59} \bar{\eta}_c(t)$.

In the model, the rate at which an agent in c meets an agent in c' is $\frac{2\times\beta_{c,c'}}{N-1}$. Therefore, in one time unit, on average $\frac{2\beta_{c,c'}}{N-1}N_cN'_c$ meetings occur between agents in c and c'. For each time unit t, the traces show $N_{c,c',uu}(t)$ meetings having occurred. Therefore:

$$\frac{2\bar{\beta}_{c,c'}(t)}{N(t)-1}N_c(t)N_{c'}(t)=N_{c,c',uu}(t)\Rightarrow\bar{\beta}_{c,c'}(t)=\frac{N_{c,c',uu}(t)}{2N(t)u_c(t)N(t)u_{c'}(t)\times\frac{1}{N(t)-1}}\Rightarrow$$

$$\bar{\beta}_{c,c'}(t)\approx\frac{N_{c,c',uu}(t)}{2\times N(t)\times u_c(t)\times u_{c'}(t)}$$

For each c and c', $\beta_{c,c'}$ can is obtained by averaging $\bar{\beta}_{c,c'}(t)$ over an hour.

Finally, in the model, the rate at which agents move from location c to c' is defined to be $\rho_{c,c'}\times N_c$. In the traces, one observes $N_{c,c',trans}(t)$ movements. Therefore: $\bar{\rho}_{c,c'}(t)N_c(t)=N_{c,c',trans}(t)\Rightarrow\bar{\rho}_{c,c'}(t)=\frac{N_{c,c',trans}(t)}{N_c(t)}$. The same averaging is applied to $\bar{\rho}_{c,c'}(t)$ to find $\rho_{c,c'}$.

The parameters obtained from the contact traces were used to build a fully parametrized model. The model was then simulated and the stochastic behaviour obtained was compared against the traces. The authors show that the model is sufficiently detailed to capture the stochastic behaviour of the real system.

The last step of the validation is checking whether the mean-field method is an appropriate method for the analysis of this model. The authors show that for the locations which usually have reasonably large populations of agents (having at least tens of taxi cabs), there exists a close correspondence between the age distributions obtained from the mean-field analysis and the distributions derived from the contact traces. For the locations at the edges of the network, where the population of the cabs were too small, the mean-field solution has more error. Due to space limitation we skip reviewing the last sections of the validation process and the interested reader is referred to [11].

5   Model Checking Mean-Field Models

In this section we discuss model-checking approach for mean-field models. The kind of analysis we can perform through model checking is rather different from the performance studies we illustrated in previous sections. Indeed, we are able to formally prove temporal properties of the execution of these systems and have an estimate of the probability of their validity at a certain time point.

There are two possible ways of describing the properties of a large population: via studying a random individual within the whole population and via considering the whole population.

The first approach is known as a *fluid model checking* [8] and it employs a bounded fragment of the *Continuous Stochastic Logic* (CSL) for describing properties of interest. Later in this section we recall the logic CSL, and explain how these properties can be checked for an individual object.

While fluid model checking is applicable to the local model only, the second approach allows us to derive the properties of the overall mean-field model. This is done using *Mean-Field Continuous Stochastic Logic* (MF-CSL) [25], which lifts the properties of the local model to the level of the overall model via *expectation operators*. MF-CSL logics relays on the local model properties when constructing the properties of the overall model, and the timed properties can be described only on the local level (for an individual object).

Note that yet another approach to model-checking mean-field models is possible, that only makes use of the deterministic limit (occupancy vector) to reason about the timed properties on the level of the overall model.

In the following we first return our attention to the single agent and its properties in Sections 5.1-5.5. Then the model-checking procedure for the whole population is addressed in Section 5.6.

5.1 Single Agent Model

An interesting consequence of the mean-field approximation theorem is the so-called *decoupling of joint probability* (for details, please refer to [3,30]), which allows us to obtain the model of the single object within the overall model, by using *fast simulation* [13,15]. The central idea of this process is to abstract the system to its fluid approximation (to obtain mean-field model of the system) and to study the evolution of a single agent as executed in parallel with the approximation of the rest of the system. The advantage is that, rather than considering/simulating the entire system, it is sufficient to consider the abstract average behaviour of the system and observe a single agent interacting with it, by decoupling its evolution from the evolution of the remaining agents. This is a faithful approximation since the dynamics of a single agent depend on the other agents only through the overall average system state. This allows us to reason about the local model within the overall model as of a time-*inhomogeneous* continuous time Markov chain (ICTMC).

Due to the time-inhomogeneity of the local model, the existing model checking algorithms for CTMCs can not be reused. Therefore, in [8] the authors develop novel CSL model checking algorithms for ICTMC models. We denote the single object model coupled with the deterministic limit (the local ICTMC) as $Z(t)$ for ease of notation. The labelling of the states of ICTMC is done on the same way as for a time-homogeneous CTMC.

5.2 Continuous Stochastic Logic

As a single agent model is described by an ICTMC, a standard CSL logic can be used to express the properties of such model. In the following we recall the definition of *bounded* CSL as in [2]:

Definition 3. CSL Syntax. *Let $p \in [0,1]$ be a real number, $\bowtie \in \{\leq, <, >, \geq\}$ a comparison operator, $I \subseteq \mathbb{R}_{\geq 0}$ a non-empty bounded time interval, and AP a set of atomic propositions with $a \in AP$. CSL **state formulas** Φ are defined by:*

$$\Phi ::= tt \mid a \mid \neg\Phi \mid \Phi_1 \wedge \Phi_2 \mid \mathcal{P}_{\bowtie p}(\phi),$$

*where ϕ is a **path formula** defined as:*

$$\phi ::= \mathcal{X}^I \Phi \mid \Phi_1 \, U^I \, \Phi_2.$$

To define the semantics of CSL formulas we first recall the notion of a path as it was defined for the CTMCs in [2]; this notion is reused for ICTMCs. An *infinite path* σ is a sequence $s_0 \xrightarrow{t_0} s_1 \xrightarrow{t_1} s_2 \xrightarrow{t_2} ...$, for $i \in \mathbb{N}$; $s_i \in S$ and $t_i \in \mathbb{R}_{>0}$ such that the probability that starting in state s_i we reach state s_{i+1} at time $t_\sigma[i] = \sum_{j=0}^{i} t_j$ is greater than zero. A finite path σ is a sequence $s_0 \xrightarrow{t_0} s_1 \xrightarrow{t_1} ...s_{l-1} \xrightarrow{t_{l-1}} s_l$ such that s_l is absorbing, and, similarly, a probability of going from s_i to s_{i+1} is greater than zero for all $i < l$.

For a given path σ, $\sigma[i] = s_i$ denotes for $i \in \mathbb{N}$ the $(i+1)st$ state of path σ. The time spent in state s_i is denoted by $\delta(\sigma; i)$. Moreover, with i the smallest index, and with $t \leq \sum_{j=0}^{i} t_j$, let $\sigma@t = \sigma[i]$ be the state occupied at time t. For finite paths σ with length $l+1$, $\sigma[i]$ and $\delta(\sigma; i)$ are defined in the way described above for $i < l$ only and $\delta(\sigma; l) = \infty$ and $\delta@t = s_l$ for $t > \sum_{j=0}^{l-1} t_j$. $Path^{Z(t)}(s_i, t_0)$ is the set of all finite and infinite paths of the ICTMC that start in state s_i and $Path^{Z(t)}(t_0)$ includes all (finite and infinite) paths of the ICTMC. A probability measure $Pr(t_0)$ on paths can be defined as in [2].

Since the local model changes with time, the satisfaction relation for a local state or path depends on time as well, and it is defined as follows:

Definition 4. *Semantics of CSL*. *Satisfaction of state and path CSL formulas for ICTMCs is given as follows:*

$$
\begin{aligned}
s, t_0 &\models tt & &\forall s \in S, \\
s, t_0 &\models a & &\text{iff } a \in L(s), \\
s, t_0 &\models \neg \Phi & &\text{iff } s, t_0 \not\models \Phi, \\
s, t_0 &\models \Phi_1 \wedge \Phi_2 & &\text{iff } s, t_0 \models \Phi_1 \text{ and } s, t_0 \models \Phi_2, \\
s, t_0 &\models \mathcal{P}_{\bowtie p}(\phi) & &\text{iff } Prob^{Z(t)}(s, t_0, \phi) \bowtie p, \\
\sigma, t_0 &\models \mathcal{X}^I \Phi & &\text{iff } \sigma[1] \text{ is defined, and } \delta(\sigma, 0) \in I, \text{ and} \\
& & &\sigma[1], (t_0 + \delta(\sigma, 0)) \models \Phi, \\
\sigma, t_0 &\models \Phi_1 \, U^I \, \Phi_2 & &\text{iff } \exists t' \in I : (\sigma@t' \models \Phi_2) \\
& & &\wedge (\forall t'' \in [t_0, t'](\sigma@t'' \models \Phi_1)),
\end{aligned}
$$

$I \subseteq \mathbb{R}_{\geq 0}$ *is a non-empty time interval and* $Prob^{Z(t)}(s, t_0, \phi)$ *is the probability measure of all paths* $\sigma \in Path^{Z(t)}(s, t_0)$ *that satisfy* ϕ *and starting in state* s, *that is* $Prob^{Z(t)}(s, t_0, \phi) = Pr\{\sigma \in Path^{Z(t)}(s, t_0) \mid \sigma, t_0 \models \phi\}$.

Only *bounded* time intervals are used in path formulas. This is motivated by the nature of convergence theorem, which is valid only for finite-time horizons. The relaxation of this restriction is possible, but we will not discuss it this tutorial, see [8], and [25] for details.

The CSL operators can be nested according to Definition 3. Model-checking of the CSL formula is done by building the *parse tree* and computing the satisfaction set of the individual operators recursively (in a bottom-up fashion), as described in [2].

Model-checking CSL formulas for ICTMCs is similar to model-checking these formulas for CTMCs. All time-independent CSL operators can be checked using standard methods (see [2]) due to the independence of the results on time.

Therefore, model-checking these operators is not included in the following discussion.

The main challenge is in model-checking *time-dependent* operators: let us first recall how these formulas are checked for time-homogeneous models. Given an arbitrary time-homogeneous CTMC $\mathcal{A}$, the probability formula containing the interval next operator $\mathcal{P}_{\bowtie p}\mathcal{X}^{[t_1,t_2]}\Phi$ is usually checked by computing the next-state probability and by comparing it with the threshold p (see [2]). This is calculated as the probability that the next jump starts within the time interval $[t_1;t_2]$ and ends in a state that satisfies Φ.

The probability formula including interval until formula $\mathcal{P}_{\bowtie p}\Phi_1 U^{[t_1,t_2]}\Phi_2$ for an arbitrary time-homogeneous CTMC $\mathcal{A}$ is checked by computing the probability of taking a path satisfying the until formula and by comparing it to the threshold p [2]. The way to calculate this probability will be presented below. Let us denote the states satisfying Φ_2 as goal states, and the set of such a states as $\mathbb{G} = [\![\Phi_2]\!]$, a set of states satisfying Φ_1 as safe states $\mathbb{S} = [\![\Phi_1]\!]$, and, similarly, a set of the unsafe states $\mathbb{U} = [\![\neg\Phi_1]\!]$ for the ease of notation. For model-checking CSL until formula, we need to consider all possible paths, starting in a safe state $s_1 \in \mathbb{S}$ at the current time and reaching a goal state $s_2 \in \mathbb{G}$ during the time interval $[t_1,t_2]$ by only visiting safe states on the way. We can split such paths in two parts: the first part models the path from the starting state s to a state $s_1 \in \mathbb{S}$ and the second part models the path from s_1 to a state $s_2 \in \mathbb{G}$ only via safe states. In the first part of the path, we only proceed along safe states thus all unsafe states $s \in \mathbb{U}$ do not need to be considered and can be made absorbing. As we want to reach a $\mathbb{G}$ state via $\mathbb{S}$ states in the second part, we can make all unsafe and goal states absorbing, because we are done as soon as we reach such a state. We, therefore, need two transformed CTMCs: $\mathcal{A}[\mathbb{U}]$ and $\mathcal{A}[\mathbb{U}\cup\mathbb{G}]$, where $\mathcal{A}[\mathbb{U}]$ is used in the first part of the path and $\mathcal{A}[\mathbb{U}\cup\mathbb{G}]$ is used in the second.

In order to calculate the probability for such a path, we accumulate the multiplied transition probabilities for all triples (s, s_1, s_2), where $s_1 \in \mathbb{S}$ and is reached before time t_1 and $s_2 \in \mathbb{G}$ and is reached within time $t_2 - t_1$.

$$\text{Prob}^{\mathcal{A}}(s, \Phi_1 U^{[t_1,t_2]}\Phi_2) = \sum_{s_1 \models \Phi_1} \sum_{s_2 \models \Phi_2} \pi_{s,s_1}^{\mathcal{A}[\mathbb{U}]}(t_1) \cdot \pi_{s_1,s_2}^{\mathcal{A}[\mathbb{U}\cup\mathbb{G}]}(t_2 - t_1). \qquad (20)$$

Hence, CSL until formulas can be solved as a combination of two reachability problems, as shown in Equation (20), namely $\pi_{s,s_1}^{\mathcal{A}[\mathbb{U}]}(t_1)$ and $\pi_{s_1,s_2}^{\mathcal{A}[\mathbb{U}\cup\mathbb{G}]}(t_2-t_1)$ that can be computed by performing transient analysis on the transformed CTMCs.

In the following we discuss the model-checking procedures that allow us to solve the interval path formulas (until and next) for the random agent, i.e. ICTMC. The procedure for checking these operators for ICTMCs is similar to that for CTMCs discussed above. However, the probabilities to take a certain path have to be calculated differently, because the Markov chain is time-inhomogeneous.

5.3   Next State Probability

Since the local mean-field model is a ICTMC the standard model-checking procedure is not applicable, therefore in the following we explain how to calculate the
next state probability of an individual agent. This probability is also changing
with time, therefore not only the next state probability at a given time t_0 is of
interest, but also the dependency of such probability measure on time. Another
important difference between checking CSL formulas for CTMC and ICTMC is
in the fact that the set of goal states (states, which satisfy Φ) can change with
time. In the following we address these differences and explain how a bounded
CSL Next formula can be checked for the local mean-field model.

We first describe how to calculate the next state probability for a given time
t_0, i.e., the probability to jump from the state s to the state, satisfying Φ, or
goal state, within time interval $[t_1, t_2]$. This probability can be found as follows:

$$\text{Prob}^{Z(t)}(s, \mathcal{X}^{[t_1,t_2]}\Phi_2, t_0) = \int_{t_0+t_1}^{t_0+t_2} q_{s,\mathbb{G}}(t) \cdot e^{-\Lambda(s,t_0,t)} dt, \tag{21}$$

where $q_{s,\mathbb{G}}(t) = \sum_{s' \in \mathbb{G}} Q_{s,s'}(t)$ is the rate of jumping from the current state s
to the goal state s' at time t; and $\Lambda(s, t_0, t) = \int_{t_0}^{t} -Q_{s,s}(\tau) d\tau$ is the cumulative
exit rate of state s between t_0 and t. The proof is straight forward and can be
found in [8].

The next state probability can now be computed numerically in two ways:
using Equation (21) or by transformation the above formula to the differential
equations and solving them. The differential equations are more convenient and
simplify the calculations, they can be obtained as in [8], and are as follows:

$$\begin{cases} \dot{P}(t) = q_{s,\mathbb{G}}(t) \cdot e - L(t), \\ \dot{L}(t) = -q_{s,s}(t), \end{cases} \tag{22}$$

where $P(t_0 + t_1) = 0$ and $L(t_0 + t_1) = \Lambda(t_0, t_0 + t_1)$. The above ODEs have to
be integrated from time $t_0 + t_1$ to time $t_0 + t_2$.

As we discussed above, for checking CSL formulas the dependency of the next
state probability on time $\text{Prob}^{Z(t)}(s, \mathcal{X}^{[t_1,t_2]}\Phi_2, t_0, t)$ is needed to be accessed.
To find this dependency one has to either calculate integral (21) for all possible
t_0, or use the differential equations (22) to define another system of the differential equations with t_0 as a independent variable. The obtained new system of
differential equations is as follows:

$$\begin{cases} \dot{\overline{P}}_s(t) = q_{s,\mathbb{G}}(t+t_2) \cdot e - L_2(t) - q_{s,\mathbb{G}}(t+t_1) \cdot e - L_1(t) - q_{s,s}(t)\overline{P}_s(t), \\ \dot{L}_1(t) = -q_{s,s}(t) + q_{s,s}(t+t_1), \\ \dot{L}_2(t) = -q_{s,s}(t) + q_{s,s}(t+t_2), \end{cases} \tag{23}$$

where $L_1(t) = \Lambda(t, t+t_1)$ and $L_2(t) = \Lambda(t, t+t_2)$. Initial conditions at time t_0
are computed by solving Equation (22).

And finally, the set of goal states can be time-dependent $\mathbb{G}(t)$, which has to
be taken into account while calculating the next state probability. It is done by

solving the above equation piecewise. All the time points $T_1, T_2, ...T_k$ when the goal set is changing are found first, where $T_0 = t_0 + t_1$ and $T_{k+1} = t_0 + t_2$. Equation (23) is solved for each time interval $[T_i, T_{i+1}]$.

For checking next formula one has to compare next state probability with the given threshold $p \in [0,1]$, hence, equation $\text{Prob}^{Z(t)}(s, \mathcal{X}^{[t_1,t_2]}\Phi_2, t_0, t) = p$ has to have a finite number of solutions. In general, this doesn't always hold, therefore, the restrictions on the rate functions of the mean-field model have to be introduced in order to insure the finite number of such solutions. In particular, the rate functions must be *piecewise real analytical functions*, as described and proved in [8].

5.4 Until Formulas. Reachability Probability

The core idea of CSL model-checking of until formulas as explained in Section 5.2 remains unchanged for time-inhomogeneous CTMCs. However, due to time-inhomogeneity it is not enough to only consider the time duration, but the exact time at which the system is observed must be taken into account. Hence, we add time t' to the notation of a time-inhomogeneous reachability problem $\pi_{s,s_1}^{Z(t)}(t', T)$ to denote that we start in state s at time t'.

A probability for an arbitrary until formula $\Phi_1 U^{[t_1,t_2]}\Phi_2$ to hold is then again calculated by computing two reachability problems on the transformed ICTMCs $Z(t)[U]$ and $Z(t)[U \wedge G]$, respectively:

$$\text{Prob}^{Z(t)}(s, \Phi_1 U^{[t_1,t_2]}\Phi_2, t') =$$
$$\sum_{s_1,t' \models \Phi_1} \sum_{s_2,t_1 \models \Phi_2} \pi_{s,s_1}^{Z(t)[U]}(t', t_1 - t') \cdot \pi_{s_1,s_2}^{Z(t)[U \wedge G]}(t_1, t_2 - t_1). \tag{24}$$

Equation (24) is valid for $t_1 > t', t_2 > t'$. If $t_1 = t'$ the first reachability problem can be omitted.

In the following we explain here how an arbitrary reachability probability $\Pi'(t', t' + T)$ can be calculated. This method is applied to both $\pi_{s,s_1}^{Z(t)[U]}(t', t_1 - t')$ and $\pi_{s_1,s_2}^{Z(t)[U \wedge G]}(t_1, t_2 - t_1)$; and the results are combined as in (24). The standard transient analysis on the modified ICTMS is used in order to calculate the reachability probability $\Pi'(t', t' + T)$. In order to find the transient probability the forward Kolmogorov equation is solved with an identity matrix as initial condition:

$$\frac{d\Pi'(t', t' + T)}{d(T)} = \Pi'(t', t' + T) \cdot Q'(t' + T), \tag{25}$$

where $Q'(t' + T)$ is the rate matrix of the modified ICTMC.

In order to check a nested CSL formula for ICTMC the dependency of transient probability on the starting time has to be found. The later is done by combining the forward and backward Kolmogorov equations:

$$\frac{d\Pi'(t, t + T)}{dt} = -Q'(t)\Pi'(t, t + T) + \Pi'(t, t + T)Q'(t + T). \tag{26}$$

The time-dependent probability matrix $\Pi'(t, t + T)$ can be obtained by solving Equation (26) with initial condition $\Pi'(t', t' + T)$. Using Kolmogorov equations for solving reachability problems on the local models $Z(t)$ is efficient due to the fact that the state space is usually quite small (see [8]).

The goal and unsafe sets in ICTMC can vary with time (e.g., in nested formulas), which has to be taken into account while calculating reachability probability. This is done by solving Equation (26) piecewise, i.e., for each time interval, where the above mentioned sets remain unchanged. At first we find the so-called discontinuity points, i.e., the time points $T_0 = t' \leq T_1 \leq T_2 \leq \cdots \leq T_k \leq T_{k+1} = T + t'$, where at least one of the sets changes. Then we do the integration separately on each time interval $[T_i, T_{i+1}]$ for $i = 0, ..., k$.

The procedure has to be slightly adjusted to ensure that only safe states are visited before a goal state is reached. We need to modify the ICTMC $Z(t)$ for each time interval $(T_i; T_{i+1})$ as follows:

1. introduce a new goal state s^*, which remains the same for all time intervals;
2. all unsafe and goal states are made absorbing;
3. all transitions leading to goal states are readdressed to the new state s^*.

Given this modified ICTMC $\overline{Z(t)}$, the transient probability matrix $\overline{\Pi'}(T_i, T_{i+1})$ is found for each time interval using the forward Kolmogorov equation, according to Equation (25).

Upon "jumps" between time intervals $[T_{i-1}, T_i]$ and $[T_i, T_{i+1}]$ it is possible that a state that was safe in the previous time interval becomes unsafe in the next. In this case the probability mass in this state is lost, since this path does not satisfy the reachability problem any-more. In the case that a state remains safe or a safe state is turned into a goal state the probability mass has to be carried over to the next time interval. This is described by the matrix $\zeta(T_i)$ of size $(|S| + 1) \times (|S| + 1)$ constructed in the following way: for each state $s \in S$ which is safe before and after T_i it follows $\zeta(T_i)_{s,s} = 1$. For each state $s \in S$ which was safe before T_i and becomes goal after T_i we have $\zeta(T_i)_{s,s*} = 1$. For the new goal state s^* the entry always equals one ($\zeta(T_i)_{s*,s*} = 1$), and all other elements of $\zeta(T_i)$ are 0.

The probability to reach a goal state before time T has passed when starting in a safe state at time t' is given then by the matrix $\Upsilon(t', t' + T)$:

$$\Upsilon(t', t' + T) = \overline{\Pi'}(t', T_1) \cdot \zeta(T_1) \cdot \overline{\Pi'}(T_1, T_2) \cdot$$
$$\zeta(T_2) \ldots \zeta(T_k) \cdot \overline{\Pi'}(T_k, t' + T). \tag{27}$$

The probability to reach the goal state s^* is unconditioned on the starting state by adding 1 for all goal states:

$$\pi_{s,s*}^{[U \vee G]}(t', t' + T) = \Upsilon_{s,s*}(t', t' + T) +$$
$$1\{s \in Sat(\mathbb{G}, t')\}. \tag{28}$$

Similarly to the dependency on time of the reachability probability while the goal and unsafe sets are fixed (see Equation (26)), the time-dependent reachability probability for varying goal and unsafe sets can be found by again combining forward and backward Kolmogorov equations using chain rule (see [8] for more details).

The method for checking state and path CSL formulas for ICTMC was presented above in this section. The convergence of the results and decidability of the algorithms are addressed in [8]. This method is applicable for the continuous time models, as the main interest of this tutorial lies in a continuous time mean-filed models. For the similar results on the *on-the-fly* fast model-checking of the PCTL properties of the individual objects in a discrete time mean-field model we refer to [29]. As a next step we provide the example, where this method is applied to a single agent of mean-field model.

5.5   Examples

In this section couple of examples of checking CSL formulas are described. We reuse the virus spread model, described in the Examples 1 and 2 (see Figure 1). As descibed in Section 2, the system of the limit ODEs (6) for the population behaviour is as follows:

$$\begin{cases} \dot{x}_1(t) = -k_1 x_3(t) + k_2 x_2(t) + k_5 x_3(t), \\ \dot{x}_2(t) = (k_1 + k_4) x_3(t) - (k_2 + k_3) x_2(t), \\ \dot{x}_3(t) = k_3 x_2(t) - (k_4 + k_5) x_3(t). \end{cases} \tag{29}$$

The coefficients that are used in the following example are given in Setting 1 in Table 4.

Let us consider the following formula

$$\Phi = \mathcal{P}_{<0.3}(\text{not infected } U^{[0,1]} \text{ infected})$$

and a predefined initial occupancy vector $\bar{x} = (0.8, 0.15, 0.05)$ at time $t' = 0$.

The only time-dependent rate of the local model is $k_1^*(t) = k_1 \cdot \frac{x_3(t)}{x_1(t)}$, where $x_1(t)$ and $x_3(t)$ are the solution of the ODEs (29) with $\bar{x}(0)$ as initial condition. Therefore the transition rate matrix $\mathbf{Q}(\bar{x}(t))$ is as follows:

$$\mathbf{Q}(\bar{x}(t)) = \begin{pmatrix} -k_1 \cdot \frac{x_3(t)}{x_1(t)} & k_1 \cdot \frac{x_3(t)}{x_1(t)} & 0 \\ k_2 & -k_2 - k_3 & k_3 \\ k_5 & k_4 & -k_5 - k_4 \end{pmatrix}.$$

To find $Prob^{Z(t)}(s, \text{not infected } U^{[0,1]} \text{ infected}, t')$ only one reachability problem $\pi_{s,s_1}^{Z(t)[\neg \text{not infected} \vee \text{infected}]}(0,1) = \pi_{s,s_1}^{Z(t)[\text{infected}]}(0,1)$ has to be solved according to the algorithm described earlier in Section 5.4. The local model $Z(t)$ is modified and all *infected* states are made absorbing. The Kolmogorov equation is used to calculate the transient probability matrix of the modified model, which consists of the reachability probabilities:

$$\Pi'(0,1) = \begin{pmatrix} 0.91 & 0.09 & 0 \\ 0 & 1 & 0 \\ 0 & 0 & 1 \end{pmatrix}.$$

Table 4. Parameter settings

Parameter		Setting 1	Setting 2
Attack	k_1	0.9	5
Inactive computer recovery	k_2	0.1	0.02
Inactive computers getting active	k_3	0.01	0.01
Active computer returns to inactive	k_4	0.3	0.5
Active computer recovery	k_5	0.3	0.5

The probability of the until formula

$$\phi = \text{not infected } U^{[0,1]} \text{ infected}$$

to hold for each starting state is as follows:

$Prob^{Z(t)}(s_1, \phi, t') = \pi_{s_1,s_2}^{Z(t)[\text{infected}]}(0,1) + \pi_{s_1,s_3}^{Z(t)[\text{infected}]}(0,1) = 0.09;$

$Prob^{Z(t)}(s_2, \phi, t')) = 0;$

$Prob^{Z(t)}(s_3, \phi, t')) = 0.$

The found above probabilities are compared with 0.3, and as one can see the formula $\mathcal{P}_{<0.3}(\text{not infected } U^{[0,1]} \text{ infected})$ holds for all states s_1, s_2, and s_3.

As was discussed earlier, the satisfaction on the CSL formula may change with time. Let us consider the same formula $\mathcal{P}_{<0.3}(\text{not infected } U^{[0,1]} \text{ infected})$ and initial occupancy vector $\bar{x} = (0.8, 0.15, 0.05)$. In the following we calculate the time-dependent probability on the predefined time interval $[0, 20]$.

The calculation of the time-dependent probabilities $Prob^{Z(t)}(s, \text{not infected } U^{[0,1]}\text{infected}, t', t)$ is done as described earlier in this section:

1. the model $Z(t)$ is modified so the infected states are made absorbing;
2. the transient probability $\Pi'(0,1)$ is calculated as described in the example above;
3. forward and backward Kolmogorov equations are used in order to construct the ODEs, describing the time-dependent transient probability of the modified model (see Equation (26));
4. These ODEs are solved using $\Pi'(0,1)$ as initial condition. The solution of the ODEs defines the required reachability probabilities.

The time-dependent probability $Prob^{Z(t)}(s_1, \text{not infected } U^{[0,1]} \text{ infected}, t', t)$ is depicted in Figure 7. Starting at states s_2 and s_3 this probability equals zero at all times, since these states do not satisfy *not infected*. In order to find the satisfaction set of this formula the following equation $Prob^{Z(t)}(s_1, \text{not infected } U^{[0,1]} \text{ infected}, t', t) = 0.3$ is solved and $t = 13.42$ is found. The satisfaction set depends on time and includes all three states s_1, s_2, and s_3 for $t \in [0, 13.42)$; and only two states s_2 and s_3 for $t \in [13.42, 20]$.

In the following we discuss a more involved example, which includes a nested until formula. The parameters of the model used in this example are given in the column Setting 2 in Table 4, the initial conditions at $t = 0$ is $\bar{x} = (0.85; 0.1; 0.05)$.

We check the following satisfaction relation:

$$\mathcal{P}_{>0.9}(\text{infected } U^{[0,15]}(\mathcal{P}_{>0.8} \; tt \; U^{[0,0.5]} \text{ infected})).$$

The formula is split into sub-formulas and the time-dependent satisfaction set of the sub-formula $\Phi_1 = (\mathcal{P}_{>0.8} tt \; U^{[0,0.5]} \text{ infected})$ is calculated first. Similarly to the previous example, the probability $Prob^{Z(t)}(s, tt \; U^{[0,0.5]} \text{ infected}, t', t)$ is calculated for all states $s \in S^o$. In Figure 7 this probability at state s_1 is depicted; the probabilities at states s_2 and s_3 equal to one, since these states are already *infected*. Similarly to the previous example, the time-dependent satisfaction set is found and equals to $Sat(\Phi_1, t', t) = \{s_2, s_3\}$ for all $t \in [0, 10.443]$ and $Sat(\Phi_1, t', t) = \{s_1, s_2, s_3\}$ for all $t \in (10.443, 15]$.

The next task is calculating the probability

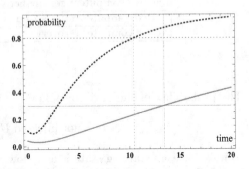

Fig. 7. The green solid line shows $Prob^{Z(t)}(s_1, \text{not infected} \quad U^{[0,1]} \text{ infected}, t', t)$. The time-dependent probability $Prob^{Z(t)}(s_1, tt \quad U^{[0,0.5]} \text{ infected}, t', t)$ is presented by the blue dotted line.

$$Prob^{Z(t)}(s, \text{infected } U^{[0,15]}\Phi_1, t', t).$$

The reachability probability for the time-varying satisfaction set of Φ_1 is calculated following the algorithm described above in this section. We first calculate all discontinuity points $T_0 = 0$, $T_1 = 10.443$ and $T_2 = 15$. An extra state s^* is added and an indicator matrix $\zeta(T_1)$ is constructed: $\zeta(T_1)_{s^*,s^*} = 1$, $\zeta(T_1)_{s_1,s_2} = 0$ for all $s_1 \neq s^*, s_2 \neq s^*$. The transient probabilities on time intervals $[0, 10.443)$ and $(10.443, 15]$ are calculated using the forward Kolmogorov equation:

$$\Pi'(0, 10.443) = \begin{pmatrix} 0.53 & 0 & 0 & 0.47 \\ 0 & 1 & 0 & 0 \\ 0 & 0 & 1 & 0 \\ 0 & 0 & 0 & 1 \end{pmatrix},$$

$$\Pi'(10.443, 15 - 10.443) = \begin{pmatrix} 1 & 0 & 0 & 0 \\ 0 & 1 & 0 & 0 \\ 0 & 0 & 1 & 0 \\ 0 & 0 & 0 & 1 \end{pmatrix}.$$

Equation (27) is used to calculate $\Upsilon(0, 15)$:

$$\Upsilon(0, 15) = \begin{pmatrix} 0 & 0 & 0 & 0.47 \\ 0 & 0 & 0 & 0 \\ 0 & 0 & 0 & 0 \\ 0 & 0 & 0 & 1 \end{pmatrix}.$$

Equation (28) is used in order to calculate the reachability probability for each state $s \in S^o$: $\pi_{s_1,s^*}^{Z(t)[\neg \text{infected} \vee \Phi_1]}(0,15) = 0.47$; $\pi_{s_2,s^*}^{Z(t)[\neg \text{infected} \vee \Phi_1]}(0,15) = 1$; $\pi_{s_3,s^*}^{Z(t)[\neg \text{infected} \vee \Phi_1]}(0,15) = 1$. The probability $Prob^{Z(t)}(s, \text{infected } U^{[0,15]}\Phi_1, t')$ is calculated according to Equation (24), and equals to 0, 1, and 1 for states s_1, s_2, and s_3 respectively. Therefore only states s_2 and s_3 satisfying the formula

$$\mathcal{P}_{>0.9}(\text{infected } U^{[0,15]}(\mathcal{P}_{>0.8} \text{ tt } U^{[0,0.5]} \text{ infected})).$$

In this section we have illustrated how the properties of a single agent in a large communication network (system of interacting objects) can be checked. Next to the fluid model checking reader might be interested in the techniques for calculation *fluid passage time*, as discussed in [18]. In the following model-checking the overall mean-field model is discussed.

5.6   On Model-Checking Overall Mean-Field Models. MF-CSL.

The properties of interest of the overall mean-field model differ from the properties which can be described by CSL. Therefore, in order to reason at the level of the overall model in terms of fractions of objects an extra layer "on top of CSL" that defines the logic MF-CSL was introduced in [25]. The latter is able to describe the behaviour of the overall system in terms of the behaviour of random local objects.

Definition 5. *Syntax of MF-CSL.* *Let $p \in [0,1]$ be a real number, and $\bowtie \in \{\leq, <, >, \geq\}$ a comparison operator. MF-CSL formulas Ψ are defined as follows:*

$$\Psi ::= tt \mid \neg \Psi \mid \Psi_1 \wedge \Psi_2 \mid \mathbb{E}_{\bowtie p}(\Phi) \mid \mathbb{ES}_{\bowtie p}(\Phi) \mid \mathbb{EP}_{\bowtie p}(\phi),$$

*where Φ is a **CSL state formula** and ϕ is a **CSL path formula**.*

□

In this definition three expectation operators were introduced: $\mathbb{E}_{\bowtie p}(\Phi)$, $\mathbb{ES}_{\bowtie p}(\Phi)$ and $\mathbb{EP}_{\bowtie p}(\phi)$, with the following interpretation:

- $\mathbb{E}_{\bowtie p}(\Phi)$ denotes whether the fraction of objects that are in a (local) state satisfying a general CSL state formula Φ fulfills $\bowtie p$;
- $\mathbb{ES}_{\bowtie p}(\Phi)$ denotes whether the fraction of objects that satisfy Φ in steady state, fulfills $\bowtie p$;
- $\mathbb{EP}_{\bowtie p}(\phi)$ denotes whether the probability of a random object to satisfy path-formula ϕ fulfills $\bowtie p$.

The formal definition of the MF-CSL semantics is as follows:

Definition 6. *Semantics of MF-CSL.* *The satisfaction relation $\models$ for MF-CSL formulas and states $\overline{x} = (x_1, x_2, \ldots, x_K)$ at time t_0 of the overall mean-field model is defined by:*

$$\overline{x} \models tt \qquad \forall \ \overline{x} \in X,$$
$$\overline{x} \models \neg\Psi \qquad \text{iff } \overline{x} \not\models \Psi,$$
$$\overline{x} \models \Psi_1 \wedge \Psi_2 \quad \text{iff } \overline{x} \models \Psi_1 \wedge \overline{x} \models \Psi_2,$$
$$\overline{x} \models \mathbb{E}_{\bowtie p}(\Phi) \quad \text{iff } \left(\sum_{j=1}^{K} x_j \cdot Ind_{(s_j, t_0 \models \Phi)} \right) \bowtie p,$$
$$\overline{x} \models \mathbb{ES}_{\bowtie p}(\Phi) \text{ iff } \left(\sum_{j=1}^{K} x_j \cdot \pi^{Z(t)}(s_j, Sat(\Phi, t_0)) \right) \bowtie p,$$
$$\overline{x} \models \mathbb{EP}_{\bowtie p}(\phi) \text{ iff } \left(\sum_{j=1}^{K} x_j \cdot Prob^{Z(t)}(s_j, \phi, t_0) \right) \bowtie p,$$

where $Sat(\Phi, t_0)$ is a satisfaction set of the CSL formula Φ at t_0, $\pi^{Z(t)}(s, Sat(\Phi, t_0))$ is a steady-state probability, $Prob^{Z(t)}(s, \phi, t_0)$ is defined as in Definition 4; and $Ind_{(s_j, t_0 \models \Phi)}$ is an indicator function, which shows whether a local state $s_j \in S$ satisfies formula Φ for a given overall state $\overline{x}$ at time t_0:

$$Ind_{(s_j, t_0 \models \Phi)} = \begin{cases} 1, & if \ s_j, t_0 \models \Phi, \\ 0, & if \ s_j, t_0 \not\models \Phi. \end{cases}$$

□

To check an MF-CSL formula at the global level (overall model), the local CSL formula has to be checked first, and the results are then used at the global level. The first step, namely CSL model-checking was explained in the previous sections, and for the algorithms for MF-CSL model-checking we refer to [25]. In the following we provide the example, which first shows the expressivity of the MF-CSL logic, and then provides the intuition behind the model-checking procedure.

Example 3. Let us consider the virus spread example to illustrate the expressive power of MF-CSL for mean-field models. In order to express the property that not more than 5% of the computers in the system are infected the following formula is used:

$$\mathbb{E}_{\leq 0,05} \text{ infected.}$$

The property "The percentage of all computers, which happen to have a probability lower than 10% of going from not infected to active infected state within 3 hours, is greater than 40%" is expressed as

$$\mathbb{E}_{>0,4}(\mathcal{P}_{<0.1}(\text{not infected } U^{[0,3]} \text{ active})).$$

If one wants to ensure that the probability of a computer to be infected within two hours from now is less than 50%, the following property has to hold:

$$\mathbb{EP}_{<0.5}(tt \ U^{[0,2]} \text{ infected}).$$

Note that in the formula above the current state of the individual is not taken into account. If the percentage of not infected computers which will become

infected within next two hours is of interest the formula has to be changed accordingly:
$$\Psi = \mathbb{EP}_{<0.5}(\text{not infected } U^{[0,2]} \text{ infected}).$$

If in a long run the system has to have a low probability (less then 2%) for a random computer to be infected the formula:

$$\mathbb{ES}_{<0.02} \text{ infected},$$

has to hold.

Let us consider the following MF-CSL formula:

$$\Psi = \mathbb{EP}_{<0.3}(\text{not infected } U^{[0,1]} \text{ infected}).$$

To check this formula against the occupancy vector $\overline{x}(0) = (0.8, 0.15, 0.05)$ we first have to check the CSL formula $\phi = (\text{not infected } U^{[0,1]} \text{ infected})$, then we have to find the expected probability for the whole formula Ψ to hold according to the semantics of the MF-CSL, and, finally, compare it with the treashhold $p = 0.3$.

The probabilities $Prob^{Z(t)}(s, \text{not infected } U^{[0,1]} \text{ infected}, 0)$ that the underlying CSL formula holds for initial condition $\overline{x}(0) = (0.8, 0.15, 0.05)$ was found earlier in Section 5.5. It equals to 0.09, 0, and 0 for states s_1, s_2, and s_3, respectively.

According to Definition 6, the weighted sum of the entries of the occupancy vector $\overline{x}(0)$ and the respective probabilities in the local model define the expected probability $\mathbb{EP}(\phi)$:
$$\sum_{j=1}^{K} x_j \cdot Prob^{Z(t)}(s_j, \phi, 0) = 0.8 \cdot 0.09 + 0.15 \cdot 0 + 0.05 \cdot 0 = 0.072 < 0.3.$$
As one can see, the occupancy vector $\overline{x}(0) = (0.8, 0.15, 0.05)$ satisfies the MF-CSL formula $\mathbb{EP}_{<0.3}(\text{not infected } U^{[0,1]} \text{ infected})$.

In this section we provided the insides for both fluid model-checking and MF-CSL model-checking on the overall model. We showed how these two approaches are related and what kind of properties can be expressed and checked using both CSL and MF-CSL logics.

6   Conclusions

This paper illustrates several aspects of applying mean-field approximations for efficient analysis of large scale stochastic models. The purpose is to provide a self-contained, example-guided and accessible tutorial for researches that are interested in the area of mean-field.

The main idea of mean-field is to provide an approximation for a large number of interacting similar objects. In contrast to existing tutorials [9] this presentation starts from the single agent model and than abstracts to a large number of these objects using the mean-field, in addition, the single agent model within the whole population an inhomogeneous CTMC.

This paper features two case study, one on the analysis of Botnets, where indeed the distribution of objects is assumed to be uniform, and one on the analysis of gossip to show how the location of objects can be taken into account using spatial mean-field models.

The performance measures that are traditionally derived from such model are mainly steady-state and transient state distributions. However, exploiting the difference between the local object and the overall mean-field model allows to apply model checking techniques to derive more complex measures of interest. Section 5 repeats the main idea of fluid model checking, that can be used to check the single agent model and hints at a new logic, called MF-CSL that can be used to specify properties on the overall model. Note that we do not focus on all the details of these techniques, but aim to show how they can be used to analyse different aspects of the system.

Mean-field approximation cannot be considered as a ready solution to the state-space explosion problem. Indeed, it is an approximation technique that must be applied carefully [33] and it provides a satisfactory first approximation of a system dynamics which requires, then, to be studied in further details to obtain a more precise analysis, as discussed in Section 1. To support the user in the correct application of these techniques, there are frameworks that allow for systematic application of mean-field techniques [9,21,36].

While the use of mean-field models in computer science already started in 1980 [28], still several open problems remain. For example, mean-field results are only reliable if the population is large enough, however it is still unclear whether and if so how this can be judged from the model at hand. Another interesting research topic would be to analyse the mean-field of models that include non-determinism.

References

1. Baccelli, F., Karpelevich, F.I., Kelbert, M.Y., Puhalskii, A.A., Rybko, A.N., Suhov, Y.M.: A mean-field limit for a class of queueing networks. Journal of Statistical Physics 66, 803–825 (1992)
2. Baier, C., Haverkort, B.R., Hermanns, H., Katoen, J.P.: Model-checking algorithms for continuous-time Markov chains. IEEE Trans. Softw. Eng. 29(7), 524–541 (2003)
3. Benaïm, M., Le Boudec, J.Y.: A class of mean field interaction models for computer and communication systems. Perform. Eval. 65(11-12), 823–838 (2008)
4. Benaïm, M., Weibull, J.W.: Deterministic approximation of stochastic evolution in games. Econometrica 71(3), 873–903 (2003)
5. Billingsley, P.: Probability and Measure, 3rd edn. Wiley-Interscience (1995)
6. Bobbio, A., Gribaudo, M., Telek, M.: Analysis of large scale interacting systems by mean field method. In: QEST, pp. 215–224 (2008)
7. Bortolussi, L.: Hybrid limits of continuous time Markov chains. In: QEST, pp. 3–12. IEEE Computer Society (2011)
8. Bortolussi, L., Hillston, J.: Fluid model checking. In: Koutny, M., Ulidowski, I. (eds.) CONCUR 2012. LNCS, vol. 7454, pp. 333–347. Springer, Heidelberg (2012)
9. Bortolussi, L., Hillston, J., Latella, D., Massink, M.: Continuous approximation of collective systems behaviour: A tutorial. Performance Evaluation 70(5), 317–349 (2013)

10. Cabspotting, http://stamen.com/clients/cabspotting
11. Chaintreau, A., Le Boudec, J.Y., Ristanovic, N.: The age of gossip: spatial mean field regime. In: SIGMETRICS/Performance, pp. 109–120. ACM (2009)
12. Ciocchetta, F., Hillston, J.: Bio-pepa: A framework for the modelling and analysis of biological systems. Theoretical Computer Science 410(33-34), 3065–3084 (2009)
13. Darling, R.W.R., Norris, J.R.: Differential equation approximations for Markov chains. Probability Surveys 5, 37–79 (2008)
14. Deavours, D.D., Clark, G., Courtney, T., Daly, D., Derisavi, S., Doyle, J.M., Sanders, W.H., Webster, P.G.: The Mobius framework and its implementation. IEEE Transactions on Software Engineering 28(10), 956–969 (2002)
15. Gast, N., Gaujal, B.: A mean field model of work stealing in large-scale systems. In: SIGMETRICS, pp. 13–24. ACM (2010)
16. Gillespie, C.S.: Moment closure approximations for mass-action models. IET Systems Biology 3, 52–58 (2009)
17. Gillespie, D.T.: Exact stochastic simulation of coupled chemical reactions. J. Phys. Chem. 81(25), 2340–2361 (1977)
18. Hayden, R., Stefanek, A., Bradley, J.T.: Fluid computation of passage time distributions in large Markov models. Theoretical Computer Science 413(1), 106–141 (2012)
19. Hayden, R.A., Bradley, J.T.: A fluid analysis framework for a markovian process algebra. Theoretical Computer Science 411(22-24), 2260–2297 (2010)
20. Hillston, J.: A compositional approach to performance modelling. Cambridge University Press (1996)
21. Hillston, J.: Fluid flow approximation of pepa models. In: QEST, pp. 33–43. IEEE Computer Society (2005)
22. Hillston, J., Tribastone, M., Gilmore, S.: Stochastic process algebras: From individuals to populations. The Computer Journal (2011)
23. Kadanoff, L.P.: More is the Same; Phase Transitions and Mean Field Theories. Journal of Statistical Physics 137, 777–797 (2009)
24. Kleczkowski, A., Grenfell, B.T.: Mean-field-type equations for spread of epidemics: the small world model. Physica A: Statistical Mechanics and its Applications 274(12), 355–360 (1999)
25. Kolesnichenko, A., de Boer, P.T., Remke, A.K.I., Haverkort, B.R.: A logic for model-checking mean-field models. In: DSN/PDF, pp. 1–12. IEEE Computer Society (2013)
26. Kolesnichenko, A., Remke, A., de Boer, P.-T., Haverkort, B.R.: Comparison of the mean-field approach and simulation in a peer-to-peer botnet case study. In: Thomas, N. (ed.) EPEW 2011. LNCS, vol. 6977, pp. 133–147. Springer, Heidelberg (2011)
27. Kurtz, T.G.: Solutions of ordinary differential equations as limits of pure jump Markov processes. Journal of Applied Probability 7(1), 49–58 (1970)
28. Kurtz, T.G.: Approximation of population processes, vol. 36. Society for Industrial Mathematics (1981)
29. Latella, D., Loreti, M., Massink, M.: On-the-fly Fast Mean-Field Model-Checking: Extended Version. Technical report (2013)
30. Le Boudec, J.Y., McDonald, D., Mundinger, J.: A generic mean field convergence result for systems of interacting objects. In: QEST, pp. 3–18. IEEE Computer Society (2007)
31. McComb, W.D.: Renormalization Methods: A Guide For Beginners. OUP, Oxford (2004)

32. Mitzenmacher, M.: The power of two choices in randomized load balancing. IEEE Trans. Parallel Distrib. Syst. 12(10), 1094–1104 (2001)
33. Pourranjbar, A., Hillston, J., Bortolussi, L.: Dont Just Go with the Flow: Cautionary Tales of Fluid Flow Approximation. In: Tribastone, M., Gilmore, S. (eds.) EPEW/UKPEW 2012. LNCS, vol. 7587, pp. 156–171. Springer, Heidelberg (2013)
34. Silva, M., Recalde, L.: On fluidification of petri nets: from discrete to hybrid and continuous models. Annual Reviews in Control 28(2), 253–266 (2004)
35. Tribastone, M.: Relating layered queueing networks and process algebra models. In: WOSP/SIPEW, pp. 183–194 (2010)
36. Tribastone, M., Gilmore, S., Hillston, J.: Scalable differential analysis of process algebra models. IEEE Trans. Software Eng. 38(1), 205–219 (2012)
37. Van Kampen, N.G.: Stochastic Processes in Physics and Chemistry. North-Holland Personal Library. Elsevier Science (2011)
38. van Ruitenbeek, E., Sanders, W.H.: Modeling peer-to-peer botnets. In: QEST, pp. 307–316. IEEE CS Press (2008)
39. Wolfram Research, Inc. Mathematica tutorial (2010), http://reference.wolfram.com/mathematica/tutorial/IntroductionToManipulate.html

Author Index